Agrotechniques and Uses of

Medicinal Plants

The Authors

Dr. Rameshwar Dass Gupta is a leading soil scientist and noted environmentalist, having specialized in soil Microbiology and Pedology. He did his M.Sc. Agriculture (Agricultural Chemistry) in 1968 from Ranchi University, India and Ph. D Agriculture (Soil Science and Water Management) in 1980 from Himachal Pradesh Krishi Vishvavidyalaya, Palampur, India. He also holds a Postgraduate Diploma in Ecology and Environment from Indian Institute of Ecology and Environment, New Delhi.

He started his professional career as an Assistant Extension Specialist (Soil Science) in Punjab Agricultural University, Ludhiana, in 1968 and has served in various capacities as Assistant Scientist, Associate Professor, Deputy Director Extension Education (Training) and Chief Scientist, Krishi Vigyan Kendra, SKUAST-Jammu. He has served as Chief Scientist and Head, Regional Agricultural Research Station, SKUAST-Jammu, and Division of Agricultural Chemistry and Soil Sciences and Agroforestry. He has over 150 peered publications like research papers, review and research papers as various book chapters and four books to his credit, namely *Problems and Management of Soil and Forest Resources of Northwest Himalayas* (1991); *Environmental Degradation of Jammu and Kashmir Himalayas and Their control* (2005) *and Environment Pollution: Hazards and Control* (2006) *and Glimpses of North-West Himalayas (2009).* He has also contributed a number of extension papers in *Intensive Agriculture, Agriculture Today, Farmers' forum, Indian Farmers' Digest, Farmers and Parliament, Agrobios Newsletter, Gram Vikas Jyoti* and daily newspapers.

Prof. Sushil Kumar Gupta has 22 years experience of teaching, research and extension in the discipline of agroforestry. He graduated in agriculture from HKKVV, Palampur (HP) in 1988 with forestry as elective subject and post graduate and doctorate in Forestry with specialization in Silviculture from Dr. Y.S. Parmar University of Horticulture and Forestry, Nauni, Solan (HP). He started his career in 1991 as Sr. Technical Assistant, Agroforestry and remained associated with research on forage production and pasture improvement in the intermediate zone of J&K. He has been teaching courses on forestry, agroforestry and environment and ecology for the past 20 years. Presently he is working as Professor, Division of Agroforestry, at Sher-e-Kashmir University of Agricultural Sciences and Technology of Jammu (J&K) and handling All India Coordinated Research Project on Agroforestry.

He has authored one book, 12 book chapters and more than 30 research publications in referred national and international journals of repute, edited training compendiums, popular articles, and project reports besides presentations in national/international conferences/seminars/symposia.

Dr. S. D. Bhardwaj has the 35 years of teaching, research and extension experience in the discipline of forestry. During his professional carrier he served on important positions like Dean, College of Forestry, Director Research, and Head, Department of Silviculture and Agroforestry at Dr YS Parmar University of Horticulture and Forestry, Solan. He obtained his Doctorate Degree in Agronomy with specialization in Medicinal and Aromatic Plants from HPKVV, Palampur (HP). He also obtained Post Graduate Diploma in Forestry from Utah, USA. He was awarded World 2000 Millennium Award in 8th Environment Congress, New Delhi, Man of the Year, 1999 by American Biographical Institute Inc., USA and Social Forestry Award by the Environment Society of India. He has authored 05 books, over 150 peered research publications including research papers, reviews and book chapters besides popular/extension articles and bulletins/reports.

Agrotechniques and Uses of
Medicinal Plants

R.D. Gupta
Ex-Associate Dean-cum-Chief Scientist, KVK, SKUAST-Jammu
Sushil Kumar Gupta
Professor, Agroforestry, SKUAST-Jammu
S.D. Bhardwaj
Ex-Dean, College of Forestry, DYPUHF, Solan

2016
Associated Publishing Company®
A Division of
Astral International Pvt. Ltd.
New Delhi – 110 002

Cataloging in Publication Data--DK
Courtesy: D.K. Agencies (P) Ltd. <docinfo@dkagencies.com>

Gupta, R. D. (Rameshwar Dass), 1941- **author.**
Agrotechniques and uses of medicinal plants / R.D. Gupta, Sushil Kumar Gupta, S.D. Bhardwaj.
pages cm

ISBN 978-93-5130-994-9 (International Edition)

1. Medicinal plants--India. 2. Materia medica, Vegetable--India. I. Gupta, Sushil Kumar, 1967- author. II. Bhardwaj, S. D., author. III. Title. IV. Title: Medicinal plants.

SB294.I4G87 2016 DDC 633.880954 23

Published by : **Associated Publishing Company®**
A Division of
Astral International Pvt. Ltd.
– ISO 9001:2008 Certified Company –
4760-61/23, Ansari Road, Darya Ganj
New Delhi-110 002
Ph. 011-43549197, 23278134
E-mail: info@astralint.com
Website: www.astralint.com

Laser Typesetting : **Classic Computer Services,** Delhi - 110 035

Printed at : **Replika Press Pvt. Ltd.**

— Dedicated to —

Late Sh. Karam Chand Mahajan (Thubwale)
a well known Hakim and Grand Father (Nana ji)
of Dr. R.D. Gupta

Acknowledgement

We express our gratitude to Prof. Pradeep K. Sharma, Vice-Chancellor, Sher-e-Kashmir University of Agricultural Sciences and Technology of Jammu and former ADG (Education), ICAR for accepting to write the foreword for this book after going through the contents of this book.

The authors would greatly be indebted to the scientists, authors/publishers, traditional folk healers, local practioners and the institutions working on medicinal plants whose knowledge and work have been referred to and incorporated in the text of this book. Some of the information contained in the extension bulletins, reports of the different departments have been included in the preparation of the manuscript and the authors would like to acknowledge all such sources.

It would be grateful to appreciate all friends and colleagues who have assisted the authors in providing the relevant information such as photographs etc. especially those from SKUAST-Jammu, the teachers/scientists from the Divisions of Agroforestry, Vegetable Science, Fruit Science and others whose names have not been mentioned.

Senior author is indebted to Late Smt. Gian Devi, his revered mother for her continuous inspirations in writing work during her life. Prof. Sushil Kumar Gupta would like to acknowledge the constant inspirations and support of his family members and children without which the documentation would have never been possible.

The authors would like to sincerely thank M/s Astral International Pvt. Ltd. for handling the manuscript expeditiously and giving the manuscript the present outlook. Prof. Sushil Kumar Gupta would like to offer sincere thanks to Mr. Sunil for his untiring assistance and editorial changes in the text.

R.D. Gupta

Sushil Kumar Gupta

S.D. Bhardwaj

Prof. Pradeep K. Sharma
Vice-Chancellor

Sher-e-Kashmir University of Agricultural Science and Technology of Jammu

Foreword

Medicinal flora is an integral part of the general vegetation of the forests and outside the forests. Out of about 8000 medicinal plants listed in different classical and modern texts, nearly 200 species are described in Indian Systems of Medicines, and about 960 medicinal plants are in active trade in India. Nearly 4000 plant species are utilized by the rural folk in curing local health problems. According to All India Ethno-botanical Survey, more than 7500 plant species are used by 4635 ethnic communities for human and animal health care across the country. These are particularly used against chronic disorders such as breast cancer, liver diseases, asthma and rheumatologic disorders. Herbal preparations are cheap and easy to prepare.

Cultivation of medicinal plants not only provides viable alternative to traditional agricultural crops but also plays a pivotal role in the improvement of socio-economic conditions of the farming communities. Cultivation of these plants has brought prosperity to young entrepreneurs in Punjab and other states. Governments of Chhattisgarh and Uttrakhand have taken lead in this direction by declaring their states as "Herbal States". Himachal Pradesh Government drew an ambitious plan in July 3, 2008 to develop the state as "Herbal Garden" of the country with active involvement of rural people. Other states like Jharkhand, North-Eastern hill states and Jammu and Kashmir, being rich in medicinal plants wealth, should also work on similar lines. Before World War-II, the J&K state used to collect about 100 tonnes each of roots of the composite *Saussurea lappa* (kuth) and *Artemisia* spp. annually, besides the cultivation of medicinal plants like belladona, podophyllum, ephedra, digitalis, pyrethrum etc. Recently the forest department undertook the development and marketing of these plants on a large scale.

After 1947, with the enactment of Jammu and Kashmir Kuth Act, medicinal plants were extracted from the forest areas for marketing in national and international markets, as cultivation of medicinal plants was not allowed. It badly destroyed forest wealth and raised long term sustainability issues. Deforestation and overgrazing of the pastures, and destruction of drug fields, further shrunk medicinal plants in forests. Same story was repeated in the forests of the Siwaliks of Jammu, Punjab, Haryana, Himachal Pradesh, Uttar Pradesh and Uttrakhand; they were once covered with dense forests and a variety of medicinal plants. These medicinal plants mostly consisted of amla, behera, arjun, baelpatar, amaltas, neem, jamun, swanjana, lassura, phalsa, karunda, girna etc. Some underutilized fruits from these forests having medicinal values, which used to flood the markets, are now not available. Such fruits are local mango, shatoot, aloocha, garna etc.

Government of India showed concern and launched a scheme "National Watershed Development Project for Rainfed Area (NWDPRA)", with one of the objectives of strengthening the cultivation of medicinal plants and small nurseries for growing indigenous species, like phalsa (*Grewia asiatica*), jamun (*Syzygium cuminii*), and dhiu (*Atrocarpus lakoocha*) were established. Herbal gardens of harar (*Terminalia chebula*), aonla (*Phyllanthus emblica*), behara (*Terminalia belerica*), neem (*Azadirachta indica*), ashwagandhya (*Withania sominifera*) were also set up in various areas depending on climatic and soil conditions. The Government of India sanctioned a scheme for Kashmir in 2010 to give boost to the cultivation of medicinal plants on commercial lines. Nevertheless, it is being realized that still there is lack of information pertaining to uses and cultivation of various medicinal plants. In view of this, I welcome publication of this book entitled "*Agrotechniques and Uses of Medicinal Plants*", by R.D. Gupta, Sushil Kumar Gupta and S. D. Bhardwaj. I congratulate the authors for the attempt. I am sure that the book will be of great value to the teachers, researchers, policy planners, practitioners, professionals, students and medicinal plant growers.

Prof. Pradeep K. Sharma

Vice-Chancellor,
SKUAST of Jammu

Preface

The term, medicinal plants, represents all those plants, which are used for medicinal purposes. They comprise both of aromatic and non-aromatic plants. Besides, medicinal purposes, these plants are also utilized for perfumery, dyes and cosmetics. All medicinal plants are rich in secondary metabolites which act as potential sources of drugs. The secondary metabolites include alkaloids, glycosides, caumarins, flavonoids, carotenoids, terroids, saponins and many others. Generally, whole plant, fruits, flowers, stems, bark, leaves, roots, gums and oleo- resins are used as medicines. There are hundreds of medicinal plants that have a long history of curative properties against various diseases and ailments. India and China are the two major producers of medicinal plants, sharing more than 40 per cent of the global production. The global herbal market is worth 60 billion USD which is growing @ 7 per cent annually. The trade in medicinal plants in the world for plant derived drugs accounts for about Rs. 2,000,000 million, whereas, in India it has been estimated to the tune of Rs. 5,500 million per annum. Total turnover of ayurvedic and herbal products is around Rs.23,000 million per annum.

There is no doubt that since time immemorial man was in search of cures for certain ailments which affected him badly. However, the scientific and experimental investigation on plants used in indigenous medicines in India commenced in 1921 in the modern laboratory of the School of Tropical Medicine, Calcutta (now Kolkata). Sir R. N. Chopra, Professor of Pharmacology was the pioneer who initiated the research work in collaboration with other experts there. Later on he moved to Jammu in 1941 as the founder Director of Drug Research Laboratory established by J&K Government. Thereafter, Council of Scientific and Industrial Research (CSIR) established Central Drug Research Institute and Central Institute of Medicinal and Aromatic Plants at Lucknow, National Chemical Laboratory at Pune, Regional Research Laboratory at Jorhat, Bhubaneswar, Hyderabad and Jammu. Multi-disciplinary research programmes for the development of modern drugs from the plant resources are going on in these central institutes using sophisticated and advanced instruments like Mass Spectrometers, Gas Liquid Chromatography (GLC),

High Pressure Liquid Chromatography (HPLC), X-ray diffraction etc. Some other parameters used to detect as well as to isolate active biochemical compounds from plants are micro-chemical tests, florescence microscopy and phyto-chemical studies. Modern tissue culture, callus culture for micro-organisms and induced mutation techniques are also used to screen, isolate and for mass propagation of selected drug yielding plants. More than 4,000 plant samples have already been screened in a systematic manner for a wide range of biochemical substances and drugs. It is not surprising that about 1000 new drugs have been introduced during the last 25 years which are mostly of plants origin including spices. To further provide impetus in this direction, Government of India is making all concerted efforts to promote cultivation of medicinal plants on a large scale as per variability of agro-climatic conditions. This compilation thus contains information collected from several publications on the subject and related topics. The present book will thus prove a boon not only to the medicinal plants growers but also to the researchers, herbal practioners, folk healers and students of Ayurveda. The title would also surely prove to be indispensable to the undergraduate and post graduate students studying courses on medicinal plants in the SAU's. This voluminous compilation is likely to boost the cause of development and promotion of usage of medicinal plants in the health care and encouraging the farmers to take up their cultivation to earn their livelihood.

We feel pleasure to have compiled the information on the traditional knowledge on medicinal plants and aspects related to their cultivation. Reader's views and suggestions on this edition are highly appreciable and would be a guiding force for future pursuits and improvement of subsequent editions.

R.D. Gupta

Sushil Kumar Gupta

S.D. Bhardwaj

Contents

Fruits Trees

Shrubs

Climbers

Herbs

Introduction

Worldwide swing back to the Indian Systems of Medicine, the demand for medicinal plants in the international market, both in the developed and developing countries has swollen. According to World Health Organization (WHO) estimate, about 80 per cent of population in developing world relies on the traditional medicines for their heath care (Mohanty *et al.*, 2011 b). It is because of its being healthy source of life cure for people especially the believers in Ayurveda, Unani and Sidha medicines. These plants and their products not only serve as valuable source of income for small and marginal farmers but also earn foreign exchequer through export. The traditional knowledge base of the Indian ethnic communities is perhaps the richest and unique in the world, particularly the traditional medicines and health care system. The classical Indian texts like 'Vedas' describe them on the use of plant wealth of the area (Mohanty *et al.*, 2011 a).

Ayurveda is a complete science based on the laws of nature and comprise not only of the use of herbs and roots but the entire plant also. Ayurveda deals not only with the agent and the host but the environment also. It means the knowledge of life or means by which the life may be prolonged. As such, Ayurveda concerns with the aspects of life which bring about a complete physical, mental, social and spiritual health. One must take herbal medicines and seasonal foods grown in the area of one's habitat (Aggarwal, 2014).

The Ayurvedic System of Medicine stands well documented in the ancient scriptures such as the *'Atharv-Veda'* (1200 BC), 'Charak Samhita' and 'Sushruta Samhita' (1000 BC) and is still being practised in India (Singh, 2010). This system is of great significance due to the wide variation in climatic conditions from Kanayakumari to Kashmir and from deserts of Rajasthan to the rain forests of Assam. Such a great biodiversity had allowed ayurvedic practioners to resort to experimentation with plants thought to possess the therapeutic properties (Singh, 2010).

In India, the history of medicinal plants can be traced back to the Vedic period (4500 BC to 1500 BC). 'Atharveda' contains detailed information on approximately 2,000 medicinal plants and their uses (Sharma and Mir, 2000). As a matter of fact, India can be called a "Botanical Garden" of the world as it is the largest producer of medicinal herbs, with more than 35,000 medicinal species. Herbal medicines have achieved proven milestones in the health care industry (Fotedar, 2007). Many of our modern drugs have been developed from plants. Immuno-modulatory, hepato-protective and anti-diabetics are some of the activities which have made the globe to think seriously regarding herbal medicines. Though each herb has its own specific use from becoming one's passion to blasting cholesterol, yet there are top 10 plants which can easily be incorporated into daily life to stay healthy, happy and muscular. Notable among them are; garlic, ashvagandha, turmeric, amla, tulsi, ginger, gloe, guggal, ajwain and vidari (Anonymous, 2009).

Herbal formulations are safe and would overcome the resistance created by the pathogens as they exist in the combined form in the protoplasm (Tapsell *et al.*, 2006). Even with the advent of allopathic drugs, a variety of drug formulations have been derived from plants used by ethanic communities. The father of medicine, Hippocrates once said, "*Let healthy food be your medicine rather than medicine be your food*". In old days, ayurvedic practioners used to prescribe the consumption of dry fruits to keep the skin glowing and also to ameliorate digestion system. The almond oil "*Badam Rogan*" was used to be administered as brain tonic and improve memory. But consumption of dry fruit decreased when allopathic practioners classified them as fattening agent and causing hypertension and diabetes. However, studies have revealed that use of dry fruits can reduce the incidence of heart disorders and number of other ailments if consumed regularly.

Many plant species (trees including fruit trees, shrubs, vegetables and herbs) are the source of natural antibiotics (secondary metabolites) belonging to alkaloids, flavonoids, coumarins, steroids, carotenoids, saponins, anthraquinones, glycosides, anti-microbial, anti- bacterial, antioxidants, anti-cancerous groups. Primary metabolites in higher plants concentrated in seeds and vegetative storage organs used commercially, are high volume-low value bulk chemicals (*e.g.* vegetable oils, fatty acids, carbohydrates etc.).The secondary metabolites directly or indirectly play an important role in the human society to combat diseases (Wink *et al.*, 2005). Rural areas of many developing countries still rely on traditional medicines for their primary health care needs and thus have found a place in day-to-day life. These medicines are relatively safer and cheaper than synthetic or modern medicines (Iwu *et al.*, 1999; Idu *et al.*, 2007; Mann *et al.*, 2008; Ammara *et al.*, 2009). People living in rural areas from their personal experience have established that these traditional remedies are the valuable source of natural products to maintain human health, but they may not understand the science behind them. They also know that some medicinal plants are highly effective only when used at therapeutic doses (Maheshwari *et al.*, 1986; Van Wyk *et al.*, 2000) and effective against many diseases like jaundice, asthama, diabetes etc.

Many new compounds including 'reserpine' (*Rauwolfia serpentina*) and 'psoralene' (*Psorylia corylifolia*), have been isolated using modern methodology of isolation techniques which have shown to display medicinal properties (Singh, 2010). Some of the herbs like Dioscorea, Artemesia, Brahmi, Berberis, Ashwagandha, Euphorbia, Alfalfa, are now being widely used (Khashu, 2012).

Traditional use of medicine is recognized as a way to learn about potential future medicines. Researchers have identified variety of compounds used in popular medicines derived from "ethnomedical" plant sources (Fabricant and Farnsworth, 2001). Plants are used medicinally in different countries and are a source of many potent and powerful drugs (Srivastava, *et al.*, 1996; Mahesh and Sathish, 2008).

India has been the store house of many herbs and spices, spread over vast geographical situations and widely useful in Ayurvedic, Unani, Siddha medicines but their chemical and pharmacological potential have not been studied to optimum (Gupta *et al.*, 2005; Sandhu and Heinrich, 2005). At present about 6,000 plant species are in traditional use with 960 species being traded and efforts for conservation and cultivation of only 178 such species is being directed to those being traded in quantities exceeding 100 metric tonnes per year (Table 1)

Table 1: Commercially Traded Species of Medicinal Plants in India

Sl.No.	*No. of Species*	*Percentage*	*Source*
1.	21	12	Temperate forests
2.	70	40	Tropical forests
3.	36	20	Cultivation/plantations
4.	46	25	Roadsides/degraded lands/wastelands
5.	5	3	Imports

Source: Ved and Goraya, 2007.

Annual demand in India for raw drugs has been estimated to be 3, 19,500 MT for 2005-06 including consumption by herbal industries, traditional health care and export. To sustain about 9500 registered herbal industries and un-registered cottage herbal units in India, continuous supply of raw material is a major threat. Data on acreage on cultivation on medicinal plants and their productivity from cultivated sources is not available. Only information on certain species such as isabgol, senna, jojoba, henna, aswagandha and milk thistle indicated that entire supply is made available through cultivation. These species account for about 1,21,400 MT production from an area of 1,18,000 ha (Singh *et al.*, 2008).

Though global interest in herbal formulations, cosmetics and food supplements is increasing but the raw material for the herbal industry is obtained either legally or by illicit means from the wild sources. This has resulted in over exploitation of resources for their botanicals leading to the extinction of many vital species. Future strategies for conservation of medicinal plants (Gupta, 2014) include ban on illicit cutting and lopping of trees, awareness among growers regarding the importance of medicinal plants, provision of incentives to the growers, setting up of national gene banks etc. To bridge the gap between demand and supply, to meet the non-commercial needs, systematic and concerted efforts towards promotion of their cultivation in home gardens, community and herbal gardens is necessary. The future line of action must be directed to following important issues:

- ✰ *In situ* conservation for the resource availability of high volume species which are largely obtained from forest habitats.
- ✰ Assessment of production potential of wild populations and adequate management strategy to ensure sustained supply in future.
- ✰ Ensuring and protecting the rights of the communities associated with the traditional health care.
- ✰ Zonalization of plant resources for development of raw drugs.
- ✰ Establishment of regional repositories of raw drugs for authenticity of trade and consumption.
- ✰ Convergence of different centrally sponsored schemes aiming at the promotion of cultivation of medicinal species to bridge the gap between demand and supply.
- ✰ Adequate market support system and strengthening of financial institutions and effective supply chain mechanism.

In light of the immense importance of medicinal plants, the benefits for the people in general and the Indian ethnic communities in particular (Mohanty *et al.,* 2011 a), it becomes imperative to compile medicinal properties and usage of important ones in the form of a book. Thus, the present pursuit *"Agrotechniques and Uses of Medicinal Plants"* deals with high value medicinal species. All these medicinal plant species have been classified in five sections namely; forest trees, fruit trees, shrubs, climbers and herbs. An attempt has been made by the authors to present each species by describing its botanical name, common name, family, introduction, medicinal values, botanical description and cultivation aspects. Cultivation technology included its climatic and soil requirements, method of planting/sowing, propagation, intercultural operations, harvesting and yield. Various terms often used in different sections of the text have also been explained/ defined in the end as glossary.

References

Aggarwal, K.K. 2014. Ayurveda: a way of life (Personal communication).

Ammara, H., Salma, R., Farah, D. and Shahid, M. 2009. Anti microbial activity of some plant extracts having hepatoprotective effects. *Journal of Medicinal Plant Research.* 3(1): 20-23.

Anonymous. 2009. Ayurved, Reinterpreted. The State Times. 14 (219): 5 (Saturday, August, 8).

Fabricant, Daniel S. and Farnsworth, Norman R. 2001. The Value of Plants Used in Traditional Medicine for Drug Discovery. *Environmental Health Perspectives.* 109 (Suppl 1) : 69-75.

Fotedar, S. 2007. Herbal medicines. The Daily Excelsior. 43(14): 3 (*Sunday Magazine,* January 14).

Gupta, M. P., Solis, P.N., Calderon, A. J., Guinneau - Sinclair, F., Correa, M. Gladames, C., Guerra, C., Espinosa, A., Alvenda, G.L., Robles, G. and Olampo, R. 2005. Medical ethnobotany of the tribes of Bocas del Toro, Panama. *Journal of Ethnopharmacology*. 96: 389-401.

Gupta, R. D. 2014. Conservation of medicinal plants. The Daily Excelsior. 50 (62): 10(Tuesday, March 4).

Idu, M., Omogbai, E. K. I., Aghimine, G. E. Amaechina, F., Timothy, O. and Omonigho, S. E. 2007. Preliminary phytochemistry, antimicrobial properties and acute toxicity of *Stachytarpheta jamaicensis* (L.) Vahl. leaves. *Trends in Medical Research* 2: 193- 198.

Iwu, M. W., Duncan, A. R. and Okon, C. O. 1999. New Antimicrobial of plant origin. In: *Perspective on New crops and New uses*. Janick, J. (Ed.), Alexandra Press, VA.

Khashu, V. 2012. Role of medicinal plants in J&K. The Daily Excelsior. 48(345): 6 (Wednesday, December, 2012.

Mahesh, B. and Satish, S. 2008. Antimicrobial activity of some important medicinal plant extract against plant and human pathogens. *World J. of Agri.Sci.* 4(S): 839-843.

Maheshwari, J.K., Singh, K.K. and Saha, S. 1986. Ethnobotany of tribals of Mirzapur District, Uttar Pradesh, Economic Botany information Service. NBRI, Lucknow.

Mann, A., Yahaya, Y., Banso, A. and John, F. 2008. Phytochemical and antimicrobial activity of *Terminalia avicennioides* extracts against some bacteria pathogens associated with patients suffering from complicated respiratory tract diseases. *Journal of Medicinal Plants Research*. 2(5): 94-97.

Mohanty, R.B. Tripathy, B.K. and Panda, T. 2011 a. Traditional knowledge Digital Library and its role in conserving medicinal plants in India. *Agrobios Newsletter*.9(12): 28-29.

Mohanty, R.B. Tripathy, B.K. and Panda, T. 2011 b. Plant based drugs in modern health care: An overview. *Agrobios Newsletter*.9(12): 27.

Sandhu, D. S. and Heinrich, M. 2005. The use of health foods, spices and other botanicals in the Sikh community in London, *Phytotherapy Research*. 19: 633-642.

Sharma, J.P and Mir, A.A. 2000. Dynamics of cold Arid Agriculture. Kalayani Publishers Ludhiana- New Delhi- Noida (UP)-Hyderabad- Chennai-Calcutta-Cuttack pp.147-148.

Singh, R.V., Singh, P., Hansen, L.A. and Graudal, L. 2008. Medicinal Plants, their Conservation, Use and Production in Southern India. Development and Environment No. 11-2008. Forest and Landscape, Denmark.

Singh, S.P. 2010. Medicinal plants: New opportunities. *The Tribune*. 130 (179): 11 (Wednesday, June 30).

Srivastava, J., Lambert, J. and Vietmeyer, N. 1996. Medicinal plants: An expanding role in development. *World Bank Technical Paper*. No. 320.

Tapsell, L.C., Hemphill, I., Cobiac, L., Patch, C.S., Sullivan, D.R., Fenech, M., Roodenrys, M., Keogh, J.B., Clifton, P.M., Williams, P.G., Fazio, V.A. and Inge, K.E. 2006. Health benefits of herbs and spices: the past, the present, the future. *Medical Journal of Australia*. 185 (4 Suppl): S 4-24.

Van Wyk, B.E and Gericke, N. 2000. Peoples Plants: A Guide to Useful Plants of Southern Africa. Briza Publications, Pretoria, South Africa, 351pp. ISBN 1–875093–19–2.

Ved, D.K. and G.S. Goraya, 2007. Demand and Supply of Medicinal Plants in India. NMPB, New Delhi and FRLHT, Bangalore, India.

Wink, M., Alfermann, A.W., Franke, R., Wetterauer, B., Distl, M., Windhovel, J., Krohn, O., Fuss, E., Garden, H., Mohagheghzaden, A., Wildi, E. and Ripplinger, P. 2005. Sustainable bioproduction of phytochemicals by plant in vitro cultures, anticancer agents. *Plant genetic resources*. 3: 90-100.

Forest Trees

1

Arjun Tree (*Terminalia arjuna* Roxb. W.&A.)

Common Name: Arjun, Arjuna

Family: Combretaceae

Introduction

Arjun is not found naturally as a component of forest vegetation. It is planted as an avenue tree and is easy to take up its cultivation from nursery stocks. It is a magnificent tree of great economic and environmental values. It can be grown up to 1000m in sub-mountainous areas of Siwaliks of Jammu, Punjab, Haryana, Himachal Pradesh and Uttrakhand. Its cultivation is picking up privately and officially.

Arjun tree is also found in parts of Madhya Pradesh, Maharashtra and Tamil Nadu chiefly along water channels, stream banks and moist places. In Himachal Pradesh, it is common in Nurpur and Nahan forest divisions. All parts of Arjun tree have many curative properties which are used in different kinds of medicine systems like Unani and Siddha as well as in Ayurveda.

Medicinal Values

Bark of the tree is astringent, tonic, febrifuge, cardiac stimulant and valuable for heart diseases. It is also found useful in diarrhoea, dysentery and ulcers. In additions to bark, its fruits and roots are also used medicinally. Bark is usually given in powdered form. Beside serving as acrid stypic, anti-dysenterial and febrifugal, it is also given to the patients suffering from high blood pressure and in urine problems (Singh, 2005). It is effective in cirrhosis of liver. Bark decoction is used in washing ulcers. Twigs are used as tooth brush (Datoon), to cure ulcers and blisters of mouth.

Bark of the arjuna tree is already being tried by Ayurveda physicians for treating cardiac patients. However, the department of biotechnology wanted to check the efficacy of these claims, "said by Dr S.K. Maulik of the Department of Pharmacology at All India Institute of Medical Sciences, and the Principal Investigator in this clinical trial (Anonymous, 2012).

Prepartions from the bark and the fruit are also used as tonic, astringent, febrifuge and deobstruent. These are used in bilious effections and for sores (Chauhan, 2006). 'Arjuna ishta', 'Arjubkwath', 'Arjunatwak churana', 'Hridayarnava Rassa', are some of the preparations which are commonly used.

Botanical Description

Arjun is a large tree having height up to 25m. It has smooth and thick bark, greenish white in colour and flaking off in large flat pieces. This is an exotic tree in Jammu (Jammu and Kashmir State) and elsewhere. It produces enormous quantities of fruit annually which goes waste. Fruit is winged, edible and sour in taste. Leaves of the tree are opposite, oblong-elliptic. Flowers of arjun tree are sessile, axillary spikes or in terminal panicles. Drupe is 2.5 to 5cm long, ovoid, fibrous, woody, glabrous and dark brown in colour with 5 hard projecting wings.

Cultural Practices

Climatic requirements: The tree is found throughout the greater parts of India having sub-tropical to tropical climate.

Soil requirements: Arjun tree prefers to grow in sandy loam to loam soils. The soil should have fine structure *i.e.,* fine granular to loose, well fertile and moist.

Cultivation: Arjun tree can be raised by means of seeds, stumps or entire transplant (Chauhan, 2006).

The fruits of the tree are soaked in water for about 24 hours or dumped in cowdung. These are then sown during month of March or April in raised beds at 20X5 cm distance. Germination is reported between 35-70 per cent and seedlings get ready for transplanting after 1.0 to 1.5 years.

The seedlings are sensitive to drought and frost, so they should not be planted in open and dry hill sides. It would be much better if prior to transplanting the seedlings, the fields are slightly irrigated.

Harvesting time of the barks is April to May. About 15-20 kg bark is obtained per tree in dry form. Usually, 3 kg fresh bark provides 1 kg dry bark.

References

Anonymous. 2012. Medicinal tree Arjuna's bark to treat cardiac patients. *The Kashmir Times*. 69 (60):12 (February 29, Wednesday)

Chauhan, N.S. 2006. Medicinal and Aromatic Plants of Himachal Pradesh. Indus Publishing Company, ES-5, Tagore Garden, New Delhi-110027

Singh, S.S. 2005. Fate of Medicinal Plant Resources of Jammu and Kashmir. Directorate of Environment and Remote Sensing Jammu, J&K State.

2

Bamboo (*Bambusa arundinacea* Retz.)

Common name: Bans, Wans

Family : Gramineae

Introduction

Bamboo the "Maggar Bamboo" is an exotic. It is a versatile, strong renewable and environment-friendly member of the grass family and the fastest growing woody species on the earth. There are more than 10 million hectares, both natural and cultivated with some 1225 species (Tewari *et al.*, 2007). About 80 per cent of these species are found in India and Asia Pacific region, including Japan.

Most of the bamboo species produce mature fibre in 3 years sooner than any other tree species. Some bamboos grow up to 1 m a day, with many culms reaching lengths of 25m or more. The world's annual production of bamboo has been estimated to be more than 20 million tonnes, with wide annual fluctuations because of large scale periodic death of bamboo clumps from gregarious flowering. The real benefits from value added products, handicrafts (mats, baskets, tools, toys and utensils) and furniture are established possibilities produced in finished form or supplied as components to small enterprises for further processing (supply of mats for production of bamboo mat board). There are emerging industrial and large scale applications too in the manufacture of wood substitutes and composites, energy, charcoal and activated carbon. Bamboo also possesses some medicinal uses.

Medicinal Values

Medicinally, bamboo is used locally on many ailments. Roots and seed sprouts are rich source of vitamins and minerals. Seeds are narcotic and given in piles also.

Bamboo shoots are a great delicacy in the Chinese cuisine. They are also eaten by the people who live in north-east India. Japanese also relish tender bamboo shoots and consumption of bamboo shoots is known to ameliorate the digestive system of the humans. Infact, the wonderful property of tasty tender bamboo shoots is that they have been shown to actually lower body fat and prevent constipation (Ponnuswami *et al.,* 2009). The juice of the pressed bamboo shoots possesses protease activity that helps digestion of proteins. The boiled bamboo shoot or pickle is served as an appetizer. Decoction of tender shoots is also used for cleaning wounds and maggot affected sores, ulcers etc. Mixed with palms jaggery, it is used to cause abortions or to induce labour and expulsion of placenta after child birth and to prevent excessive loss of blood.

The powdered hardened secretion from bamboo is used internally to treat asthma, coughs, bile fever, leprosy, swellings, cuts, wounds and laxative (Ralte *et al.,* 2013). In China, ingredients from the roots of the black bamboo help kidney disease. Phytosterols in bamboo shoots have nutraceutical properties. Bamboo beer prepared from bamboo leaf extract lowers blood lipids, strengthens the heart, slows aging and prolongs the life span.

Roots and leaves of the bamboo have also been used to treat intestinal worms, diarrhoea, veneral diseases and cancer. 'Tabasheer or banslochan' is a siliceous secretion found in the culms of various species of bamboos (Singh and Singh, 2007). It has enjoyed considerable reputation in medicine as a cooling tonic, aphrodisiac cure to asthma, cough and female disorders.

Botanical Description

Bamboo is botanically a tree like grass belonging to the sub-family Bambusoideae. Although bamboo is a grass, many larger bamboo are tree like and sometimes called 'bamboo tree'. Bamboo grows as tall as 120 m but unlike trees they do not have secondary growth. Bamboo, a woody stemmed, evergreen woody perennial, has jointed stems with regular nodes above. Culm height ranges from few meters to 40 meters with stem diameter ranging from 1 mm to 30 cm. Internodes are hollow; buds appear on the nodes which develop into branches.

Rhizomes are sub-terrainean which send off new culms. The rhizomes of certain bamboo species may spread for several feet (a few meters). There is an alternative type of rhizome growth that results in the rhizome being very short, and sometimes, almost bulbous.

Bamboo does not flower annually but only once in lifetime. Flowering cycle in bamboo varies from 7 to 60 years depending on species. Flowering can either be sporadic or gregarious. Bamboo fruit is indehiscent, one seeded caryopsis. Seeds are also used as grains during the periods of scarcity. The spreading or running bamboos are generally the species for temperate climate and clumping bamboos for tropical climate. The spreading bamboo needs plenty of space to form a groove.

Cultural Practices

Climatic requirements: Bamboo can be grown throughout the sub-tropical and sub temperate zones of the world. Bamboo is a drought tolerant and can tolerate diverse climate and moisture regimes. Bamboo require sparsely 1270 mm to as high as 6350 mm of rainfall or sometimes even more.

Soil requirements: Bamboo is a crop which should be grown in a well drained sandy loam to loamy clay soil at a soil pH of 7 or a bit less. Although is prefers well drained soil but some species can flourish well along the river banks, beds of streams and other moist places (Fartyal *et al.*, 2004). It can flourish well even on degraded land or non-cropped areas where no other kind of cultivation is feasible. Bamboo can be grown in any type of soil but the loamy soil is the best suited for its cultivation (Singh and Singh, 2007).

Bamboo plantation on farmland: As bamboo is a long term cropping option, therefore, it cannot be replaced even after several harvestings. Thus, bamboo tree should be planted in the fields where it can be placed on long term basis. For this purpose, it can be grown in the field in the following pattern:

(a) **Bund plantation:** Bamboo is sown in furrows or bunds either to limit the boundary of the plots or to act as a wind break. For bund purpose, the cultivars should be either thorny or shrubby in nature like *Bambusa bambos* or *Dendrocalamus strictus*. In this method, the distance maintained between two plants is 1-3m, and within 1-2 years they form a thick fence.

(b) **Block plantation:** In this method, bamboo is planted in degraded land and there is no direct contact between the planted bamboo and other crops grown.

(c) **Inter-cropping:** In inter-cropping method, bamboo can be grown in direct association with other crops hence competition for growth is more. Distance between two rows and plants should not be kept less than 8-10m or even more. According to Ponnuswami *et al.* (2009) propagation is made with rhizome division, culm segments or seed but generally rhizomes are preferred. Propagation is almost entirely by vegetative means using cuttings made from the underground rhizomes. Make cuttings 30 cm long and then plant them end to end 15 cm deep. The cuttings should have two or three nodes.

Manures and fertilizers: The pit where bamboo is to be planted, should be filled with 5kg compost or FYM, 100g urea, 100g single superphosphate and 50g muriate of potash.

Yield: The total annual production of bamboo in India has been estimated to be around 5 Mt. The annual yield of bamboos ha^{-1}, depending upon the species varies from about 0.2 to 4.0 tonnes.

References

Fartyal, S., Tewari, S and Kumar, V. 2004. Bamboo based agroforestry: A new turn in farming. *Indian Farmers Digest.* 37 (12): 15-16.

Ponnuswami, V, Prabhu, M. and Ramesh Kumar, A. 2009. Bamboo shoots- A nutritious and delicious vegetable. *Agrobios New letter.* 8(5): 28-31.

Ralte, Z., Lalmuanzovl and Agami, T. 2013. Bamboo shoot: opportunities and processing. *Indian Farmers Digest.* 46 (9): 18-20.

Singh, A.K and Singh, B.2007. Bamboo as an ornamental plant. *Indian Farmers Digest.* 40(11): 37-39.

Tewari, S., Tewari, S.K. and Shukla, A.N 2007. Bamboo, for improving rural economy. *Indian Farmers Digest.* 40 (12): 17-18.

3

Bastard Myrobalan (*Terminalia belerica* Gaertn. Roxb.)

Common name: Bahera

Family: Combretaceae

Introduction

Bahera or Baheda is a forest tree and also found outside the forests. This is the most useful tree of Ayurvedic System of Medicine and constitutes one of three components of "*Triphala*". Its leaves are used as fodder and wood as timber and fuel wood.

Bahera is a large sized tree found in deciduous forests throughout the greater part of India except arid regions of the country (Anonymous, 2002). Bahera is a common associate of sal (*Shorea robusta*) and teak (*Tectona grandis*) occurring more or less scattered. In the Indian peninsula, bahera is mostly found in the moist valleys. In Siwalik hills it rises upto 800 m altitude. Because of over-exploitation of bahera for fodder, timber and medicinal uses, it is disappearing from the forests areas (Singh, 2005). Owing to its declining population, the trees are now found scattered in Siwalik hills of Jammu, Himachal Pradesh, Uttrakhand, Haryana and Punjab. As all parts of bahera are used for medicinal purposes, so it becomes an imperative to know fully about its utility in different diseases (Anonymous, 1992, 2001).

Medicinal Uses

Fruit is the constitutent of famous Ayurvedic formulation "*Triphala*" and is used in innumerable diseases and body disorders. Fruit kernel possesses narcotic properties and is very effective in dyspepsia, dropsy and piles. Fruit pulp is used in diarrhoea, leprosy and in fever. Half-ripe fruit is a purgative. Fruit extract has anti-bacterial properties and used in hair problems. It is demulcent also and is effective in rheumatism.

The fruit of bahera is quite bitter and pungent. It has proved good for respiratory disorders and cures cough. It is also laxative and is effective in piles. The bark of bahera tree has been found diuretic and is very effective in urinary problems besides use in anaemia and leucoderma. The bahera has anti-mutagenic effects and is used in herbal digestive tonic for ruminants alongwith harar. Oil of bahera has been proved very effective in children suffering from oozing ears.

Other Uses

In the leather industry, bahera is used to give flexibility to the leather. Dyeing and ink production units in India use bahera tree. The timber is mainly used in house building as rafters, boards and also for packing cases. The timber is also found suitable for the manufacture of plywood and tea-chests.

Botanical Description

Bahera is a large deciduous forest tree species but is also found outside the forests throughout India. It reaches up to a height of 30-50 m with rounded crown and brownish grey bark. Leaves are alternate, large and broad elliptical. Flowers are small, simple, and solitary, in spikes, with offensive odor and appear in May; appear along with new leaves. The fruits are light-yellow, ovoid called drupes and the kernels are sweet (Orwa *et al.*, 2009). The nut is stony. Animals eat fruits and cause the seed dispersal. The fruit is collected for local use only and not used commercially. Its propagation is easy from seed, directly sown in the field or from nurseries stock.

Cultural Practices

Climate requirements: In the natural environment, where it shows its luxuriant growth, maximum temperature lie between 36 to 46° Celsius and minimum temperature varies from –1° to 15°. The normal rainfall varies from about 1000 mm to 3000 mm. In Jammu Siwaliks, it grows under sub-tropical climatic conditions especially in *Kandi* belt at an elevation between 300 to 800 m.

Soil requirements: Bahera can thrive in variety of soils having texture from loamy sand/sandy loam to loam/silty loam, provided moisture retention capacity is adequate. The best growth, however, has been noticed in sandy loam soils having adequate moisture and sub-soil drainage. Moderate to deep soils with medium to high organic matter and high water holding capacity are preferred.

Nursery Raising: Pre-sowing seed treatment has become an imperative to produce better results regarding seed germination and growth of seedlings (Anonymous, 2002). For this purpose, seeds are soaked in concentrated sulphuric acid (H_2SO_4) for 12 minutes.

Nursery sowing is done in the third and last week of March. For this, seed collected in winter is preferably used. Optimum spacing in the nursery is 20 cm 20 cm. Weeding and irrigation of nursery beds is necessary. The germination is almost completed in 30 to 60 days.

Planting techniques: Planting of bahera seedlings can be done in June-July, when the seedlings are about 3-4 month old to ensure good field survival. Planting of the seedlings is done with a ball of earth in pits of size 30 cm 30 cm 30 cm. Spacing adopted is normally 3 m 3 m.

Plant protection measures: The heart rot disease recorded on the bahera tree is caused by *Fomes fastuosus*. This disease is also caused by *F. senex* and *Ganoderma applanatum*. *Corticium salmonicolor* and *C. repens* cause canker and thread blight. *Puccinia terminaliae* causes woody galls on branches and leaves.

Harvesting and collection: The fruits ripen from November to March and fall soon after ripening. These are collected in second week of February to first week of March from the ground. The fruits are then dried before storage.

References

Anonymous. 1992. Forest Research Division, Jammu (Annual Report). Jammu and Kashmir Forest Department, India.

Anonymous. 2001. Social Forestry in Jammu and Kashmir. Directorate of Social Forestry, Jammu and Kashmir.

Anonymous. 2002. Package of Practices for Agroforestry Species. Directorate of Extension Education. Dr. Y.S. Parmar University of Horticulture and Forestry, Nauni, Solan (HP), India.

Orwa C, A. Mutua, Kindt, R., Jamnadass, R. and S. Anthony. 2009 Agroforestry Tree Data base: a tree reference and selection guide version 4.0.

http: //www.worldagroforestry.org/sites/treedbs/treedatabases.asp

Singh, S.S. 2005. Fate of the Medicinal Plant Resources of Jammu and Kashmir. Directorate of Environment and Remote Sensing, Jammu.

4

Bengal Quince (*Aegle marmelos* L. Corr. Serr)

Common name: Bil, bel, bael

Family: Rutaceae

Introduction

Bael is one of the indigenous fruit trees of India which is found throughout the country in each town and village (Jamwal, 2010). In West Bengal, Bael is known as Bengal Quince. Besides high nutritional value fruits, it also possesses medicinal values. All the parts of the Bael tree *viz.*, stem, bark, leaves, roots and fruits have important uses. The leaves of the Bael tree without any kind of damage are used to worship the Lord Shiva. It is believed that any person, offering leaves on the Lord Shiva's *Ling* (*Shivling*) is relieved from the sins which he has committed or commits.

Due to its hardy nature, it grows naturally under very adverse agro-climatic and soil conditions. The fruits of the tree can be kept for a longer period and can tolerate transport and marketing hazards. However, inspite of its so much nutritional, medicinal and religious importance, Bael has not been given due attention. Rather, it has been even neglected by many of the people. If this tree is cultivated scientifically it can pay good dividend to the farmers.

Medicinal Value

The bael tree is an important source of Ayurvedic medicines (Rikhye, 2013). Pulp of the Bael fruits is used as medicine against diarrhoea and dysentery as *sharbat*. It is also a good source of vitamin-A due to the presence of carotene (precursor of vitamin-A) and vitamin-C. As such, it is useful in curing night blindness and scurvy diseases, particularly in children.

Pulp also constitutes a rich source of vitamin-B_2 (riboflavin), the consumption of which improves loss of appetite and weight. Its use is also beneficial to the patients suffering from sore throat, cataract, swollen nose (Sharma and Pareek, 1993) as well as those affected with osteoporosis (loss of calcium). Fruit is valuable in habitual constipation, chronic dysentery and dyspepsia. Fruit rind is given in amoebic dysentery and unripe fruit is used in treating piles. Rind of the fruit is effective in vomiting. Recent studies indicate (Rikhye, 2013) that root of bael is part of the *Dasmula i.e.* 10 roots. The *Dasmula* denotes the roots of ten different trees which are integral to the formulation of any ayurvedic preparation and proves that its properties are widely used in any ayurvedic medicines.

Chemical Composition

Average nutrients present in 100 g^{-1} of edible portion of Bael, are presented in Table 4.1. which indicate that it is the richest source of carbohydrates having 31.8 g 100 g^{-1} of edible portion of Bael followed by protein 1.80 g, minerals 1.70 and fat 0.40 g 100 g^{-1}. Carotene, a precursor of vitamin-A has been found to the extent of 55 mg and vitamin C (8 mg 100 g^{-1}) of edible portion of Bael. Thiamine, riboflavin, niacin, are the other vitamins, present in the edible portion of the Bael.

Table 4.1: Average Nutrients Contents in Edible Portion of *Bael*

Sl.No.	*Constituents*	*Amount (per 100g)*
1.	Water	61.50 g
2.	Carbohydrates	31.80 g
3.	Proteins	1.80 g
4.	Fats	0.40 g
5.	Minerals	1.70 g
6.	Carotene	55.00 mg
7.	Thiamine	0.13 mg
8.	Riboflavin	1.19 mg
9.	Niacin	1.10 mg
10.	Vitamin-C	8.00 mg

Source: Khan *et al.*, 2001 and Mittal, 2005.

Religious Values

It is believed devotee of Lord Shiva, who mounts two or three offerings of leaves of Bael with strong and deep prayer, can get salvation from his sins.

Botanical Description

Bael is a food medicine from time immemorial in India. In Sanskrit it is called *Bilva* or *Shrifal*. It is the name of the fruit that appears yearly on bael tree. Full grown tree is large in size with thorny branches which are full of leaves. Every leaf is characterized by having three leaflets. Fruits are green in colour and after ripening they become yellow in colour and are very hard. Owing to the hard nature of the fruit, it is also called stone apple. Generally, bael tree has two varieties. One is wild with natural growth, and the second one is planted for cultivation under direct supervision. Its fruit has few seeds of golden colour with comparatively small size. Fully ripe fruit contains pulp that is sweet in taste and has its own delicacy for preparation of syrup with good flavour. Some people prepare preserve (Murba) out of pulp of the fruit (Mittal, 2005).

Cultural Practices

Climatic requirements: The bael tree is indigenous to the drier regions of India, extending to Burma (Rikhye, 2013). Tropical climate with hot and dry summer are ideal for its growth though it can grow well at very low temperature also *i.e.* to an extent of minus 70 degree Celsius (Khan *et al.*, 2001). Arid and semi-arid regions of the country with low rainfall, low atmospheric humidity, high vapour deficient, saline soil, poor water holding capacity have too been found suitable for the growth of Bael.

Soil requirements: Bael tree grows in a variety of soils. However, a well drained sandy loam soil is considered to be the best for Bael cultivation. But it can also grow in swampy, alkaline and stony soils, acidic and saline soils with pH ranging from 5 to 10. In lower Siwaliks of Jammu, Himachal Pradesh, Punjab and Haryana it occurs naturally in soils having pH range from 7.0 to 8.5, with free calcium carbonate content from 5 to 10 per cent.

Propagation: Bael is generally propagated by sowing the seeds (freshly extracted from the fruits) during the month of June-July. The seedlings which get ready after a year, are transplanted. However, plants raised by seedlings, start bearing fruits after 8 to 10 years, seldom produce a true to type plant. Hence, propagation should be done by budding. Patch budding has been found to be the best and highly successful when carried out in the month of June or July (Pushkar *et al.*, 2011).

Out Planting: The best time for out planting the seedlings is June to July but with assured irrigation facilities, planting can also be done during February to March.

Pits of 1 m 1 m 1 m size are dug one month prior to planting at a distance of 10 m apart. After exposing to direct sunlight for one month, these pits are refilled with 50 : 50 FYM and top soil to about 10 cm above ground level. Addition of 250 g single super phosphate and 100 g *neem* cake is made in each pits. Later on, pits are irrigated and when the soil becomes workable, the plants are placed in the pits and watered immediately. In alkali soils, the hard layer, if any, should be removed before planting and soil of each pit should be mixed with 5 kg of gypsum. For saline soils, the pit soil should be replaced with good fertile soil. For usar soil, 20-25 kg sand and 5-8 kg gypsum or pyrite should be mixed in the pit (Pushkar *et al.*, 2011)

Varieties: Mirzapuri, Rampuri, Khamaria Kagzi, Gonda and Faizabadi, are the few varieties recommended for cultivation. Varieties recently developed by GB Pant University of Agriculture and Technology, Pantnagar and ND University of Agriculture and Technology, Faizabad, are detailed in the Table 4.2.

Table 4.2: A Few Standard Varieties of Bael Evolved by GBPUAT, Pantnagar and NDUAT, Faizabad

Sl.No.	*Variety*	*Main Characteristics*
1.	Pant Shivani	It is a mid season, tall, vigrous and heavy bearing variety. The colour of fruit and pulp is lemon yellow with pleasant flavour and medium thin rind. Keeping quality of fruit is best.
2.	Pant Aparna	It is dwarf, precocious, almost thornless and heavy bearer. Fruits are very tasty with yellow in colour, globoid and thin rind with low ascorbic acid and is good for processing.
3.	Pant Urvashi	It is mid season, tall, vigrous and heavy bearing type. Fruits are ovoid-oblong.
4.	Pant Sujata	It is early mid season variety having a problem of splitting. Trees are medium dwarf, dense and heavy bearer. Fruits are globoid depressed at both ends.
5.	Narender Bael-5	It is a dwarf and spreading type of variety. Fruits are roundish, medium sized, thin rind with yellowish pulp and low seed content. Fruit is excellent in taste and flavour.
6.	Narender Bael-9	Trees of this variety are medium and spreading type, starts bearing at early stage. Fruits are elongated roundish with thin rind and pinkish colour. They are good in taste and are slightly acidic.

Source: Khan *et al.*, 2001 and Pushkar *et al.*, 2011

Manures and fertilizers: To get better growth of the Bael tree and fruits, 10 kg well decomposed FYM or compost (black-brown in colour and amorphous type) should be added along with urea (125 g), single superphosphate (150 g) and muriate of potash (80 g) per plant (one year old). The dose of urea, single superphosphate and muriate of potash should be increased with the same rate up to the age of 10 years (Table 4.3) and then at constant rate every year.

Half of the urea and full dose of single superphosphate and muriate of potash should be applied during March and the remaining half of the urea during July. Sufficient amount of FYM or compost should also be added during March in the tree basins.

Irrigation: Young plants require frequent irrigation but after being established irrigation is not needed. It is because Bael trees shed their leaves in summer and escape drought.

Training and pruning: To keep the plants straight and in a proper shape, they require to be supported with bamboo sticks. At a height of 1 meter, branches are permitted to develop freely. Any sprout below 1 meter and/or weak-diseased branch should be pruned-off. Old Bael trees can also be top worked in the month of June or July for good fruiting.

Table 4.3: Fertilizer Schedule for *Bael* Tree

Age of the Tree (Years)	*Amount of Fertilizers (g)*		
	Ureau Urea	*Single Superphosphate*	*Muriate of Potash*
1.	125	150	80
2.	250	300	160
3.	375	450	240
4.	500	600	320
5.	625	750	400
6.	750	900	480
7.	875	1050	560
8.	1000	1200	640
9.	1125	1350	720
10.	1250	1500	800

Source: Khan *et al.*, 2001

Intercultural operations: Young plants should be kept free from weeds through hand weeding. No additional interculture is required to the planted trees, even if inter-cropping has been followed. Any legume or forage crop can be easily taken as intercrop in the plantation during rainy season.

Plant protection: Bael trees being very hardy, are not susceptible to an assault of insect pests except lemon butter fly larvae and diseases. Lemon butter fly larvae can be controlled by spraying of botanical pesticide – *neem*.

Harvesting and yield: The fruits get ready for harvesting in the month of April to May. The fruits should be picked individually (when become yellow in colour) with a portion of stalk and should not be allowed to drop. Fruit stalk protects from infection, renders easy handling and gives an indication about the ripening as the stalks get easily detached from the fruits after ripening. A 10 to 15 years old tree may produce 200-400 fruits. However, a crop of 1000 to 1500 fruits per trees old trees, of 30-35 years, is not uncommon.

References

Jamwal, J. 2010. Uncommon but useful fruit plant. *Kashmir Times*. 67 (267): 12 (September 25)

Khan, F.A., Ahmad, I., Ram, D. and Khan, F.U. 2001. Grow Bael for more money. *Intensive Agriculture*. 39 (5-6) : 10-11.

Mittal, S. R. 2005. Bel-the great medicinal Food. *Ayurveda*. August Issue, pp.31.

Prasad, R.N. and Rathore, S.S. 1988. Fruit trees for agroforestry in arid region. *Farmers and Parliament*. 23 (10) : 11-12 and 24.

Pushkar, N. C., Chandra, R. and Singh, H. K. 2011. Bael- a medicine tree. *Indian Farmers Digest*.44(5): 16-18

Rikhye, D. 2013. Boundless benevolence of bael. *The Sunday Tribune Spectrum*. 133 9262): 5 (22 September)

Sharma, S. and Pareek, O.P. 1993. Horticulture paves the way to better days for them. *Intensive Agriculture*. 30 (11-12) : 8-13.

5

Birch
(*Betula utilis* D. Don)

Common name: Bhojpatar

Family: Betulaceae

Introduction

The tree is native to Himalayas growing at an elevation up to 4,500 m. In Jammu and Kashmir Himalayas, it grows at even high altitudes. It most commonly occurs in the Lidder Valley, Gurez and Kishan Ganga valleys, Amarnath Kolohoi and also in Leh and Kargil areas of Ladakh region of the Jammu and Kashmir. Besides having medicinal uses, it gives an excellent fuel, its dried leaves can be used while preparing *rice pulow* both sweet and salty.

Medicinal Values

Astringent in taste, high in action bhojpatar is hot in effect. Its taste after digestion in pungent and facifies all the three morbidities. Its bark is used as medicines in form of powder or decoction. In *Charak Samhita*, its use as local application in skin diseases and its preparation in ear diseases is indicated. When there is pus in ear, oozing out of the ear and deafness, oil prepared from bhojpatar with drugs like 'vidanga', 'nagarmothaetic', has proved beneficial. Bhojpatar is also useful in epilepsy, insanity, hysteria, convulsions. It lowers the effect of poisons and checks obesity. It also checks bleeding from nose and is useful in treating diarrhoea and dysentery. It is endowed with properties of healing wounds. When there is difficulty in expulsion of placenta, fumes of bhojpatar helps it (Anonymous, 2003).

Other Uses

Bark is used as covering, protection from evil effects of devil etc. Sages have been using its bark as a part of their lower garment. People visiting Kedarnath, Badrinath, Gangotri tie its bark under their feet, thus, making their journey comfortable. It has been considered auspicious due to its occurrence in that divine region. Its presence is auspicious as it is known to keep evil soul away.

Till the invention of paper, its bark was used for storing information in the written form. Its timber can be used for agricultural implements, thrashing sticks and fuel wood.

Botanical Description

Betula utilis is a medium sized, the only broadleaved angiosperm in the Himalayas which dominates extensive areas at sub-alpine altitudes. It frequently grows in the wide openings of the coniferous vegetation and underground shrubs like *Rhododendron* are common. It generally attains a height of 20 m. In higher altitudes tree remains stunted due to extreme wind pressure. The bark is soft, reddish brown, or white shining like a paper which peels off in thin stripes. Its heart wood is of red and white in colour. Leaves are ovate, 5 to 10 cm long, with serrated margins, and slightly hairy.Flowers are monoecious, with a few male catkins and short, solitary (sometimes paired) female catkins. Flowers start appearing in summer (April-May) which mature into fruits in autumn season (September–October). Seed has membranous wings.

Cultural Practices

Climatic requirements: Bhojpatar requires cool temperate climate but the sub-alpine type is ideal to get good growth.

Soil requirements: The plant prefers light, sandy, medium loamy and heavy clay soils. Well drained soil conditions are good for growth and development. The plant prefers slightly acidic to neutral soil reaction. However, it grows even in slightly alkaline soil reaction *i.e.* pH ranging from 7.0 to 8.5.

Artificial regeneration: It grows slowly in the nursery and will require at least two years in the nursery for being ready to be planted. Otherwise nursery treatment is the safe as for *B. alnoides*. In trials, it has been found that its growth is slow and survival often poor. It is quite evident from the data of plantation at Kharidunga, where its mean height was observed at 74 cm after 4-5 years.

The seed ripens between August and October, and is sown then, being left to stratify naturally over winter. The flowers bloom from May to June. They are pollinated by wind. Germination of fresh seed is 32 per cent and of stratified seed is 65 per cent.

Harvesting and yield: Harvesting of ripened seed is done during October-November.

References

Anonymous, 2003. Conservation and sustainable utilization of Medicinal Plants, Jammu and Kashmir State. Plan Proposal for Support under the PBFB Scheme of the Global Environmental Facility. Facility-GEF, State Level Planning Committee. Jammu and Kashmir Forest Department.

6

Blue Gum Tree (*Eucalyptus globulus* L. Her)

Common name: Eucalyptus, gum tree

Family: Myrtaceae

Introduction

In India, Eucalyptus is more than 200 years old. It is now grown in many states of the country. In Jammu and Kashmir *Eucalyptus citriodora*, the perfume Eucalyptus had been planted in Poonch, Kathua, Jammu early during Dogra regime (Singh, 2005). Though eucalyptus planting is economically profitable, yet it is not ecologically viable. Several species like *Eucalyptus globulus, E. territicornis etc.* are to be planted on account of following reasons:

(a) They are very fast growing species which can sustain in variety of climate.

(b) Within 5 to 7 years after planting, they yield outstanding and bumper production.

(c) They can produce good quality of firewood, planking wood and raw material for plywood, paper and pulp industry.

Apart from these, eucalyptus has a number of medicinal properties.

Medicinal Values

Eucalyptus leaves, bark and roots contain active principles of many drugs. Oil from leaves is a source of numerous downstream products. Together with its richness in many drugs and ease of cultivation, eucalyptus is a tree of the future and holds out high promise. Dried leaves, gum exudation of stem and oil distilled from fresh leaves are used. Eucalyptus wood, wool and saw dust is used as deodorants. Oil is used externally in catarrh of broncho-pulmonary mucous membrane. Its other preparations are given in intermittent and septic fevers, diphtheria, whooping cough and catarrhal infections of urino-genital organs, and oil applied in surgical wounds, ulcers.

Leaves, when chewed, perfume the breath and harden spongy and bleeding gums. Inhalation of hot infusion of leaves is highly beneficial in bronchitis where cough is constant with a free water and frothy expectoration and in nasal infections. Infusion is given internally in small doses (1/2 to 1 ounce) in ulceration of mouth and throat in children. Diptheria is treated effectively by disinfecting the air of patient's room by pouring steaming hot water on fresh leaves. Eucalyptus trees control malaria in the locality where they are growing.

Tincture of eucalyptus is effective in the treatment of the infection of the urethra, bladder, vagina, in chronic bowel complaints especially chronic dysentery. Eucalyptus gum is also effective in these ailments. Powder of dry leaves like that of neem, tulsi and ajwain is found effective in the problems of alimentary canal and also in chest troubles, cough etc.

Leaves are applied externally on chronic wounds resistant to other treatments. Fluid extract is largely used in cleaning foul ulcers and offensive discharges of all kinds. This fluid is also effective in skin diseases.

Other Uses

The oil which is obtained from the leaves, is widely used in soaps, perfumery and cosmetic industries. It is also used for the manufacture of synthetic menthol, hydroxy aitrenellol from the oil of *E. citriodora* which is used for the manufacture of high grade perfumes, soaps etc. and has export market in addition to internal demand (Anonymous, 2014).

Botanical Description

Eucalyptus has been found to be very fast growing evergreen species. It is characterized by having less foliage and small leaves and as such it possesses less CO_2 absorbing capacity than those of broad leaved species. Evapo-transpiration rate in eucalyptus leaves is comparatively slow and the moisture contribution to the atmosphere by eucalyptus plantation is rather poor as compared to broad leaved forest.

Stem is erect with bluish grey. The leaves are leathery, lanceolate, alternate, 30 cm in length and often waxy or glossy. Leaf glands contain a volatile oil used for variety of purposes like medicines, expectorants or inhalants. Flowers are solitary or 3 in umbels. Fruit is cup-shaped called capsule, 2.5 cm in length and with 4 cells having large number of seeds. The leaves are rich in phenolic compounds so the decomposition of the fallen leaves cannot enrich the soil (Roy, 1995). Due to presence of phenolic compounds the leaves are not a preferred fodder.

Cultural Practices

Climatic requirements: Eucalyptus species can sustain variety of climates *i.e.,* from sea coast to an elevation of 2500 m and in the regions which experience a rainfall of 750 mm to 3500 mm.

Soil requirements: Eucalyptus can be grown on wide variety of soils but its best growth can be achieved in Inceptisols or alfisols having sandy loam/loamy sand to silty loam/clay loam texture.

Planting time: Generally, eucalyptus can be planted at any time of the year *i.e.* in the mid spring season, or in the autumn season. However, the best planting time is considered to be the beginning of the rainy season (July to August).

Planting method: Always choose a right and an appropriate eucalyptus species for the area or the site. Select about 0.30 m to 0.45 m tall seedling or 1 year old seedlings for planting. Pits of 1m^3 are dug well in advance. Care should be taken not to break any tangled roots.

After care: Eucalyptus requires relatively little care after planting. Although many species are drought tolerant, once established must not be allowed to dry after planting. Watering once a week or once a day depending upon the consistency of the soil is required.

Mulching: Mulching is recommended if eucalyptus has been planted on a site where the soil is liable to dry up or freeze or become infested with a number of weeds. If the climate is mild there is no need of mulching.

To shun the weed growth including the grasses, mulch the eucalyptus plantations heavily *i.e.* about 7.5-15.0 cm thick. Weeds and grasses should not be allowed to grow up around the base of trees as it will slow its growth by competing for nutrients and moisture.

A variety of materials can be used as suitable mulch consisting of bark of trees, wood chips, leaves and grass clippings. Compost, manure and other products containing nitrogen are generally less suitable because weeds will grow up in them more readily.

Pruning: Generally species do not require to be pruned. However, they are very responsive to pruning and will re-sprout vigorously from very large branches. A large proportion of foliage can be removed from the tree if necessary.

Harvesting and yield: Within five to seven years after plantating, it can give an outstanding and bumper production. Even if the trees are planted at 1x1m they are expected to give about 100 tonnes of dry wood ha^{-1} or even more at the age of 10 years (Roy,1995).

References

Anonymous. 2014. Agri Expo: Chemical Catalogue. Jhansi castel 7, Cooperage Road Bombay (Mumbai) – 400039, India (Personnel Communication).

Roy, B.1995. Eucalyptus and biomonopoly – A blessing to economy but a threat to ecology. *Intensive Agriculture*. 33(1-2): 1-5.

Singh, S. 2005. Fate of the Medicinal Plant Resources of Jammu and Kashmir. Directorate of Environment and Remote Sensing, Jammu, J&K State.

7

Chinar
(*Platanus orientalis* L.)

Common name: Chinar

Family: Platanaceae

Introduction

The trees whether exist inside or outside the forests play a pivotal role not only in the overall development, progress, agricultural and industrial production and prosperity of a state/ country but also maintain its salubrious climate, greenery, beauty and keep the environment free from pollution. These species occur in pure groups, occasionally in scattered, sporadic or mixed form.

The chinar is a well noted tree of Kashmir valley with great historic significance whose spell binding charm has been immortalized in exquisite prose and poetry over so many centuries. In Kashmir, during autumn, the foliage of the chinar tree turn into a crimson blaze alluring the heart of a lover but provoking imagination of a poet. Thus, chinar renders the summer refreshing experience and winter a visual happiness. "The chinar or *Boin* (Kashmiri name) is a royal tree and as a shade tree the chinar is unrivalled," writes Walter R. Lawrence about the tree - "With spreading branches, broad leaves and thin bark, which comes off in flakes." The tree lends its name from Persian language, meaning bronze colour.

The chinar tree is valued for its wood, besides medicinal values (anti-scorbutic and anti-rheumatic properties).

Medicinal Values and Symbolic Identity

The bark of the tree has anti-scorbutic and anti-rheumatic properties. It is used in diarrhoea, dysentery, hernia and tooth ache. Fresh leaves are brushed and applied to the eyes in case of ophthalmia. Chemically, the bark contains 1.5 per cent platanin, 7.9 per cent tannin and 7.3 per cent non-tannins. The roots of the chinar tree contain 60 per cent phlobanphene and the shoots and leaves have allactoin and the sap of the tree contains upto 90 per cent mannitol.

Economic Value

The species is valued for its wood which is fine-grained, moderately hard and heavy (Gupta, 2003). People utilise its twigs and bark for *hookah* smoking. Due to immense green foliage, the chinar has a good ecological value and checks pollution. The wood is also used for making small boxes, trays and similar articles which are lacquered and painted. It is used as firewood and in charcoal making from fallen leaves for use in fire pots (*kangris*) during winter (Sagwal, 2002). A kangri full of charcoal can last for 10-12 hours (Bhat, 2009). It is a cheap source of warmth to conquer cold and is a part of tradition. The wood is used for cabinet making, furniture, veneers, carving, coach-building and pulpwood. Place of chinar leaf impression in wood carving, papier-mache and earthenware holds the glory of Kashmir craft industry (Bhat, 2009). Chinar leaf in fact, has acquired a symbolic identity and genuineness for different arts and crafts of Kashmir. It is artistically printed and pasted with thread work in different garments-shawls, namdas, carpets and gabbas.

Distribution

In Kashmir valley, this majestic tree is found in whole of the Kashmir region, covering all the districts *viz.*, Srinagar, Pulwama, Anantnag, Badgam, Baramulla and Kupwara. The groves of chinars and poplars are almost present in every village, town and the capital city. The Chinar trees can also be seen along the banks of the river Jhelum, in gardens, parks, temples, compounds of shrines, mosques and also in the graveyards.

Chinar tree has been growing in other parts of the Jammu and Kashmir and some parts of India but the climate of Kashmir suits well for its growth and development. However, these days the famed chinar trees are fast dwindling in number. Nearly 20 per cent of the chinar trees have already become extinct (Gupta, 2005) in Kashmir valley. Their number which stood at 42,000 during 1970, has reduced just to 20,986 during 2007 (Anonymous, 2007). In the year 2009 the number of chinar trees further dwindled to about 17,396. Out of these chinar trees, 7,000 were 400 years old, while 10,000 were below 300 years old (Bhat, 2009).

Botanical Description

Chinar is a gigantic tree (Yusuf, 2010). Its height and circumference varies according to its age. For instance, the height of 635 years old chinar tree was found to be 14.78 m and circumference 31.85m (Bhat, 2009). The tree had replaced the previous largest one at Bijbehara which was 19.70 m wide and 13.3 m tall. The tree is very hardy as it does not call for any after care once it establishes its root system. Another safeguard is that cattle do not relish the chinar green foliage and as such no damage is rendered to the young trees. The robust tree does not freely allow the growth of other vegetation under its canopy. Being deciduous, its leaves are naturally shed during autumn at intervals. During rains it yields huge quantity of leaf mass.

Cultural Practices

Climatic requirements: The handsome chinar tree growing upto 30 m height and 12 m girth, requires sub-temperate to temperate climate for its best growth and development, though can be grown in sub-tropical and tropical climates also.

Soil requirements: It can be grown on a variety of soils but the best soil for its growth as found in Kashmir, is loam to silt loam/clay loam rich in organic matter and other nutrients. The soil must be well drained. Stagnant water will prove injurious to the roots of chinar tree. In sub-tropical or tropical climates, it can be grown even in sandy loam soils provided they are rich in organic matter including other plant nutrients.

Planting time: Being deciduous nature, the best planting time is February to March, although it can be planted during rainy season (July-August), but lot of care is required.

Protection of young plantations: For an effective management, the following impact points must be taken into consideration:-

(*a*) The soil around the tree must be kept well hoed during summer.

(*b*) Either the tree be guarded using barbed wire or thorny branches in the initial stages of its growth.

(*c*) Train the tree to a modified leader system so as to retain its majesty.

Disease and pests: The chinar tree has been subjected to several diseases and insect pests. For this purposes, the following measures need to be followed:-

(*a*) The diseases like damping-off may be severe and can be prevented by using fungicide captan.

(*b*) Defoliating insects can be controlled with lindane at 10-day intervals whenever newly hatched larvae appear.

References

Anonymous. 2007. Save Chinar. *The Daily Excelsior*. 47 (27): 6 (September, 19).

Bhat, D. 2009. Chinars of Kashmir. *The Kashmir Times*. 66 (108): 4 (April, 19).

Fazili, F.E. 1994. The charming chinar. *The Kashmir Times* [Sunday Times, December

18, 1994], pp 9.

Gupta, R.D. 2003. Chinar tree in trouble. *The Daily Excelsior*. 39 (109) : 1 [April, 20].

Gupta, R.D. 2005. *Environmental Degradation of Jammu and Kashmir Himalayas and their Control*. Associated Publishing Company, Karol Bagh, New Delhi.

Parvaiz, A. 2009. Kashmir's majestic chinars under threat. *The Kashmir Times*. 66(94): 1 (April, 19).

Sagwal, S.S. 2002. Chinar calls for care (personnel communication). Sher-e-Kashmir University of Agricultural Sciences and Technology, Srinagar, J&K (India).

Singh, J.B. 1995. The chinar of Kashmir. Haryali – Quarterly Newsletter. April-June, 1995, Inaugural Issue, pp. 22.

Yusuf, G. I. 2010. Amid decline of chinar trees, pledge made for more plantation. *The Kashmir Times*. 67(80): 7 (March, 22)

8

Cinnamon (*Cinnamomum zeylanicum* Blume)

Common name: Dalchini

Family: Lauraceae

Introduction

Cinnamon is a small, evergreen tree which is native to Sri Lanka. In spite of its exotic distant origin, it is known and widely used in ancient world. The Arabs were the first to introduce to the West and dominated the trade for centuries through their networking trading routes as far as China. Their account of where about and how cinnamon was obtained proves that ex-market techniques were very old. Cinnamon serves as a spice used in traditional medicine, by herbalists and has been a part of healing remedies for centuries (Khosla, 2013).

Medicinal Values

Cinnamon is aromatic, carminative, stimulant and anti-microbial in actions. Stimulates the urinary tract and can be used for problems of kidneys and urinary reactions. It is also used for cough and congestion of the respiratory tract. In Ayurvedic medicines, cinnamon oil is used in external applications for rheumatism, aching joints and stiffness. It is also used for toothache and sore gums like clove oil. In India, it is used at the first sign of cold to prevent it from taking hold fully (Luthra, 2006).

The essential oil of cinnamon has anti-coagulant properties which help to thin the blood and improve the blood circulation. The anti-microbial action helps to preserve food and can be used in place of food preservatives. It is a warming aromatic tonic that helps in improving the digestion and also assists in reducing cholesterol.

Recently, circulatory stimulant effects of cinnamon have been reported (Khosla, 2013) in plants and Ayurveda. It helps in the reduction of total and bad cholesterol (LDL) and increase in good cholesterol (HLD).

Cinnamon bark is widely used as a spice as a condiment and flavouring agent. Leaves contain 'eugenol' and used as a substitute for cloves. Cinnamon powder is used to enhance flavour of drinks, and sweets. Cinnamon is a warming aromatic tonic that stimulates the digestive system and can help reduce cholesterol and blood sugar levels. It makes an essential ingredient of Indian spice tea which is considered to have aphrodisiac properties.

Botanical Description

Cinnamon trees are small, tropical evergreen 10-15m tall with aromatic bark and leaves. The bark is reddish brown and smooth. Leaves have reddish appearance when young but they darken to deep green at maturity. Leaves are ovate-oblong in shape, 7-18 cm long with a pointed tip, dark green on the upper surface, lighter below. The flowers which are arranged in panicles, yellowish-white in colour have distinct odour. The fruit is a drupe, bluish purple, 1 cm long containing a single seed.

Cultural Practices

Climatic requirements: This crop can be grown up to an altitude of 800-1000 m receiving an annual rainfall of 1600 to 2500 mm.

Soil requirements: It grows on wide variety of soils. However, sandy, lateritic soils with high humus content are the most suitable for its production.

Propagation: June-December is considered to be the most optimum season for propagating this crop. Propagation can be done through seeds collected from selected mother trees and are sown immediately in nursery beds in rows of 12cm apart. July-August is the best season for sowing. Seedlings from beds, are then transplanted to polythene bags where they attain a height of 15-20 cm.

Dig pits of 60 cm at 2m x3m spacing for field planting. Fill the pits with top soil and FYM @10kg. One year old seedlings or rooted cuttings are transplanted under partial shade.

Irrigation: During summer protective watering (once or twice) has proved to be beneficial.

Manures and fertilizers: Well decomposed FYM or compost should be added @ 2 kg per plant and the dose is increased every year @ 2 kg per plant up to 10 years. Thereafter, the dose of the manure becomes constant *i.e.* 20 kg. If this process is continued, the application of chemical fertilizers can be avoided.

Harvesting and yield: Harvesting starts form 4th or 5th year after planting. The shoots are cut during May and again in November. As soon as rains are ceased, cutting of shoots for peeling of bark is started. After cutting, young shoots sprout from the stumps which are again ready for harvesting after about 18 months. The bark is peeled from the selected shoots of 18-24 months old by scraping off the outer bark, then beating the branch evenly with a hammer to loosen the inner bark. The stems must be processed immediately after harvesting while the inner bark is still wet. The outer, woody portion is discarded. About 35 kg of leaf oil is obtained ha^{-1} $year^{-1}$. Besides this, 100-200g of dried bark is also obtained.

References

Khosla, I. 2013. Spices of life. *The Tribune* 133(276): 3.

Luthra, P.K. 2006. Medicinal Properties of 50 trees of India. *Gram Vikas Jyoti*. July-September Issue, pp. 63-68.

9

Cutch Tree (*Acacia catechu* Willd.)

Common Name: Katha

Family: Mimosaceae

Introduction

Khair is one of the commonest trees of the tropical Himalayas, especially in Jammu region of Jammu and Kashmir State, Himachal Pradesh, Uttrakhand and sub-mountainous areas of Punjab, Haryana and Uttar Pradesh. However, during the past 4-5 decades, it has now become an uncommon tree, owing to over exploitation for obtaining katha. Together with grazing and browsing as well as other biotic pressure, the growing stocks are declining drastically. But khair tree is still seen everywhere up to 1300 m altitude.

Khair is a tree of Ayurveda and finds use in large number of preparations although its bulk use is in the manufacture of catechu. All parts of this tree like roots, bark, stems, fruits, have many medicinal properties.

Medicinal Values

Catechu (*Katha*) has long been used in Indian System of Medicines. It is astringent, produces cooling effect and is digestive. As a home remedy, it is used for bad throat, mouth and gum problems, and in cough and diarrhoea. Its paste/ decoction is applied externally in ulcers, boils and skin eruptions. It is a good food additive and as an excellent natural colouring material. Its gum is a good substitute for true gum obtained from *Acacia senegal* and is used in mascatories.

Botanical Description

In Hindi, it is called 'Katha', in English, 'Catechu' and in Sanskrit it is known as 'Khadira'. Under severe biotic pressure, khair tree has lost gregarious regenerations. Existing stocks have been over exploited for manufacture of katha. The traditional katha based cottage industry is exhausted. The sustained supply of khair wood for cottage industry is affected by ruthless mass felling thereby endangering their existence if the species is not given protection against biotic interference.

It is a small to moderate sized, deciduous tree reaching 3–15 m with short spines, hooked in pairs. The stem is dark brown to black, with fissured bark peeling off in long strips in mature trees (Orwa *et al.*, 2009). Leaves small (1-2 cm), bipinnate with 60-100 leaflets. Leaves are shed in Feburary -March and new flush appears in end of April or during May. Flowers appearing along with new leaves are small, pale yellow or white in cylindrical axillary spikes. Pods are straight, strap-shaped, green in colour which turn brown at maturity and usually with 4-7 seeds. Pods dehisce on maturity and seed dispersal is through winds. Tree has a strong tap root.

Cultural Practices

Climatic requirements: Like other species of *Accacia* such as *Accacia modesta* and *Accacia nilotica,* Khair also requires sub-tropical to tropical climate. Sub-mountainous areas of Siwaliks having rainfall 1000-1200 mm, such as *Kandi* belt of Jammu, Himachal Pradesh, Punjab, Haryana, Uttar Pradesh and Uttarakhand have been found to be the best for its growth.

Soil requirements: Khair prefers to grow on varied types of soils but its best growth occurs in sandy loam to loam soils provided these are well drained. It can even be grown in saline and calcareous soils.

Artificial Regeneration: Artificial reproduction is the process of renewal of the forest species by sowing, planting, or other artificial means to get the best growth of the forest crop. Planting refers to the transplanting of saplings in the area to be regenerated, through nursery raised seedlings. Khair can be raised by direct sowing and planting out stumps (Mohsin, 2004).

Direct Sowing: Seed of khair is sown in the month of June-July in patches or strips, with soil worked up to 15 cm depth in advance. Patches are about 45cm x 45cm, made at a spacing of 2.5m x 2.5m. Strips are 45 cm wide, spaced 2.5m apart. About 50 seeds are sown m^{-2} of patch or strip area. Seed rate for patch sowing is 0.5 kg ha^{-1} and 2.0 kg ha^{-1} for strip sowing.

Planting out: Another method of raising khair is by planting out seedlings raised in polybags, about 3-4 months old and 20-30cm tall. Planting is done in 30 cm^3 pits, at a spacing of 2.5 m x 2.5 m, if the plantation is along the field bunds, the spacing is increased to 4-5m.

Planting of polybag raised seedlings is preferable to direct sowing or planting out bare rooted plants. Sowing is done in February-March, usually 2 seeds in each container filled with a mixture of sieved soil and compost in the ratio of 2:1. Regular weeding and watering is done. Germination begins in about a week and is complete in a fortnight. In weeding, one seedling is ultimately retained in each container.

Stump planting: Another method is by planting stumps or root and shoot cuttings, which is least common of all methods. For making stumps, 3-4 month old nursery seedlings reach suitable size. 7 cm of shoot, 25cm of root and 2cm (minimum) thick at collar, make ideal stumps.

Yield: The yield of the cutch tree varies according to the sites where it is grown. Good sites produce 6 tonnes wood ha^{-1}, moderate 5 tonnes and poor 4 tonnes ha^{-1}.

References

Mohsin, F. 2004. Artificial reproduction of important agroforestry trees. *Indian Framers Digest*. 37 (12): 5-8.

Orwa C, A. Mutua, Kindt, R., Jamnadass, R. and S. Anthony. 2009 Agroforestry Tree Data base: a tree reference and selection guide version 4.0.

http: //www.worldagroforestry.org/sites/treedbs/treedatabases.asp.

10

Drumstick Tree (*Moringa oleifera* Lam.)

Common name: Sowanjana

Family: Moringaceae

Introduction

Drumstick is a low to medium high tenuous tree with yellowish white flowers and fernlike divided leaves. All parts of the tree *viz.*, tender leaves, roots, flowers and fruits are consumable. It is grown as a hedge or a single tree. The tree provides light shade and is useful as its sticks serve for supporting vegetable climbers.

Although the tree grows all over India but in South India it is a common tree in many of the home gardens. The Royal Commonwealth Society for the Blind, discovered that the leaves of this tree could well be used as a medication against blindness. Use of its leaves saved about 60,000 children in India from going blind owing to malnutrition, particularly those suffering from Vitamin A deficiency (Gupta, 2003).

The leaves of drumstick tree are not very popular as a market vegetable but their chemical composition reveals that these are rich source of Vitamins A and C. Not only this, crushed seeds of mature dry pods of the drumstick tree when added to a pot of water, serve as a natural coagulant and purifies the water. Central Institute of

Medicinal and Aromatic Plants, Lucknow, has recently isolated a compound from the pods called "Niaziridin" having properties to enhance absorption of drugs.

Medicinal Values

All the plant parts *i.e.* leaves, flowers, fruits, roots, seeds, bark and gum have enormous medicinal values. The chemical composition of the drumstick leaves as shown in Tables 10.1–10.3 indicate that leaves of the drumstick tree are not only a good source of calories and proteins but are also rich in ascorbic acid (Vitamin C), nicotinic acid and carotene-precursor of Vitamin A as well as calcium and iron.

Table 10.1: Composition of the Drumstick Leaves of Edible Portion (100 g^{-1})

Sl.No.	*Component*	*g/mg*
1.	Dry matter	22.4 g
2.	Calories	72.0 g
3.	Proteins	7.4 g
4.	Fibre	1.2 g
5.	Calcium	29.5 mg
6.	Iron	3.6 mg
7.	Carotene	8.9 mg
8.	Ascorbic acid	16.5 mg

Source : Melogi, 1989.

Table 10.2: Nutrient Content of Leaves and Pods (100^{-1} g leaf or pod)

Sl.No.	*Component*	*Leaves*	*Pods*
1.	Carbohydrates (g)	13.4	3.7
2.	Protein (g)	2.5	6.8
3.	Fat (g)	1.7	1.7
4.	Fibre (g)	0.9	4.8
5.	Calories (g)	92.0	26.0
6.	Calcium (mg)	440.0	30.0
7.	Magnesium (mg)	24.0	24.0
8.	Potassium (mg)	259.0	259.0
9.	Phosphorus (mg)	70.0	110.0
10.	Copper (mg)	1.1	3.1
11.	Iron (mg)	7.0	5.3
12.	Mineral salt (mg)	2.3	2.0
13.	Oxalic acid (mg)	101.0	101.0
14.	Water (per cent)	75.0	86.6

Source : Tewari *et al.*, 2004.

Table 10.3: Nutrient Content of Moringa Cake

Sl.No.	*Nutrient*	*Amount (per cent)*
1.	Crude protein	58.93
2.	Phosphoric acid	1.09
3.	Potash	0.80
4.	Lime (CaO)	0.40

Source : Tewari *et al.*, 2004.

It has been found that a cupful fresh juice of drumstick leaves provides as much as Vitamin A as one gets by eating 9 eggs or half kg butter or 80 cups of cow's milk (Melogi, 1989). Likewise, it supplies as much as Vitamin C as 6 oranges or 16 bananas or 8 apples.

A cupful of fresh leaf juice is heated on a water bath and coagulated and clear fluid obtained from this, is filtered. It is then mixed with milk and sugar and fed to infants and children once a day to develop strong bones and resistance against malnutrition.

Soup of the leaves not only prevents malnutrition but also checks anaemia, seasonal attacks of colds, asthma, loss of strength and sterility.

Fresh leaf juice, mixed with lime juice applied on pimples can cure them and keep skin fresh for a long time. If taken with honey can cure night blindness. Roasted leaves are used to reduce pain and inflammation caused by injuries.

Curry prepared from fresh leaves is good for pregnant and lactating women because it increases milk production during lactation. It acts as a very good tonic for persons suffering from allergies. The leaves are also used to reduce the infection caused by the mumps. Drumstick is used in Indian folk medicines for the treatment of various illnesses (Mallick and Kumari, 2011).

The paste of leaves is used on wounds and inflammation. The bark of the plant is good for chest diseases, bites and as cardiac circulatory stimulants.

Chemical Composition

All parts of the tree are used in the treatments of ascites, rheumatism, venomous. Estrogenic substances and pectin-esterase have also been reported to be present both in the leaves and pods. Leaf also contains essential amino acids like arginine, histidine, lysine, trypatophane, leucine, isoleucine, valine, phenylalanine, methionine and threonine. Leaf juice shows strong antibacterial properties. The roots contain an active antibiotic "pterygospermin" which inhibits the growth of many Gram positive and Gram negative bacteria. In higher concentration it is active against fungi.

The root bark contains two alkaloids *viz.*, moringine and moringinine. The former is identical to benzylamine in effects. The latter serves as symptomatic nerve endings, produce a rise in blood pressure, acceleration of heart beat and constriction of blood vessel. It also inhibits the tone and movements of involuntary muscles of gastro-intestinal tract and relaxes bronchioles. An alkaloid named spirochin in roots, in high doses, paralyses the vague nerve. The root bark contains traces of an essential oil with pungent smell. The flowers contain trace of alkaloid and the ash is rich in potassium and calcium. The seeds are considered antipyretic, acrid and bitter. Oil from seeds is traded as *Ben* or *Behen,* applied in rheumatism and gout. Oil is edible besides its use in illumination and cosmetics.

Application of 'Niaziridin' molecule isolated from the drumstick tree: As stated earlier that the Central Institute of Medicinal and Aromatic Plants, Lucknow has filed a patent application for newly discovered molecule called "Niaziridin". This molecule has been isolated from the pods of the drumstick tree and has the capacity to enhance the absorption of a number of drugs. Such drugs mostly consist of antibiotics such as rifampicin, tetracycline and ampilicin through the gastro-intestinal membrane and nutrients by humans (it augments the uptake of nutrients like Vitamin B_{12}). This molecule is a bio-enhancer that does not possess drug activity of its own but promotes the uptake in combination therapy. It can be proved useful for the treatment of cancer, tuberculosis and various other ailments.

Other Uses

The wood can be used as timber for making toys, floats and sandwich material. The timber is, however, of little use in carpentry. Bark yields coarse fibre which is used for making poor quality ropes and cordage (Tewari *et al.,* 2004). Twigs are used as fodder, leaves are palatable particularly to camel and domesticated animals.

Traditional Methods of Utilization

In rural dwellings, the tender leaves are frequently used for preparing a number of dishes *i.e.* as vegetable, mixed with pulses and seasoned. In South India, the leaves are mixed with pigeon pea, combined with other leafy vegetable to prepare *sambhar*. Soup and curry can be prepared from the leaves. The leaves can also be mixed with wheat flour to prepare *chapatis*.

In South Indian homes, the young fresh leaves of drumstick tree are used to prepare *chutney* after mixing with red chillies and tamarind (*imli*).

A number of value added products can be prepared from the seeds/pods of the drumstick plant *e.g.* dehydrated moringa, moringa powder, moringa flesh, meso-carp powder and moringa or *sohanjana* pickle. Tender drumstick leaves after chopping make an excellent garnish to use for any vegetable dishes, daals, sambars, salad etc. (Jamwal, 2010).

Botanical Description

The drumstick tree commonly known as *"Sohanjana", "Sohanjane Di Mooli", "Sohanjan Di Phali"* or *"Sohanjane Di Jadh"*, is a small sized prennial plant. *Sahajan, Muring, Muger, Mumigi, Ganigana, Shiguru, Shobanjan,* are the other local names of the drumstick tree. If it is grown through seed it becomes a short duration annual plant, but if it is raised through stem cuttings, it is a perennial tree.

This species is mainly found to occur in semi-arid, tropical and sub-tropical areas. The plant is medium in size, deciduous with sparse foliage, easily recognizable when in flowering and fruiting. It reached to a height of 10-30 m, with bole crooked and generally forked at the base. Crown usually wide and open and is a deep rooted species. Leaves compound, alternate, with opposite pinnae, bearing leaflets in pairs, dark green above and pale on the under surface. Flowers produced in loose axillary panicles in March-April are pinkish -white and produced throughout the year; very fragrant. Fruits large, with distinctive long pods of size between 25 cm to 120 cm, constricted, gradually tapering to end points. On maturity, pods split to disperse rounded blackish seeds (about 20 in number), with papery wings.

Cultural Practices

Climatic requirements: The drumstick tree requires tropical/sub-tropical climatic conditions. In sub-tropical region of Jammu and Kashmir, Punjab, Haryana, Uttrakhand and Himachal, it occurs both in irrigated and un-irrigated sub-mountain areas.

Soil requirements: In South India, the red lateritic and black cotton soils have been found suitable for the growth of drumstick trees. Although soil having pH range of 6.5 to 7.5 has been found to be the optimum for the growth but it can be grown under moderately acidic (pH 5.5 - 6.0) and alkaline (pH 8.0 - 8.5) soil conditions also.

Sowing time and method of sowing: June to July and November to December are the two best seasons for the cultivation of the drumstick. For annual crop, it is sown in pits through seeds, having size 45 cm 45 cm 45 cm at row to row and plant to plant spacing of 2.5 m 2.5 m. Seed rate is 500 g ha^{-1}. The pits are excavated about one month earlier prior to sowing and are filled with 15 kg of FYM pit^{-1}. Usually two seeds are sown in the centre of the pit at 3 cm depth. The seeds germinate with in 6 - 10 days after sowing.

For perennial crop, 1000 stem cuttings per hectare are required which are planted in the pits, having each 1 m 1 m 1 m size, with row to row and plant to plant, spacing of 3.5 m 3.5 m. One year old stems are used for propagation. Stem cuttings are taken from branches of 1 - 1.35 m long and 14 - 16 cm in girth. The cuttings are planted in pits.

Manures and fertilizers: Generally, organic manures *i.e.*, farm yard manure and vermicompost are applied at the rate of 75 kg and 25 kg plant^{-1} upto the age of 1 year or more. Before 1 year of age, *i.e.* 3 months, the plants are supplied with 100 kg of urea and 35 kg of diammonium phosphate and 50 kg of muriate of potash ha^{-1} (1000 plants).

Irrigation: It depends upon the needs of the crop. Sowing and out-planting are the most critical stages.

Intercultural operations: One weeding/hoeing is provided to the crop, one month after its sowing. This will keep the field free from weeds. However, when the plants attain a height of about 1 m, there is no need of intercultural operation.

Harvesting and yield: The plants start bearing 200 fruits year^{-1}. The ratoon crop is obtained by pruning the trees to a height of 1 m after harvesting. The ratoons begin to bear after 4 to 5 months. 3–5 ratoons can be taken depending upon the variety of the drumstick plant.

Yield: On an average, the yield of drumstick tree varies from 32-40 quintals ha^{-1}. However, with the application of organic fertilizers and organic pesticides, the yield of the tree has been increased using modern technology from 32-40 quintals to 72-80 quintals ha^{-1}.

References

Gupta, R.D. 2003. Drumstick tree disappearing in Jammu. The Daily Excelsior. The Sunday Magazine, August 10, 2003, pp. 4.

Jamwal, J. 2010. *Moringa oleifera*- an other useful plant. The Kashmir times. 67(288): 13 (Saturday, October 16).

Mallick, M.A and Kumari, B. 2011. Recent advances in Indian herbal drug research: Heptoprotective herbal drugs. *Agrobios Newsletter*. 10 95): 46-48.

Melogi, U. 1989. Leave it to drumstick leaves. *Home Science*. 27 (6 to 12) : 27 and 31.

Sharma, K.K., Sambyal, K.S., Sharma, N.K. and Gandotra, S. 2006. Agronomical practices for the cultivation of high demand plant species of medicinal and economic importance. Directorate of Common Area Development, Talab Tillo, Jammu, Jammu and Kashmir State.

Tewari, S.K., Pandey, A. and Kumar, V. 2004. Moringa : A useful agroforestry tree. *Indian Farmers Digest*. 37 (12) : 13-14.

11

Flame of the Forest (*Butea monosperma* Lam. Taub.)

Common name: Dhak, palas, palasha

Family: Leguminosae, sub-family: Fabaceae

Introduction

Palasha is considered to be the embodiment of the absolute Brahaman, and hence called the Brahmavriksha. The Skanda Purana says, "Just as the Brahman is at the apex of the caste system, the Brahmavriksha is the best among all the trees". Palasha is very dear to Lord Shiva who pervades it in his various forms and moods. He is present at its roots as Hara (The seizer), on its stem as Shuladhara (The trident –bearing one), in its leaves as Shiva (The auspicious), in its fruits as his son Ganesha or Ganapati, in its bark as the Lord of the river Ganga, in the pith morrow of the tree as the Lord of feeling and in its twig as the Lord of all Gods (Ishvara).

In light of the above, there would be no denying fact if palasha is described as the seed of the Devas or Devtas, the bestower of desires and destructive of great sins. Taking palasha leaf as food is regarded to be very auspicious and provides an advantage of thousand horse sacrifices (Ashvamedha Yagnas). Even the sight of the palasha tree early in the morning is believed to have a spiritual significance as it saves one from going to hell. The worship of the tree on Thursday in Vhaturmasya blesses with virtue and ends the miseries of life. Apart from religious significance as stated above, Palasha possesses many medicinal properties.

Medicinal Values

Palasha is medicinal plant of great importance. It is a potential source of valuable drugs. Root, bark and seed as well as flowers and leaves have medicinal uses. When externally applied, the pulp of leaves checks bleeding. Its poultice is used to cure boils, pimples, swellings and dhobi's itch; and its bark to alleviate abdominal tumors and bleeding piles. The gum of the Palasha tree contains tannins and is valuable for treatment of diarrhoea. The seeds are used as anthelmintic in the treatment of round worms and tapeworms. Their usefulness in treatment of roundworms has been confirmed through experiments (Kapoor, 2006). The root bark has been found to have some action on blood pressure.

Palasha is considered to represent immense beauty because of the yellow flowers which grow on the tree in bunches. The tree looks very attractive from a distance. Due to this quality, the tree is also called 'Flame of the Forest'. Flowers boiled in water, made into a fine paste and applied on the lower part of the belly to cure problems of stones. In case of acute pain due to the presence of stones or painful urination, this paste is highly beneficial (Mohindra, 2010).

Other Uses

The tree is valued as a host tree for lac insect. The flowers yield a yellow dye. The wood lasts well under water and is used for well curbs and water scoops. It is also used for minor construction work and as fuel. Leaves are largely used for making bowls (*dauna*) and plates (*pattals*) for serving food, and as fodder during lean periods. In Jammu region, the trees are used in several Hindu rituals.

Distribution

Tree is a native to tropical and sub-tropical parts of the Indian sub-continent and South-East Asia and across the adjoining nations of Paksitan, Nepal, Srilanka and Indonesia. Palasha is also a very common tree of sub-mountains area of Siwaliks known as *kandi* belt of Jammu, Punjab, Haryana, Himachal Pradesh and Uttrakhand. Besides pure crop it occurs in mixture with other tree species and also forms extensive patches on the slopes of Siwaliks ranges.

Botanical Description

Palasha, also called "Flame of the Forest", is found to occur in drier parts most often in open grassland areas upto 1200m. In sanskrit, Palsha is called 'Kinshuka' which means parrot like, referring to beak like keel of the flower and, therefore, this tree is also named as Parrot tree. In Dogri dialect, it is called as Palasha and name of village "Pallawnwalla" has been named after widespread occurrence of this tree in the area. Its Hindi name is 'Dhaak' or 'Palash', and in Tamil its known by 'Parasu' and in Malayali as 'Murku' or 'Chhamita' or 'Brahmavriksha'.

It is a medium sized, deciduous tree about 10-15 m in height, reddish gum-containing bark and crooked trunk. The tree with tri-foliate leaflets representing Hindu Triad with Lord Brahma on the left, Lord Vishnu in the middle and Lord Shiva on the right as has been stated in the Vedas (Vidyarthi, 2010). Flowers in racemes, bright orange-red, more rarely yellow, with dense pubescence. The tree usually starts flowering in the mid February but sometimes the flowers appear in early April especially in 'Kandi' belt of Jammu, Punjab and Himachal Pradesh (Anonymous, 2011). Fruits are large, indehiscent pods, pale yellowish-brown or grey when ripe. Fruit contains a single seed near the apex, ellipsoid, flattened, about 3 cm long. Leaves are shed during the dry season i.e. in December-January and appear after flowering.

Cultural Practices

Climatic requirements: Palasha grows very well under tropical to sub-tropical localities. It also grows in drylands as is seen in the *kandi* belt of Jammu.

Soil requirements: It grows on wide variety of soils. However, its best growth can be seen in loamy sand to sandy loam soils but it also occurs on gravelly to stony soils.

Natural regeneration: Dhak or Palasha regenerates naturally with great ease. However, presently natural regeneration of this tree has been very much affected as a result of biotic pressure, otherwise it regenerates well from seed and root suckers. Habitats of palasha are being encroached by agricultural activities and human settlements.

References

Anonymous 2011. The beauty of the bastard tree. *The Daily Excelsior* 47 (134): 4 (15th May, Sunday Magazine)

Kapoor, S. K. 2006. Palasha-The Brahmavriksha of the Puranas. *Ayurveda.* December Issue pp. 12-13.

Mohindra, K. 2010. Floral remedies. *Ayurveda.* 1(18): 14-15.

Vidyarthi, O.P. Sharma 2010. Fascinating Tree. *The Daily Excelsior* 46 (349): 4 (19th December, In *Sunday Magazine*).

12

Golden Shower Tree (*Cassia fistula* L.)

Common name: Amaltas, karangal

Family: Fabaceae

Introduction

Although amaltas is found throughout India upto an altitude of 1300m but is more common in most of the evergreen forests (Chauhan, 2006). It is also a common tree of the Siwaliks rising to 1300m in the outer Himalayas but nowhere grows gregariously or in pure patches (Singh, 2005). The tree is much favoured for planting in road side avenues and gardens. Amaltas is one of the most important medicinal trees of Ayurveda and pharmacopoeias.

Medicinal Values

Amaltas is used asemetic, febrifuge, laxative, reliever of thoracic obstructions and purgative. It is useful in boils, leprosy, ringworms, flatulence, dyspepsia, constipation, indigestion, fever and heart diseases.

Botanical Description

Amaltas is a medium sized tree nearly 15m tall. Leaves pinnately compound; leaflets are 4-8 pairs large, ovate-lanceolate, leathery, entire acute. Flowers are bright yellow appearing like Japanese lantern in long and drooping racemes. Fruit is 20-40 cm x3cm cylindrical pod, dark brown with strong wood stalk having a blunt tip and seeds are embedded in black, sweetish pulp (Chauhan, 2006).

Cultural Practices

Climatic requirements: Amaltas is found throughout India having tropical to sub-tropical climate. However, its best growth occurs in hot sub-humid tropical to sub-temperate areas of Himachal Pradesh, Jammu and Kashmir, Punjab, Uttrakhand and Uttar Pradesh.

Soil requirements: Amaltas prefers light porous soil and is immune from browsing by cattle and goat. It can also be grown in gravelly sandy loam or loamy sand soils and is found in kandi belt of Jammu having such kinds of soils.

Method of sowing: The best method of its cultivation is by sowing seed in furrows prepared 5.0-12.5cm apart and 10-15cm deep (Chauhan, 2006). The furrow is closed by another furrow ploughed alongside. Broadcast sowing by close ploughing in one direction costs about twice but gives 2.7 times more seedlings than the first method.

Intercultural operations: Thinning and weeding is done during the first season of its cultivation.

Harvesting: The tree has good coppicing ability. The coppiced buds send out large number of shoots and new harvest can be taken after a year.

Pod yield of amaltas varies from 1200-1600 kg ha^{-1}.

References

Chauhan, N.S. 2006. Medicinal and aromatic plants of Himachal Pradesh. 92nd Edition. Indus Publishing company FS-5 Tagore Garden, New Delhi-110027.

Singh, S.S. 2005. Fate of the Medicinal Plant Resources of Jammu and Kashmir. Directorate of Environment and Remote Sensing, Jammu.

13

Gum Arabic Tree [*Vachellia nilotica* (L) P.J.H. Hurter Mabb.] Syn. (i) *Acacia nilotica*, (ii) *Acacia arabica*

Common name: Kikar, babool

Family: Fabaceace

Introduction

Babul, a leguminous tree is native of western portion of the Indo-Gangetic plains and northern part of Deccan Plateau. In India, it is widely found in almost all the states and union territories except in northern-eastern states, Kashmir valley and Kerala (Deswal *et al.*, 2001).

In arid regions of Rajasthan, Haryana, Andhra Pradesh, Karnataka and other parts of the country babul is rich and excellent source of fuel wood. It is also used for charcoal making. Babul in fact is, a multipurpose tree and has many economic uses besides being a medicinal plant.

Medicinal Uses

Bark of the tree can be used for healing wounds, strengthening the teeth and treating stomach troubles. It is also used against eye troubles, throat infection and liver complaints (Singh, 2005). In stomach troubles like diarrhoea, dysentery and other stomach disorders bark powder is used. Bark of the tree is also found effective in diabetes. Decoction of bark is effective in sore throat as gargle. It is highly effective in urino-genital troubles. Seed is also found to be demulcent and possesses anti-viral properties.

It is worthwhile to mention that babul is a tree of Ayurveda and many Pharmacoepias as all its parts are medicinal in nature. Bark yields gum, used for many purposes. It is a famous tree for bark tannin and sustains tanneries though not in Jammu and Kashmir State.

Other Uses

As the charcoal of babul is superior to the charcoal from other tree species so in many parts of Punjab, Haryana and Uttar Pradesh its fuel wood is largely used in brick kilns. Leaves and pods of babul are an excellent fodder for goats and camels in Rajasthan and Haryana. Even the seeds of babul are eaten after roasting or raw at the time of acute food scarcity in Rajasthan due to its richness of food nutrients which are mentioned in Table 13.1.

Table 13.1: Nutritional Content of *Babul* Seeds

Sl.No.	*Nutrient*	*Amount per cent, Mg 100 g^{-1}*
1.	Crude Protein	26.4 per cent
2.	Crude Fibre	2.7 per cent
3.	Calcium	63.0 mg
4.	Phosphorus (P)	420.0 mg
5.	Iron (Fe)	4.9 mg
6.	Niacin	3.1 mg
7.	Asorbic acid	4.5 mg
8.	Thiamin	0.2 mg

Source : Deswal *et al.*, 2001.

Tanning material is obtained from the bark and pods of the tree when dried. Gum obtained having medicinal properties, is known as 'Indian Gum Arabic'. However, its gum is considered inferior to the gum obtained from *Acacia catechu* (Khair) or *Acacia senegal*.

Variety of dyes from yellow to black through brown can be obtained from leaves, pods, bark and wood extract depending upon their varying proportions. Branches of the babul tree have good use as fencing material around the field boundaries because they possess spines. It can be raised as live hedge and hardy avenue tree also.

Botanical Description

Babul is an evergreen, moderate sized tree (height varies from 2.5 to 25.0 m) with spreading crown. It is a spiny tree with bi-pinnate leaves and leaflets in the range of 10-20 pairs. Flowers are in globe heads. Babul is not a gregarious tree but comes

up naturally in good quality unless kept down in good browsing.

Even though its status in natural vegetation of Jammu is doubtful, the tree occurs naturally in the plains, Kandi belt and Siwaliks rising up to 800 to 1200m. It is not a very gregarious tree as Phulai (*Acacia modesta*) but it is vigorous, hardy and regenerates well naturally.

Babul is very promising species for arid region plantation on common lands, homesteads, under agroforestry and social forestry (including strip plantation, along roadsides, railway lines and canals) as it grows very fast, fixes atmospheric nitrogen, improving soil fertility and its physical and chemical proprieties by the addition of organic matter through litter fall (Deswal *et al.*2001).

Cultural Practices

Climatic requirements: The native babul is fast growing and good for natural plantations in semi arid/arid climatic regions. It grows quickly during the rainy season but can survive long drought conditions. Babul grows best in areas, where annual rainfall is more than 200 mm *i.e.* 200-1270 mm (Deswal *et al.*2001; Krishan and Toky, 1995). Babul can even survive when gets well established, with annual rainfall as low as 75 mm.

Soil requirements: Although babul can be successfully grown on marginal lands yet it grows best on the alluvial soils in riverine areas. It can also be grown successfully on salt affected soils or on black cotton soils. The existence of even saline water in the subsoil is not injurious. It has also very high resistance against oxygen deficiency in the soil and is therefore, suitable for cultivation in swampy and stiff soil.

Propagation: Natural conditions is optimum and Babul tree regenerates if seeds get sufficient moisture during rainy season, but for its artificial plantation nursery raised seedlings are the most suitable planting material.

Nursery sowing : As seeds of babul have a hard seed coat and remain dormant for a long time, so seeds must be soaked in hot water (80° Celsius) for about half an hour for better germination. These seeds are then sown in the nursery in polythene bags filled with soil and well decomposed organic manure (compost or FYM) in 1:1 ratio. Two to three seeds should be sown in each bag during February or March.

Outplanting of the seedlings: After 8-14 weeks, the seedlings grown in the polythene bags *i.e.* before their roots begin to coil, become ready for out planting. These are then planted in pits of 30 cm^3 size at spacing of 3m x 3m or 4m x 4m depending upon the purpose. Under agroforestry plantation, spacing may be increased to 5m x 5m to 6m x 6m for raising agricultural crops.

Interculture and irrigation: Proper weeding and cleaning, and 2-3 life saving irrigations are required during first year of plantation for better growth of the Babul and its establishment.

Economics and market value: Babul is a versatile tree with multifarious use, is liked by the people for planting on their field boundaries, common lands as its every part has direct use for them. It is also being planted on large scale under agroforestry/social forestry programme. This species is of utmost importance for the people of arid regions to meet their requirement for fuel wood, fodder and timber.

Annual yield of babul is commonly reported to be 35 to 39 tonnes of wood ha^{-1}.

References

Deswal, R.P.S, Dhankhar, R.S. and Singh, B. 2001. Babul-A valuable tree for farm forestry plantation in dry areas. *Intensive Agriculture* 39(3-4): 16-17

Krishan, B and Tokey, O.P. 1995. Babul-A perfect tree. *Intensive Agriculture* 33(3-4): 18-20

Singh, S.S. 2005. Fate of the Medicinal Plant Resources of Jammu and Kashmir. Directorate of Environment and Remote Sensing, Jammu, Govt. of J&K.

14

Indian Beech Tree [*Millettia pinnata* (L.) Panigrahi] Syn. *Pongamia pinnata* L.

Common name: Karanj, pongam, sukhchain

Family: Leguminoceae

Introduction

The natural distribution of pongam is along coasts and river banks in India and Burma. This species is native to Asian sub-continent and has been introduced to humid tropical lowland in the Philippines, Malaya, Australia, United States and Indonesia (Anonymous, 2011). Pongam has variety of uses. Apart from using its wood for making cart wheels, cabinet making, agricultural implements, tool handles and combs, a thick yellow orange to brown coloured oil extracted from the seeds, has lot of medicinal properties.

Medicinal Values

Karanj oil is known to have value in folk medicine for the treatment of rheumatism, as well as human and animal skin diseases. It is effective in enhancing the pigmentation of skin affected by leucoderma or scabies.

Other Uses

The leaves of the karanj are fed to cattle and readily consumed by goats. However, in some areas, it is eaten by farm animals. Its fodder value is greatest in arid regions. Its oil cake is used as a poultry feed and as organic manure.

Dried leaves are used as insect repellent in stored grains. The oil cake, when applied to the soil, has pesticidal value, particularly against nematodes besides improving soil fertility.

Botanical Description

Pongam is a medium sized tree that generally attains a height of about 8m and a trunk diameter of more than 50 cm. The trunk is generally short with thick branches, spreading into a dense hemispherical crown of dark green leaves. The bark is grey to brown and yellow inside.

The compound pinnate leaves consist of 5 or 7 leaflets which are arranged in 2 or 3 pairs. Flowers borne on racemes are pink, light purple or white. Pods are elliptical, thick walled and usually contain a single seed. Seeds are 10-20 cm long, oblong and light brown in colour.

Cultural Practices

Climatic requirements: Pongam thrives in areas having an annual rainfall ranging from 500 to 2500 mm and climate ranging from humid to sub-tropical in its natural habitat. This species grows to elevations of 1200m, but in the Himalayan foot hills it is not found above 600 m elevation.

Soil requirements: Pongam can grow on all types of soils, ranging from stony to sandy to clay including Vertisols. It does not do well on dry sandy soils but is highly tolerant to salinity.

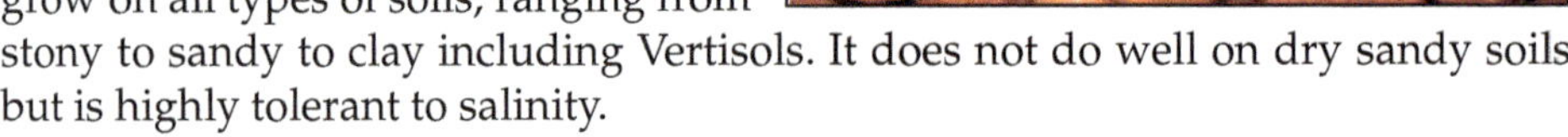

Propagation: Pongam is easily established by direct seeding or by planting nursery raised seedlings or stump cuttings of 1-2 cm root collar diameter.

In Peninsular India, the sowing season is April to June. Seedlings attain a height of 25-30 cm in their first growing season. Transplanting is done at the beginning of the next rainy season when seedlings are 60 cm in height. Seedlings have large root systems so soil should be retained around the roots during transplantation. The spacing adopted in avenue plantings is about 8 m. In block planting the spacing can range from 2x2m to 5x5m.

Yield: The seed yield per tree ranges from about 10 to 50 Kg. There are nearly 1500 to 1700 seeds per kg.

References

Anonymous. 2011. Karanj (*Pongamia pinnata*). In know your trees. *Wastelands News.* 24(3), July-September Issue, pp.16-17

15

Indian Black Berry (*Syzygium cumini* L.) Syn. *Eugenia jambolana* Lam.

Common name: Jamun

Family: Myrtaceae

Introduction

Jamun (Jamoon) is an indigenous minor fruit of Indian origin. According to Hindu tradition, Lord Rama used to subsist on this fruit along with his consort Goddess Sita and brother Lakshman during fourteen years of exile. Because of this mythological episode, many Hindus regard Jamun as a fruit of the Gods, especially in the Gujarat. It is a common tree of the Indian Siwalik hills and plains of North-Western Himalayan states *viz.*, Jammu region of Jammu and Kashmir, Punjab, Haryana, Himachal Pradesh and Uttrakhand and parts of Uttar Pradesh. It comes up naturally through dispersal of seed by wild animals and birds. It occurs along with other trees in the forests and outside but nowhere has it formed a dominant or a pure patch. Nowadays, Jamun is also planted under social forestry/farm or agroforestry.

It is remarkable to note that about half a century ago, jamun fruit was used to be sold in the market, not in the shops but on the pavement, but this practice has now vanished. It is an important ingredient of many ayurvedic preparations and has a potential for drugs; yet to be exploited (Singh, 2005; Jamwal, 2012).

Medicinal Uses

Jamun fruit is universally accepted to cure diabetes because of its effect on the pancreas. The fruit and the seed possess glucoside and the jamboline and ellagic acid which are reported to have the ability to check conversion of starch into glucose in case of excess production of sugar. Jamun fruit is also a remedy for bleeding piles and liver disorder.

Jamun is believed to prevent diarrhoea, pharyngitis and splenopathy. Processed products like jambolan vinegar can be administered in case of enlarged spleen, chronic diarrhoea and urine retention. The seed is used in various alternative systems of medicine like Ayurveda, Unani and Chinese. Bark and leaves of this plant also have various medicinal properties like control of blood pressure, gingivitis etc. The leaves of jamun have also anti-bacterial properties and as such are used for strengthening teeth and gums. Studies have also indicated that they help in the prevention of cancer, arthritis and heart disease (Anonymous, 2012).

Nutritional Values

Fruit contains a fair amount of minerals such as calcium and potassium, vitamins like vitamin-B complex, vitamin-C, free sugars-glucose, mannose and sucrose and aminoacids-alanine, arginine and asparagine. The purple colour of the fruit is due to the presence of anthocyanin pigment. The major anthocyanins are delphinidin and palunidin. Malvidin, peonidin and cyanidin are also present in small proportion. Three esters *viz.*, dihydro myricetin acetate, geranyl butyrate and terpinyl valerate have been identified which are responsible for the characteristic flavour of the jamun. Astringency of the fruit is due to the presence of the phenolic compounds. Jamun has a significant oxidant activity compared to other fruits like sapota, papaya, banana and guava. The anti-oxidant activity is attributed to the occurrence of phenolic acids and anthocyanins. The other nutritional constituents of the fruit are resin, albumin, gallic acid, tannic acid and essential oil. These compounds prevent heart disease, cancer and combat aging. It also prevents age related memory loss. Apart from Vitamin-C, the blue berries also carry a good amount of iron, potassium and foliate (Anjali *et al.*, 2014).

Jamun (blackberry) is a good source of carotene and folic acid. Apart from being rich in mineral salts, it has many medicinal properties. When consumed regularly it helps to control diabetes. Its bark is used to cure urinary disorders, de-worming and also in herbal medicines by traditional healers. It is rich in calcium, magnesium, phosphorus, sodium and vitamin C.

Ripe fruits are consumed fresh with salt or processed into a refreshing drink for hot summers. Special character of the blue berries is that they make the urine acidic and destroy the harmful bacteria if present in the gall bladder. If the blue berries are eaten in excess, they darken the stool colour (Sawhney, 2005). It should not be confused with intestinal bleeding. The blue colour of the berries is because of flavonoides besides anthocyanins.

Botanical Description

Jamun or Jamoon is known as Blueberry or Blackberry in English and Jambu in Sanskrit. It is a fast growing, tall, evergreen tree (25-30 m), with coarse bark near the base and is grown as a shady tree/and or as wind break on road sides and as avenues. Crown is irregular or globular with dense foliage. Leaves are opposite, simple, entire, elliptic to broadly oblong, smooth and glossy. Flowers are white to pinkish, appear in branched clusters at stem tips. The tree produces large quantity of fruit during May to July. The fruit is oblong, ovoid, one seeded berry, starts greening and turns purplish red, pink to shining crimson black as it matures.

A variant of the tree produces white coloured fruits.

Cultural Practices

Climatic requirements: In India, the maximum number of Jamun trees is found scattered throughout the tropical and sub-tropical climatic regions. It also occurs in the lower ranges of the Himalayas upto an elevation of 1300m and in the Kumaon hills up to 1600m (Yumnam and Swamy, 2010)

Soil requirements: Jamun can be grown on wide variety of soils. But its best growth is seen in soils having sandy loam and loam texture. The soils must be well drained and pH in the range of neutral to slightly alkaline.

Propagation: It is naturally propagated by seeds. Since Jamun is highly cross pollinated crop, so wide variability exists in this species, though it is propagated through seeds. Jamun offers opportunity for establishing clones by simple seedling selection. Plus trees or elite trees selected for total yield, size and quality of fruit etc. are the starting point of improvement programme. Selection of a large number of elite trees serves as immediate seed source with fairly high genetic gain.

Jamun is propagated vegetatively by softwood grafting, wedge grafting and T-budding etc. It has been found that propagation of jamun by softwood grafting gave 97.5 per cent success during June and 50.0 per cent during August.

Harvesting and Yield: The ripe fruits are available during summer and before the onset of the rainy season. During ripening fruits change their colour from green to purple or bluish black. Fully ripe fruits are harvested daity either by hand picking or by shaking the branches and collecting them on polythene sheets or large sized cloth piece.

A fully grown tree yields about 80-100 kg of fruits.

References

Anjali, K.B., Akshay, K.R. and Sudharani, N. 2014. Medicinal importance of Jamun. *Agrobios Newsletter*, (4): 38-9.

Annonymous. 2012. Why black berry is good for you. *The Kashmir Times*. 69(117): 12 (27th April).

Jamwal, J. 2012. Jamun the medicinal plant. *The Kashmir Times*. 69 (22): 9 (Sunday Magazine, Jammu

Sawhney, D. 2005. Flavours of the mouth. *Ayurveda*. August Issue pp., 15-16.

Singh, S. S. 2005. Fate of the Medicinal Plant Resources of Jammu and Kashmir Directorate of Environment and Remote Sensing, Jammu.

Yumnam, S.S. and Swamy, G.S.K. 2010. The Jamun: A versatile under exploited fruit tree. *Agrobios News-letter*. 9(6): 24.

16

Indian Cherry (*Cordia dichotoma* L.)

Common name: Lasura

Family: Boraginaceae

Introduction

Lasura is commercially cultivated in Chittorgarh, Ajmer, Sirohi, Bhilwara, Alwar, Pali and Jaipur districts of Rajasthan; Guna, Datia, Neemach, districts of Madhya Pradesh; Rewari in Haryana and Vadodara of Gujarat. It is a income generator to the low and mid hill farmers of Himachal Pradesh, where it is a popular vegetable, especially in Kangra, Una, Bilaspur, Hamirpur and Mandi districts and sold on high rates during summer-rainy season. The sub-mountain areas of Siwaliks of Jammu, locally known as *kandi belt,* are very popular for cultivation of lasura as well as local mango, garna, phalsa etc. The well known areas famous for cultivation of lasura were Chinor, Raipur, Kote, Bhalwal, Amb Ghrota in Bhalwal block of Jammu district. Similarly lasura also grow a in Raya, Raya Suchani, Patti, areas of Samba. But now days, lasura has almost got vanished owing to over exploitation. Over exploitation of the lasura is attributed to use of all its parts for various purposes.

The tender fruits of lasura are mostly used as vegetable. It also yields high quality wood and possesses antifungal and larvicidal compound cordiaquinones. Unripe fresh fruits are acrid and used for pickle and curry besides as vegetable. Leaves are also used for preparing vegetable in India. The trees serve as a source of fuel wood and fruits have medicinal values.

Medicinal Values

The fruits of the plant contain saponins, flavonoids, proteins, carbohydrates including gum, oleic acid, linoleic acid, palmitic and stearic acids vis-à-vis arachidic and behenic acids. Its fruits are used as cooling, astringent, diuretic, aphrodisiac, emollient, expectorant, anti-worm, anti-diarrhoeal, anti-helmenthic, purgative and also as a liver tonic.

Seeds of the species are anti-inflammatory. Two compounds; alpha-amyrin and 5- dirhamnoside have been isolated from the seeds. The bark is medicinal and several chemicals have been identified such as allantoin, beta-sitosterol and 3, 5-dihydroxy-4 methoxy flavonones-7-o-alpha-L-rhamnopyranoside. The fruit is eaten to suppress cough and chest complaints and to treat soar throat, as it has demulcent properties. The pulp is also applied as an emollient to mature abscesses, to calm rheumatic pain and as an anthelmintic. In Tanzania, the fruit pulp is applied on ring worm. Its leaves are applied to wounds and ulcers. Several preparations of *lasura* have been used in traditional medicines and oesto-articular diseases.

The nutritive value of lasura indicates that it is very good source of carbohydrates, mineral matter and proteins.

Botanical Description

Lasura, also known as cherry of the desert, is a small to moderate sized deciduous tree available throughout India and other warmer regions of the world. It is a tree with a short trunk having spreading crown. Flowers are short stalked, bisexual and white in colour, appear in loose corymbose cymes. Leaves of lasura are simple, broad, ovate, alternate and 3-5 nerved. Young leaves tend to be hairy. The fruits are drupaceous, 2.5 to 5.0 cm long having round or pear shape, which attain yellowish brown to light pink colour on ripening. When fully ripe, central stony part is surrounded by sweetish mucilaginous pulp which is fully enjoyed by the young ones especially children. The average weight of the fruit is 5 g and pulp content is 69.9 per cent. The hard stone is 1- 4 seeded (Singh *et al.*, 2012).

Cultural Practices

Climatic requirements: Lasura thrives well under tropical and sub-tropical climate up to an elevation of 200 to 1500 m. It requires warmer climate and is susceptible to frost. But can tolerate temperature as low as 1 degree Celsius for few days. It can tolerate temperature as high as 48 to 50 degree Celsius during summer months, particularly in arid region of Rajasthan. It thrives well in areas receiving an average annual rainfall of 250-300 mm.

Soil requirements: Plant are very hardy and can grow in almost all kinds of soils. It can thrive well on neglected, less fertile, sandy, swampy, saline and alkaline soils. However, water logged soil is not suitable for its cultivation.

Propagation: Propagation is through freshly harvested seeds sown directly into containers, beds or trays and pricked out when the first pair of true leaves have formed. Germination is epigeous. One or two seedling may appear from single stone. Sowing is done in June to July at a depth of 2 cm in lines spaced about 20 cm apart. A seed rate of 80 g m^{-2} of nursery area is required. Germination starts in about 3-4 weeks and completes in 6 weeks.

Seedlings are ready for planting in the field after one year with the commencement of monsoon rains. Flowering takes place from March to May with the new leaves. The old leaves are shed during winter and the trees are leafless for short period in early summer. Fruits are formed soon after flowering, develop quickly and ripen from June to August in North India and normally before May in South India.

Manure and fertilizers: Application of well rotten FYM before flowering and nitrogenous, phosphatic and potassic fertilizers after pruning and fruit set in split doses promote better vegetative growth and fruiting of the trees. To achieve this goal, application of 100 g N, 125g P_2O_5 and 50 g K_2O per tree should be done besides 10 kg FYM per tree during monsoon season. Application of 20-30 kg FYM and 1 kg diammonium phosphate to 5 years old plant is found beneficial to increase fruit yield in semi-arid areas.

Irrigation: Through, lasura is a hardy plant but watering is necessary to young plants after planting at an interval of about 15-20 days during winter and 8-12 days during summer. Monsoon planting requires very little water and depends on weather conditions. When plantation becomes one or two years old generally watering is not required but if watered, plants show vigorous growth and gives higher yield. Irrigation should be provided during active growth period of plants, particularly during the period of fruit growth. Hoeing and weeding of pits and provision of mulching is recommended to check the loss of water through evaporation from the soil.

Pests and diseases: A large number of insect pests are reported, defoliators being the most important, which must be controlled using neem based organic insecticides.

Harvesting and yield: Tree starts production after 4-5 years of planting. Fruits are ready for harvesting by the middle of May and they should be harvested at a tender stage when fruit colour is still green. Maturity indices for ripe fruit are decided by turning of colour from green to yellow pink, sweetness of pulp and hardening of seed. Fruits mostly ripen between May- July in arid/semi-arid areas but in tropical condition the crop is comparatively early than sub-tropical conditions. Yield varies with the age of the tree, climate and management practice. Young plants produce 5-10 kg green fruit per tree while a full grown tree yields nearly 50 kg fruits per tree.

References

Singh, Y., Gopal, G., Sharma, M.and Bhardwaj, A. 2012. Lasura (*Cordia dichotoma*)-A multipurpose tree vegetable for nutrition, health and economy. *Agrobios News letter*. 11(4) : 27 – 28.

17

Indian Liliac
(*Azadirachta indica* J. Juss.)

Common name: Nim, neem

Family: Meliaceae

Introduction

A multipurpose neem tree endowed with power to control pests is the base for various medicines extensively used in the indigenous systems of medicines like Ayurveda and Unani. The name *Azadiractha indica* is derived from the Persian, *"Azad darakht-i-Hind"* literally meaning free tree of India, evincing that neem is a well noted tree of India and rather a native to the Indian sub-continent. In India and across the world, it is called by various names (Table 17.1). Neem has a history which goes back thousands of years and has been identified on 5000 years old seals of the Indus Valley civilization (Husain, 2013).

The medicinal, insecticidal and religious significance of neem tree is well enshrined in Hindu mythology. In fact, neem was the most celebrated medicinal tree of India and its mention has been found in a number of *Puranic* texts like Atharava Veda, Amartkosha, Ghrysutra etc. They all have mentioned the outstanding qualities of the neem tree as a source of medicines and natural pesticide. In Sanskrit, neem is called 'sarvaroga nibarini', which means it is the cure of all ailments.

In India, millions of people have been cleansing their teeth with neem twigs for centuries, curing skin disorders with neem leaf juice, taking neem tea as a tonic, and keeping neem leaves in their beds if suffering from small pox and grain bins, cupboards and silk/warm clothes to do away with various kinds of insects.

Villagers too use this tree to get relief from body pains, fevers, infections and other complaints and they call it "Village Pharmacy".

Table 17.1: Common Names of *Neem*

Sl.No.	Language	Name
	India	
1.	Sanskrit	*Nimba, Nimbou, Arishtha* (reliever of sickness)
2.	Hindi	*Neem* or *Nimb*
3.	Urdu	*Nim* or *Neem*
4.	Tamil/Telugu/Malayam	*Vepa, Veppan, Vembu*
5.	Bengali	*Nim*
6.	Marathi/Gujarati	*Limba, Limbro, Nimba*
7.	Punjab	*Neem*
8.	Dogri	*Nim*
9.	Oriya	*Nim*
10.	Kannada	*Bevu*
	World	
1.	Engish	*Neem,* Indian Lilac
2.	French	*Azadira-d-Inde, Margousier, Azadira*
3.	Portugese	*Margosa* (Goa)
4.	Spanish	*Nim Margosa*
5.	Germany or German	*Niembaum*
6.	Burmese	*Tamar, Tamarkha*
7.	Sindhi	*Nimmi*
8.	Sri Lanka	*Kohomba*
9.	Indonesia	*Mindi*
10.	Nigeria	*Dongoyaro*
11.	Persian	*Nib, Azad Darakht-i-Hind*
12.	Malaya	*Veppa*
13.	Singapore	*Kohumba, Nimba*

Source: Koul, 1993 and Koul, 2002, Tikoo, 2004.

Medicinal Values

Almost all the parts of the tree, such as leaves, flowers, fruits, seeds, twigs, bark and roots are utilized as medicines, pesticides, mosquito repellent, fertilizer, diabetic food or animal feed.

Ayurveda doctors prescribe numerous therapeutic values/applications of neem for skin disorders, antidotes of poisons, anti-dermatatic and antiseptic, healing of wounds and sores. Use of mixture of neem seeds and sesame for ulcers, neem juice with salt for intestinal worms and neem juice with honey for jaundice is a common practice.

Decoction of neem leaves alone or blended with honey or other soothing herbs are used to cure eczema and other skin rashes besides dermatitis. Neem concentrate is used to treat fungus infections.

Bath in neem water protects against heat and controls rashes and boils especially those suffering from measles/chicken pox. Dried leaves are mixed in food grains to control storage pests. This was the common practice in the villages of Jammu and Kathua districts of Jammu and Kashmir State before 1947 but this simple method slowly disappeared on the advent of chemical insecticides. Dried leaves used to be mixed with mud for making grain storage bins and burning, dried leaves drive away the mosquitoes. Leaves can also be used to cure vaginal itches, uterine inflammations and white discharge in case of women.

In villages, and now even in towns and cities, the people use fresh neem twigs (*datoon*) to cleanse their teeth, particularly in pyorrhea. Neem based tooth pastes are also available in the market.

Neem's bark is a bitter tonic, astringent and antipyretic. It also breaks the periodic sequence of fevers like malaria. The gum yielded by the tree is effective in curing chronic skin infections, leprosy and general debility.

Grind the flowers and make a fine paste, apply this on boils and wounds to reduce burning sensation and heat (Mohindra, 2010).

Oil (margosa) extracted from seeds is anti-helminthic and effective against ring worms, thread worms and scabies. It is best for lice and dandruff. Neem seed oil contains oleic, palmitic and stearic acids and also nimbosterol and tannin. Margosa oil has amazing antiseptic properties as well. Nimbosterol (dried) mixed with honey and black pepper powder can cure colds, stop bleeding and help a patient suffering from piles. *Neembola* capsules from neem oil have been recommended for curing diabetes. The neem oil is said to assist longevity, protection against heart diseases, high blood pressure and arthritis. Besides, it has ingredients which lower down the cholesterol and clear the arteries of fat. Neem oil is also useful for vaginal itches and uterine inflammations. Intra-vaginal application of 1 ml of purified neem oil is claimed to be an effective contraceptive and is the base of an Indian product called "Sensal". Common medicinal uses of neem have been listed in Table 17.2.

A recent study conducted in New Jersey has indicated that extracts from neem tree, profuse in tropical and sub-tropical areas, may thwart the HIV virus from multiplying (Anonymous, 2012).

Insecticidal Values

Although almost all parts of the neem tree account for biologically active compounds yet the maximum insecticidal activity has been found in the seed kernel. These products are systemic in action and have been found to be effective in controlling over 300 insect species. Apart from these, its various products are also responsible for controlling host of nematodes and other plant pathogens. Neem derivatives exert behavioural, physiological and biological effects on insects.

Table 17.2: Common Medicinal Uses of *Neem* in Health Care

Disease	*Part(s) of Neem to be Used for Curing the Disease*
Hypertension and diabetes	Chewing of leaves (8-12) in the morning for about a month protects the body from these diseases.
Dermatitis and eczema	Decoction of leaves alone or blended with honey cures eczema, skin rashes and dermatitis.
Pyorrhea and other dental disorders	Brushing with neem twigs makes gums strong and prevents pyorrhea. Also protects teeth from decay and relieves of pain.
Vaginal itches, uterine inflammations and white discharge in women	Neem oil is useful for curing these diseases. Intra-vaginal application of 1 ml of purified neem oil is claimed to be an effective contraceptive. Leaves can also be used to cure vaginal itches and uterine inflammations.
Controlling of cough troubles	Gargles with neem leave juice mixed with a few drops of honey relieves from cough trouble.
Preventing of rash and boils	Bath taken in neem water acts as a preventive measure to control rashes and boils.
Curing of constipation	Rinsing mouth with hot water containing 10-15 neem leaves early in the morning helps in curing constipation.
Hair care	A decoction prepared from neem leaves is used as wash to prevent premature graying of hair, lice infestation and dandruff.
Curing of conjunctivitis	Fresh juice of tender neem leaves after slightly warming and cooling can be applied to get relief from the disease.
Controlling of ring worms and thread worms	The juice of neem leaves (about 10 ml) alongwith a teaspoon of honey kills these worms. Oil extracted from seed is equally effective in controlling these worms.
Curing of dysentery	Taking one tea spoonful juice of leaves in the morning helps in curing dysentery.
Falling of hair	Application of neem oil on hair not only stops the falling hair but also helps to grow hair on the bald portion.
Curing of piles	The paste of neem bark with turmeric and rock salt applied externally over the anal region cures piles.
Controlling of excessive sweating	Application of neem oil on palm and sole reduces excessive sweating.
Relieving in asthma	Taking 20-30 drops of neem oil with betel leaf gives relief in asthma.
Preventing further spreading of chickenpox	Leaves are used to soothe the patients suffering from chickenpox and prevention of further spread. The patient is given bath with water in which leaves are soaked. Keeping leaves on the patient's bed and fanning him on these leaves, is an old aged practice.
Curing of heart disease	Neem delays the coagulation of blood, calms heart beats and helps to reduce elevated heart beat and high blood pressure.
Reducing of cancers	Presence of limonoides and other compounds in the bark, leaves and seed can help in reducing tumors and cancers.
Healing/protection of ulcers	Neem extract give significant protection from the discomfort of gastric and duodenal ulcers.

Source : Sen, 2006; International Research Council, 1992, Pal, 2001 and Gupta, 2006.

The seeds of neem are rich source of Tetranor – triterpenoids and non-isopperenoids. However, the most biological active component of neem is highly oxygenated *Azadirachtin*. Neem products with or without *Azadirachtin* have shown excellent results in controlling about 300 insects belonging to orders like Coleoptera, Diptera, Homoptera, Isoptera, Lepdioptera, Orthoptera, Dictyoptera, Thysanoptera and Siphanoptera.

Various products of neem have now been recommended for controlling the pests of rice, pulses, cotton and vegetables as repellent, anti-feedant, ovipositor deterrent, growth regulator or sterilant.

Neem has also been found effective for the control of certain plant diseases such as fruit rot of banana, wilt of cotton and rice tungro virus. Nematicidal value of neem has been well recognized. Medical neem cake when mixed with soil has been found effective against nematodes in tomato, okra, eggplant, guava, citrus fruit, cole crops, etc.

Neem Based Insecticidal Formulations

A number of commercial formulations are now available in India, which have been enlisted in Table 17.3. Their use in agriculture is as follows:

- ✰ It even disrupts mating, sexual communication and causes sterility in insects.
- ✰ After oil extraction, neem cake can be used for controlling nematodes and soil pests of cardamom, vegetables, sugarcane and citrus. Neem cake is traditionally added in rice fields as a fertilizer.

Table 17.3: *Neem* Based Insecticides and Fungicides

Name	*Crop in which Used*	*Pest Controlled*	*Dose (l ha^{-1})*
Margocide	Cotton, tomato and cabbage	Jassid white fly and Aphids	5.0
Margocide	Cotton	Boll worms	2.5
Margocide	Tomato	Fruit borer	2.5
Neembecidine	Paddy	Gall midge	0.4 – 0.5
Neembecidine	Cotton	Aphids, Leaf roller	0.4 – 0.7
Bioneem	Cotton	Aphid, Jassid, Boll worm	2.5 – 3.0
Neembasol	Cotton	White fly, Boll worm,Leaf roller	1.5
Neembazal F	Cotton	Aphid, Hopper, White fly	1.0
Neem Gold	Tobacco	*Spodoptera litura*	1.25
Neem Gold	Coffee	Stem borer	1.25

Source : Saxena, 1996.

- ✰ Neem leaves and kernel liberally mixed with compost can make good eco-friendly pesticide, which the nature has bestowed upon the man.
- ✰ Neem is an air purifier and protects people from number of pollution problems (Gupta, 2005a and Gupta, 2005b). It helps in checking the vehicular pollution and industrial emissions. It has been observed that under its shade, temperature never goes beyond 37 degree Celsius.

- ✰ A number of neem based products have been in the Indian market for some years now. Among them is "Margo" soap, neem toothpaste.
- ✰ Neem coated urea is used in rice soils as a slow release nitrogenous fertilizer.

Recently, traditional uses of neem tree and its legends have been mentioned (Gupta, 2013). There is a myth that the neem tree is associated with Shitala Mata, the Goddess of small pox, who along with seven sisters resides in the tree and is propitiated on specific days named after her on Shitala Shastahi (sixth day of Magha in Bengal); Shitala Saptami (seventh day of Shravana in Gujrat) and Shitala Ashtami (eighth day of Phalguna in North India). It is customary to place neem tree leaves near the patients suffering from small pox or to fan them with fresh leaves or the branches of the tree for early recovery. There is another belief or myth that nectar, obtained from churning ocean by Demons and Gods, when was brought to heaven, its few drops fell on the neem tree. Hence, the tree is considered an auspicious, sacred and emblem of prosperity in Hindu mythology.

Botanical Description

Neem is a large sized evergreen tree having height ranging from 15 to 30 m with an average height of 25 m and 2.5 m in girth. It can survive upto 200-300 years. It has a large canopy, its spreading branches constitute round crown as much as 20 m wide. The tree remains green in leaves round the year except during extreme drought. The short, usually straight bole 30 to 80 cm in diameter has moderately thick, strong bark with furrow.

The roots penetrate deep into the soil, however, upon injury; they produce suckers especially in dry localities.

The small, white, bisexual flowers are borne in axillary clusters. They have honey like scent and attract bees. The flowers appear during March-April in India and honey produced by neem is much prized.

The fruits are smooth, ellipsoidal drupe and about 2 cm long. They are green in the beginning but become white, pale yellow and/or yellow to greenish yellow when ripe and have sweet pulp and brown seed kernel. It is the kernel which is mostly used for pest control. The annual seed yield of a mature tree may range from 50 to 100 kg depending upon the age, soil, rainfall and variety of the tree. Although a neem tree begins to fruit in 3-5 years but becomes fully productive after 10 years. About 50 kg of fresh fruit yields approximately 30 kg seed from which about 6 kg oil and 24 kg neem cake are obtained.

Distribution

Although neem is thought to have originated in Assam and Burma yet its exact origin is uncertain. Some believe that neem is native to the whole of Indian sub-continent; others attribute it to dry forest areas throughout South and South-east Asia including Pakistan, Sri Lanka, Thailand, Malaysia and Indonesia. It is, however, in India that neem tree is most widely used. The tree is grown from southern tip of Kerala to the Himalayan hills, sub-tropical to tropical, semi-arid to wet tropical regions and upto 1000 m elevation.

Now neem tree is being grown in some 30 countries particularly those in the regions along the Sahara's Southern fringe where the tree provides precious fuel and lumber and is helping to halt the march of the deserts. Over the last century or so, the tree has also been established in Fiji, Mauritius, the Caribbean islands and many countries of South and Central America. Countries like Australia, Philippines, Kenya, Germany, and Italy have also raised large scale neem plantations for commercial purposes (Elanchezhyn *et al.*, 2006 and Sen, 2006).

Cultural Practices

Climate requirements: Neem tree grows almost everywhere in the lowland tropics. However, it thrives best in areas with annual rainfall of 400 to 1200 mm. It thrives even in areas under extreme hot conditions where temperature may go upto 50° Celsius but never tolerates freezing or extended cold. It does well near the equator also. The tree grows faster even without irrigation, in arid and semi-arid regions with rainfall of 500 mm or less. It can tolerate high temperature, low rainfall and long dry spells.

Soil requirements: Neem tree can be grown in a wide range of soils, varying from dry infertile soils to sterile or barren, stony shallow or even where there is a hardpan near the surface. It can thrive even in nutrient poor dry soil. In fact, the neem leaves which are slightly alkaline (pH 8.2) are good for neutralizing acidity in soil. However, it cannot tolerate excessive wet sites and fails if grown in water-logged soils.

The eroded land can be treated faster by growing neem trees on it. Since neem has extensive root system which extracts nutrients from highly leached sandy soils and improve their fertility. Hence it is an ideal tree for afforestation of dry and refractory terrain. It is tolerant to soil alkalinity, salinity and acidity; hence it thrives best on all types of soils including shallow rocky/sandy or clayey soils.

Propagation: The tree can be easily propagated both sexually and vegetatively. It can be grown using seeds, seedlings and root suckers. It is, however, normally grown from seeds, either planted directly on the site or planted through seedlings raised in the nursery. The common season of planting is rainy season.

Harvesting and yield: After ripening, fruit becomes yellow or greenish yellow. On an average, a normal and mature tree yields 30 kg seed, though yield varies from 50 to 80 kg.

References

Anonymous. 2012. Neem may hold clues for HIV treatment. The Kashmir Times.

69(117): 2 (27th of April).

Elanchezhyan, C., Harisudan, C. and Sudhanan, F. 2006. *Neem* -a multipurpose tree. *Agrobios Newsletter*. 5 (1) : 20.

Gupta, R. D. 2013. Neem Tree: Traditional uses and legends. The State Times. 18(44): 8

Gupta, R.D. 2005a. Environmental degradation of Jammu and Kashmir Himalayas and their control. Associated Publishing Company, Karol Bagh, New Delhi, India.

Gupta, R.D. 2005b. *Neem* : A potential organic fertilizer. CAD's Subzar. 3 (1) : 5 [May, 2005].

Gupta, R.D. 2006. Environmental pollution: Hazards and control. Concept Publishing Company, New Delhi, India.

Husain, S. 2013. Home care with neem. *The Daily Excelsior* 49 (261): 4

International Research Council. 1992. Neem – A tree for solving global problems. National Academy Press, Washington DC.

Jain, S.K. 1991. Bitter sweet portions. Nature-cum-Nim and Emblica. *The Indian Express* [Magazine, March 24, 1991].

Koul, M.K. 1993. Know your *neem*. *The Kashmir Times* [*The Sunday Magazine*, 29th August, 1993].

Koul, M.K. 2002. Don't neglect your neighbourhood *neem*. *The State Times*, 27th October issue.

Mohindra, K. 2010. Floral remedies. *Ayurveda*. 1(18): 14-15

Pal, M.S. 2001. Indigenous technologies for safe storage. *Intensive Agriculture*. 39 (1-2) : 17-19.

Saxena, R.D. 1996. Report on awareness programme on pesticide and sustainable agriculture. 288, Sidarth Enclave, New Delhi, India.

Sen, R. 2006. Sarbarogo-Nibarini, "*Neem*". *Farmer's Forum*. 6 (8) : 17-18.

Tikoo, R. K. 2004. Neem tree – Natural insecticide (unpublished). *Personnel communication*.

18

Jatropha
(*Jatropha curcas* Linn.)

Common name: Ratanjot, jaffolata, jamalgota

Family: Euphorbiaceae

Introduction

Potentially limited reserve of fossil fuels, their continued higher price and threats from pollution have led to conservation efforts and the development and use of alternative environment friendly energy sources. Bio-diesel from plant sources is gaining popularity owing to its multitude benefits ranging from energy security, employment generation and keeping our environment pollution free. Mixing of only five per cent bio-diesel to the present diesel availability in India, can save crores of rupees, every year and the Government of India is taking keen interest in promoting plantation of bio-diesel plant *Jatropha curcas* on waste/ degraded un-used lands.

It derives its name from Greek word 'Jatros' (doctor) and 'Tropha' (food) which implies medicinal uses. It is native of Mexico and Tropical South America but naturalized now in many Latin American, African and Asian countries (Pandey *et al.*, 2005). In India, it was introduced by Portuguese seafarers in the 16th century as oil yielding plant.

Now it is found in semi-wild conditions throughout the country including Andaman Islands. It is known as 'Jangli arandi', 'Safed arand', 'Dravanti', 'Bagherenda' and 'Rattanmanaku' in different states of India. In English, it is known as Physic nut or Purging nut. It is because that all plant parts including seed, leaves and bark, fresh or decoction are used in traditional medicines and for veterinary purposes (Sengar and Choudhary, 2011).

Farmers in the Jammu region of Jammu and Kashmir generally plant this species as a live fence for protecting their agricultural fields. Besides, hedge fence, the dried stems are used as a poor quality fuelwood. The plant establishes easily, grows rapidly and is drought resistant. Once established, it can give economic returns upto 15-20 years. It has yield potential of 12 tonnes ha^{-1} year^{-1} after 5 years of growth.

Medicinal Uses

Green twigs are used as tooth brush to cure toothache. Leaves and young branches are bound on the affected portion for joint pains. Its roots are also used as an antidote for snakebite. In addition to medicinal uses, leaves and twigs are incorporated into or laid on top of soil for green manuring and weed control. Jatropha oil is also used as lubricant for bullock and horse carts axels and other agriculture implements. The conditions for its plantation on degraded barren and unproductive lands are quite suitable in low altitude, sub-tropical belt commonly known as Siwaliks. Large scale cultivation of *Jatropha curcas* (Ratanjyot) will open up employment opportunities in kandi (rainfed) belt of Jammu division and help farmers to upgrade their life standard.

Botanical Description

Jatropha curcas is a small tree or large shrub which grows between 3 to 5m but can attain a height upto 8 to 10 m under favourable conditions (Singh *et al.*, 2010). The plant shows articulated growth with a morphological discontinuity at each increment. The branches are glabrous, ascending and stout. The bark is pale brown, papery peeling which exudes whitish coloured watery latex when cut. The leaves are large, green to pale green, alternate to opposite three to five lobed with a spiral phyllotaxis. Inflorescences are formed terminally on branches and are complex possessing main and co-florescences with paracladia. The plant is monoecious and flowers are unisexual, occasionally hermaphrodite. Ten stamens are arranged in two distinct whorls of five each in a single column in the androecium and in close proximity to each other. In the gynoecium, the three slender styles are connate to about two-thirds of their length, dilating to massive bifurcate stigmata. The female flowers are 4-5 times more than the male ones. Pollination is by insects. The fruits are ellipsoid capsules approximately 2.5-3 cm long, 2-3 cm in diameter, green, turn yellow to black on maturity. The exocarp remains fleshy until the seed are matured. The seeds are black, ellipsoid, triangular-convex and approximately 1.5-2.0 cm long and 1-1.2 cm thick and resemble with castor seed.

The plant flowers during rainy season and sometimes flowering is also noticed during September-October. It has nearly 422 fruits and 1587 seeds/kg. The 100 seed weight ranges between 48.82 to 71.88 g.

Cultural Practices

Climatic requirements: Since jatropha is hardy species and can be grown in wide range of agro-climatic conditions including moderate to high temperature areas. Dry and semi -arid conditions with annual rainfall ranging from 200- 1200 mm are suitable for its growth. It can withstand dry weather conditions and grows well on drained soils with good aeration. It is adapted naturally to nutritionally poor and dried seasonal streams. It can withstand long periods of drought by shedding most of its leaves to reduce transpiration loss. It can be grown in areas of low rainfall upto 250 mm per annum. It can thrive on poorest stony, gravelly, sandy, eroded lands and even in the crevices of rocks. It can tolerate low frost and frost burned plants re-grow.

Soil requirements: It is suitable for enrichment of soil, reduction of wind and water erosion. It can also be grown on problematic soils such as saline, alkaline and sodic soils. Jatropha plantations cannot be established on soil with prolonged flooding or that have very high water table.

Propagation: Jatropha can be propagated from seed as well as cuttings.

Seed sowing, sowing time and seed rate: The seedlings can be raised in nursery beds as well as in poly bags. For raising bare-rooted seedlings, the beds should be properly prepared and mixed with well rotten farm yard manure. Fresh, well dried, insect free and fungus free seed are sown in the raised nursery beds at a spacing of 20x20 cm and a depth of 2-3 cm. Before sowing, seeds are given overnight soaking in water for uniform germination. 5-6 kg of seed is required for one hectare plantation (two seeds dibbled per hole). After sowing beds are to be mulched with paddy straw, fresh adhatoda leaves or any other suitable material (Anonymous, 2012). The seed germination starts after 5-7 days and continues for 10-15 days. Weak and diseased seedlings are pricked out when they are about one month old and only one seedling is maintained per hole. The beds are watered as and when required to keep soil loose and prevent cracking. The nursery is raised during the month of February- March so that plants are ready by the month of July for field planting. Plants are carefully uprooted from nursery and should be wrapped in wet gunny bags during transportation.

Seedlings can also be raised in poly bags. Poly bags of 20x20 cm size should be filled with a mixture containing soil, sand and farm yard manure (FYM) in the ratio of 2: 1: 1. Two seeds are sown in each bag and watered regularly. After one month, weaker of the two germinated seedlings should be removed which can be used for gap filling. Seedlings are ready for planting in about three months and grow so fast that one or two weeding is sufficient. The roots of seedlings raised more than 6-8 weeks come out of the polybags. For this region, seedling should not be kept too long in the nursery and given regular root pruning.

Propagation through cuttings: The cuttings of 40-50 cm length and 1.5-2.0 cm diameter should be collected from one year old terminal branches. The cuttings treated with 100 ppm IBA for 24 hours have been found to have highest survival rate. For quick establishment of hedges and plantations for soil erosion control, cuttings can be directly planted at the site. Plants propagated by cuttings will normally produce seed within one year of planting and growth is rapid. A tap root is not usually formed by vegetatively propagated plants. For long lived plantations, plants propagated by seed are better.

Planting techniques: Direct seeding -The seed should be sown with the onset of the monsoon. The pit size should be 45x45x45 cm and filled with soil mixed with farm yard manure. The planting spacing is kept on the basis of soil condition, rainfall, irrigation and method of plantation. In each pit two seeds are sown 4-6 cm deep and covered with any mulching material. The excess number of germinated seedlings can be used in gap filling and only one seedling per pit is maintained. Proper weeding is necessary if raised through direct seeding. After seedling stage, growth is more rapid and seedling can effectively compete with weeds.

Transplanting: Outplanting of nursery grown stock give higher survival rate than direct seeding because of weed competition, excess rains or long dry spell. However, raising seedling require extra labour in the nursery and in transplanting.

Planting Systems

(i) *Block plantation-* These can be done at a spacing of 2x2 m apart with plant population of 2,500 per hectare. High density plantation can also be raised at a spacing of 1.5 x 1.5 m or 2 x 1.5 m to accommodate 5000 or 4440 plants per hectare. There may be a possibility of thinning these trees to wider spacing as they mature.

(ii) *Living-fences*-Living fences protect crops from wild life, domestic animals and can be managed for seed production. Cutting/seedlings can be planted densely at a spacing of 30-50 cm in single or double row system surrounding an orchard/farm.

(iii) *Bund plantation-* It can be planted on field bunds 1x 1 m spacing. This plantation can be managed as shrubs by proper pruning and training methods. Alley farming at the early growth stages. Vegetables, medicinal and aromatic plants, oilseeds, and pulses can be grown in the interspaces. However, required intercropping can be done when planted at a wider spacing of 4x4 m which provides enough space for intercrops to flourish as well as ease in agricultural operations. For alley cropping spacing within the rows depends upon the slope of the land. Usually closer spacing is recommended for better erosion control on steeper slopes. Following combinations with Jatropha may be practised to augment additional farm income.

1. **Vegetables**: Turmeric, Ginger, Potato, Chillies, Colocacia, Peas, cowpea etc.
2. **Medicinal and Aromatic Plants**: Ashwagandha, Kalihari, Sarpagandha, Aloe, Safed musli, Aspargus, Lemongrass, Mucuna etc.
3. **Pulses and Oilseeds**: Gram, Mung, Mustard, Sunflower etc.

Inter-cultural practices: The plantations should be kept free from weeds. Light harrowing and weeding is necessary for proper growth of plants.

Irrigation: *Jatropha curcas* requires less water, but during the first year of plantation, light irrigation is required depending upon the soil type and agro-climatic conditions. Usually, second year onwards, the irrigation is not required and it can withstand long drought periods.

Manures and fertilizers: Apply 2-4 kg of well decomposed organic manure per pit or 2 kg organic manure plus 100-200 g neem cake per pit to provide nutrition. Application of biofertilizer especially *Azospirillium* 10 g + 10 g mycorrhizal fungi per pit at the time of planting has improved the uptake of plant nutrients and maintenance of fertility.

Pruning: Pruning is done after 2 years of plantation at a height of 30 cm to develop more number of branches. This will induce a bushy stature to plant rather than tree like vertical woody growth. During the 3rd year 2/3 top portion of each side branch should be pruned thereby retaining 1/3 of branch on the plant. Periodical pruning can be carried out depending upon the vegetative growth of plants. Pruning should be done when the tree sheds leaves and enters into a period of dormancy (December to February). Deformed, dead and weak branches should also be removed. A sharp secateur is used for pruning to avoid peeling of the bark.

Pinching: The terminal growing twig is to be pinched to induce secondary branches. Likewise, secondary and tertiary branches are to be pinched or pruned at the end of the first year to induce a minimum of 20-25 branches at the end of second year. Once in ten years, the plant may be coppiced at height of 30 cm from ground level for rejuvenation. The growth is quick and the plant will start yielding in about a years' period. This will be useful to induce new growth and yield stabilization there on.

Harvesting and yield: Jatropha plantation when raised through seed starts flowering after two years of plantation. The flowering takes place during the month of April-May and sometimes flowering is also noticed in the month of September-October. Fruits mature after about 90 days of flowering and harvesting of fruit is practised in December-January when capsule turns pale yellow to black. The jatropha plants start yielding from the second year of planting but only a limited quantity of seed can be harvested. On an average 5 tonnes of seed under un-irrigated and 12 tonnes per hectare under irrigated conditions can be harvested from 6th year onwards with good care and management practices. Economics of jatropha seed is worked out on the basis of seed yield at different age of plantation and current market rate of seed. The net returns per ha will be around Rs. 25,000-30,000 under normal conditions. In addition to this, it has additional benefits like de-oiled cake, intercrop, glycerine etc.

Oil content and extraction: The oil content in seed ranges from 30 per cent to 40 per cent of seed weight and 50 per cent to 60 per cent of the kernel under quality seed plantation. Jatropha oil can be extracted by expellers which already exist at village level. Solvent extraction gives maximum oil, but cannot be achieved on small scale. Mechanical pressing technique is simple and feasible for farmers.

References

Anonymous, 2012. Establishing Jatropha plantations in Jammu Shivaliks. Bulletin published by Division of Agroforestry, SKUAST-Jammu under National Oilseeds and Vegetable Oils Development Board project.

Pandey, A., Pandey, S., Mishra, A., Koushal, R. Tewari, S and Chilana, K. 2005. Jatropha- A panacea for wastelands. *Indian Farmers Digest*. 38 (11): 36-38

Sengar, R.S. and Choudhary, R. 2011. Challenges and opportunities in growing biodiesel plant in India. *Indian Farmers Digest*. 44(11): 20-23.

Singh N., Srivastava, A and Srivastava, N. 2010. *Jatropha curcas* – The biodiesel crop. *Indian Farmers Digest*. 43 (7): 26-27 and 44.

19

Mulberry Tree (*Morus alba* L.)

Common name: Tut or shatut

Family: Moraceae

Introduction

In India, mulberry is a very old, exotic tree, which is not commonly associated with forest vegetation though occurs occasionally from seed dispersed by birds and animals. It is cultivated extensively for silk and has been hybridized to produce innumerable varieties (Singh, 2005). Mulberry in fact, is a multipurpose tree as every part of it is directly or indirectly used for one or other purposes. For instance, mulberry leaves are utilized for feeding domestic animals and for rearing silk worms. Feeding on mulberry leaves have been well known to ameliorate milk yield. It is because mulberry leaves contain lot of crude protein, crude fat, crude fibre and vitamins. Even leaf stock after feeding can also be used for feeding cattle without any adverse effect on their health (Gupta, 2007). About 30 per cent of mulberry waste could be available after silk worm feeding, which is being used as cattle feed. 'Silk and Milk' concept which depends upon the symbiotic relationship between dairying and sericulture has now become an established practice in some parts of Karnataka and other states. Apart from role of mulberry in cattle and silkworm feed, it is a potential source of drugs because of the presence of non-active constituents the anthocyanins (Rangaswami *et al.*, 1976).

Medicinal Values

Mulberry tree contains different chemicals of medicinal value in leaf, fruit, stem and root portions. Mulberry leaf juice has a special quality of moisturizing the skin and keeps the skin smooth and healthy. It also prevents throat infection, irritation and inflammation. Traditionally fruit also benefit kidneys, general weakness, fatigue, anemia, constipation and graying of hair.

Fruit juice is used for preventing high fever, diarrhoea and cold/cough. Fruit is used in preparing syrup which has cooling and laxative properties. Roots of black mulberry are useful for diabetic patients. Fruits of mulberry indeed, are called as esteemed desert fruit. It is eaten as fresh and can be used as pickle, squash and beverage. It is a good source of vitamins besides malic acid, pectin, mucilage and colouring matter and gives lot of strength to the body. Analysis of fruit juice is presented in Table 19.1.

Table 19.1: Analysis of Fruit Juice

Constituent	*Percentage*
Water	85-88 per cent
Carbohydrate (sugars, mainly glucose and fructose, producing the sweet taste)	7.8-9.2 per cent
Protein	0.4-1.5 per cent
Fat (mainly fatty acids, such as linoleic, stearic, and oleic acids in the seeds)	0.4-0.5 per cent
Free acids (mainly malic acid, producing the sour taste)	1.1-1.9 per cent
Fiber	0.9-1.4 per cent
Minerals	0.7-0.9 per cent

Source: Dharmananda, 2003.

Leaves of quality mulberry are sometimes cooked as vegetable. Due to high protein content, young leaves constitute an excellent substitute for protein deficient diet. The growing leaves contain 6.6 to 6.8 per cent protein, 4 per cent soluble carbohydrates, 0.6 per cent fat and several minerals-Ca, P, Mg, Zn, Cu, Fe, Mn and vitamins; A,B, C and D.

Rural people in South India prefer mulberry leaves equally good for consumption like amaranths, pea, cucumber, and drumstick leaves. Mulberry curry is prepared from its tender leaves.

Other Uses

Chlorophyll extracted from leaves is used for cosmetics *viz.*, soap, hair oil, toothpaste and medicines. In China serving mulberry juice in the villages to the visitors is considered the most common hospitality.

Thick stems can be utilized as fencing material. Dried stems can be used as medium grade fuel wood. Thin and soft twigs can be used for making baskets and handicraft items. Very small and soft twigs along with dry leaves are used to prepare compost.

Bark stripped of from waste branches is used for making paper pulp, which is very common in China and some European countries. A white and soft fibre suitable for use in textile industry is also extracted from the bark, present in phloem and is composed of cellulose, lignin and pentosan. On account of its elasticity and flexibility, the wood is of immense importance for the manufacture of sport goods like hockey sticks, tennis rackets, badminton rackets, cricket stumps. Wood can also be used for constructing houses and making agricultural implements. Tannin of the wood is considered to be suitable for tanning and colouring purpose. Yellow dye is obtained from the roots and used in dying process.

Botanical Description

Mulberry is a fast growing, deciduous, woody perennial tree species occurring in wild or as cultivated (10-20 m in height). Older bark is grey with narrow irregular, vertical fissures which splits easily, stems with pubescence but free from thorns. It has a deep root system. The leaves are simple, alternate, stipulate, petiolate, entire or lobed. Plants are generally dioecious. Inflorescence is catkin with pendent or drooping peduncle bearing unisexual flowers. Inflorescence is always auxiliary. Male catkins are usually longer than the female catkins. The ovary is one-celled and the stigma is bi-fid. The chief pollinating agent in mulberry is wind. Mulberry fruit is a multiple fruit, 2-3 cm long, mainly violet black in colour on ripening in June to August and is eaten by birds and wild animals.

Cultural Practices

Climatic requirements: Mulberry tree is known to grow both in tropical/ sub-tropical and temperate climatic condition. 22–30°C temperature, 1000-2000 mm rainfall and 65-80 percent humidity are optimum for optimum growth. It can also be seen growing in desert climate.

Soil requirements: As white mulberry is somewhat drought tolerant so it prefers evenly moist, well drained soil. Red mulberry grows best in rich soil, whereas, black mulberry prefers warm moist, well drained soil. Flat or slightly slopy land is suitable for mulberry cultivation. The optimum range of soil pH is 6.5 to 6.8.

On an average mulberry likes warm, well drained soil, preferably a deep loam. Mulberry is a deep-rooted plant and the top soil should be sufficiently deep (60 to 90 cm). Shallow soils such as those found on chalk or gravel are not recommended. For soils too acidic or alkaline, corrective measures like application of lime or gypsum are required.

Land preparation: Land should be ploughed deep before the onset of monsoon, followed by 2 - 3 light ploughings to bring the soil to a fine tilth and leveled.

Planting: The best time to plant the mulberry trees is during winter season when the tree is in dormant condition. Prior to planting the fields should be well prepared and plants should be planted in appropriate spacing. The distance between trees should be maintained at least 4-5 m. For planting in rainfed areas wider spacing needs to be adopted.

Generally, mulberry is planted through cuttings but planting of saplings gives better results. Cuttings of 20 -22 cm length with 3 to 4 buds are planted. Cuttings are placed deep, leaving one bud on top, 2 to 3 cm above the ground and the soil around each cutting should be pressed well to keep the cuttings firm in the soil. Direct planting of cuttings under rainfed conditions results in failures and under such situations 4 months old saplings are recommended. The planting during the monsoon period have the advantage of receiving fairly distributed rain from June to October.

Manures and fertilizers: 20 tonnes of farm yard manure (FYM) per hectare is added and mixed well in soil at the time of land preparation. 28 kg of nitrogen, 12 kg of phosphorus and 12 kg of potassium is recommended per acre or 70 KgN and 30 kg each of P and K ha^{-1} after every shoot harvesting. Use of biofertilizers especially that of *Azotobacter* has proved beneficial not only to cut the dose of nitrogenous fertilizer but has also increased the yield of mulberry leaves (Srivastava, 2015).

Inter-culture and irrigation: After about 1 month of planting, a light weeding/ hoeing must be done. Second weeding must be done around 3 months after planting. Weeding operations must be thorough and should be regularly carried out. Although somewhat drought tolerant mulberry require to be watered during dry seasons. If the roots become too dry during drought, the fruits are likely to dry before they have fully ripened. Adequate irrigation has to be given immediately and in the event of failure of rain.

Harvesting and yield: Leaves are harvested either individually or shoot harvesting. The latter is more economical and commonly used. The expected leaf yield in temperate climate for variety 'Chak Majra' is 20-22 tonnes/ha/year. In case of S-1635 leaf yield ranged from 35-40 MT ha^{-1} $year^{-1}$.

References

Dharmananda, Subhuti 2003. Fruit as medicine : *Morus* fruit (mulberry). Institute for Traditional Medicine, Portland, Oregon

Gupta, R. D. 2007. Mulberry: a tree of multipurpose uses. The Daily Excelsior 43 (222): 3 (The Sunday Magazine)

Rangaswami, G. Narasimhanna, M.N., Kasi Viswanathan K. and Sastry, C.R. 1976. *Manual on Sericulture Vol.1*, Rome, FAO.

Singh, S. S. 2005. Fate of the Medicinal Plant Resources of Jammu and Kashmir. Directorate of Environment and Remote Sensing, Jammu, Govt. of J&K.

Srivastava, R. P., Ramanagouda, S.H and Basera, A. 2015. Role of biofertilizers and micronutrients in mulberry cultivation. *Agrobios Newsletter*. 49 (1): 10-11.

20

Orchid Tree (*Bauhinia variegata* Linn.)

Common Name: Kachnar, katrer

Family: Leguminosae

Introduction

Kachnar, a native of India, is popularly known as the orchid tree or mountain ebony. It has its own remarkable brand of hardiness and the striking beauty of orchid like flowers. The tree is often lopped for fodder and its flowers are taken as vegetable, raw or cooked. Generally the flowers of the tree, after boiling in water are mixed with curd. The dish so prepared is called "Dehi Raita" which is used by many of the rural people. Almost all the parts of the tree are medicinal in nature.

Medicinal Values

Bark of the tree is astringent, roots and seeds are digestive. Bark is anti inflammatory, useful in skin diseases, ulcers and scrofula. Dried buds are used in dysentery, piles and killing worms. Roots are used as antidote to snake poisoning. Decoction of roots is used in dyspepsia (Singh, 2005).

Other Uses

Kachnar is a good species whose value as fodder and fuel wood is known to everybody (Pachnanda, 1999).

Botanical Description

Kachnar is a fairly medium sized to a large deciduous tree reaching up to a height of 6-15 m, with smooth and dark brown bark. The tree comes into leaves, towards the end of summer. Leaves are deeply cordate 11-15 nerved, sub-coriaceous, cleft through about a quarter, lobes obtuse. Flowers are large, fragrant, sessile or shortly pedicellated in few flowered lateral corymbs. Calyx tube is slender, the limb is spathaceous, petals ovate or oblong. Flowers are 3-5 across usually in clusters at the branch tips, appear in February to April. Pods are 10-15 cm long, narrow, flattened, containing 5 to 10 seeds.

Cultural Practices

Climatic requirements: Sub-tropical to humid sub-tropical is the best climate for the growth of Katrer, though can be grown in the Sub-Himalayan tract also from the Indus eastwards, dry forests of Eastern, Central and South India (Chauhan, 2006).

Soil requirements: Katrer can be grown on variety of soils but the most suitable is sandy loam or loamy sand soil having good amount of organic matter. In sub-mountain areas of Siwaliks, it can be grown even in the gravelly sandy loam soils.

Cultivation: Natural reproduction occurs from the seeds. The seeds are dispersed on dehiscence of the pods on the trees and germinate rapidly when furnished with favourable conditions.

Artificial propagation is carried out by direct sowing of the seeds and through stump planting. Sowing is done in lines by maintaining spacing about 3m apart. Germination starts in about a week. To raise entire plants for planting in July-August, sowing is done during April-May or March-April using previous year's seed.

The entire plants have to be planted out with the ball of soil. Stumps or container grown plants can be preferred for transplanting.

New Growth of the Tree

The orchid tree comes into leaves towards the end of summer. By the time the monsoons commence, it has already developed a lush green canopy with beautiful flowers and fruits which serve as source of food with the retreat of the monsoon, the large leaves of the tree begin to dry and by the end of November, a cracking carpet of dry leaves covers the ground. On decomposition by various groups of micro organisms the carpet of the leaves reduces to organic matter – a source of nutrients for the growth of the tree.

Meanwhile, the tree remains dormant during winter but prevents desiccation and withstands the cold during the dry winter months are completed. Soon it begins to form the buds that will flower and later bear fruits in the coming season.

By the end of February or first fortnight of March, depending upon the weather/ agro-climatic conditions this work is nearly at an end and the tree waits for the right moment to send forth its first new growth. This moment actually consists of certain combination of atmospheric humidity and warmth coinciding with the emergence of some insects which will pollinate the orchid like flowers. No sooner the time is right the tree rapidly begins to bristle with spikes of 2.5 cm long buds.

Precautions

Fresh seeds should be sown at the onset of monsoon. They come up very well in places where soil depth is very good.

References

Chauhan, N.S. 2006. Medicinal and Aromatic plants of Himachal Pradesh, Second Edition. Indus Publishing Company FS-5, Tagore Garden New Delhi-110027.

Pachnanda, U.1999. Choice of species for degraded lands. *Haryali. A quarterly Newsletter*. 11-13: 19-21.

Singh, S.S. 2005. Fate of the Medicinal Plant Resources of Jammu and Kashmir. Directorate of Environment and Remote Sensing, Jammu, Govt. of J&K.

21

Rudraksh Tree (*Elaeocarpus gantirus* Roxb.) Syn. *E. sphaericus* Gaertn.

Common name: Rudraksh

Family: Elaeocarpaceae

Introduction

Rudraksh tree inhabits area from Manila, Philippines and passing through Myanmar to entire North-East India, Bangladesh, Nepal, Bhutan. However, presently, it is located in Eastern Nepal. About 70 per cent of Rudrakhsh trees are found in Indonesia, twenty five per cent in Nepal and five per cent in India (Sood and Jaswal, 2013).

Seeds of rudraksh tree known popularly as 'Rudraksha', have fascinated people across the world due to their mystical properties. It is rather amazing that such seeds have gone on to receive so much attention and reverence worldwide. Primarily rudraksh is a symbol of spirituality and worn singly or in multiples in the form of rosary (*malas*). Many people wear Rudraksh for its good omen, which may be as a talisman or just for good health. Health benefits of rudraksh have been known since long. It is because it is useful as a medicinal plant. Apart from its medicinal value, a recent study revealed that rudraksh is capable of neutralising all the harmful radio and communication frequencies on the human body.

Medicinal Values

Rudraksh is useful as a medicinal tree, specifically for cardiac problems, blood pressure and mental disorders. (Seetha, 2013). Three-mukhi (facets) rudraksh is also used in disorders such as diabetes, ailments of the stomach, fever, defects of eyes and even cancer. A child suffering from repeated occurrence of fever or is very week, three mukhi Rudraksh is prescribed to be worn around the neck to get relief. Four mukhi rudraksh has proved to be effective in curing diseases related to mind (loss of memory and epilepsy etc.) and controlling diseases of gall bladder (Sharma, 2014).

From medical point of view, eight mukhi Rudraksh is believed to cure diseases of nervous system, prostrate and gall bladder. Eleven mukhi rudraksh is considered good for stomach disorders, acidity, liver and breast diseases. Thirteen mukhi rudraksh is used for psychiatric illness, while the fourteen mukhi is believed to be effective against diseases of the heart, eyes, skin and ulcers.

Botanical Description

Rudraksh is a large, evergreen broad leaved tree. It generally grows up to 15-30 m height or more up to 60 m depending upon the area, with trunks up to 1.2 m in diameter. It has leaves like mango trees, light green in colour, which changes to yellowish red, and then grey before falling. Flowers have a mild aroma similar to 'Rat ki Rani' flowers and are white in colour. Flowering takes place in April-May.

The fruits are 2-4 cm in diameter and green in colour. After maturity, the colour of the fruits begins to turn into bluish and violet, then deep brown and finally black. Inside the fruit is the stony endocarp or bead which is attached to the stem from the base of the stone. The bead has very hard rough surface with uneven groves and a long cavity in the centre from the point where it remains attached by the stem.

Cultural Practices

Climatic requirements: Rudraksh grows best in tropical and sub-tropical climatic conditions with temperature range from 25-30° Celsius.

Soil requirements: Seeds of Rudraksh require well drained sandy loam or loamy sand soils for their good germination. Good supply of moisture and air in the soil is must. However, it does not grow well in areas of abundant sand or water logged soils.

Planting: It can be propagated by planting the mature beads in soil which take a year to sprout into seedlings. Special care is required like daily watering, shelter from cold, direct sun light and protection from insects. Seed quality affects the seedling growth. Tree starts giving fruit after 7 years and thereafter for long time.

Manures and fertilizers: Bio-fertilizer, cow dung manure and compost is found to be the most suitable for Rudraksh.

Harvesting: The trees start giving fruits about 7 years after their planting. Fruit setting takes place the month of June and get ripened during 2-4 months *i.e.* by August to October. Two thousand fruits are obtained per tree.

References

Seetha, K.N. 2013. The power of Rudraksh. In *Tattvaloka*. 36(1): 59-78

Sharma, O.P. 2014. 151 Medicinal Plants of Jammu and Kashmir. Directorate of Social Forestry Project, Jammu and Kashmir Govt. (Personnel Communication).

Sood, S and Jaswal, D. 2013. Mystic Medicine Rudraksh. *Ayurveda*. 1 (22) : 4-5.

22

Sandal Wood Tree (*Santalum album* Linn.)

Common name: Chandan

Family: Santalaceae

Introduction

Sandal tree is found all over India (Bangalore, 2007) with more than 90 per cent area in Karnatka and Tamil Nadu. Other states where sandal tree grows consist of Andhra Pradesh, Kerala, Maharashtra, Madhya Pradesh, Orissa, Uttar Pradesh, Bihar and Rajasthan. In Himachal Pradesh it grows naturally at Bilaspur and Jawalaji area of Dehra Forest Division of Kangra district. It flourishes well from 250 to 1,000 m on different types of soils like sandy loam, loam and clay loam soils in Himachal Pradesh (Bhardwaj, 2014). Sandal wood tree on steam distillation yields an essential oil, which is used in perfumery as well as in medicines.

Sandalwood Oil in Perfumery and Medicines

Indian sandalwood consists of the highest oil content. Sandalwood oil contains santalol *i.e.*, santalol 'a' and santalol 'b' as compared to other species of genus *Santalum*. Both wood and oil find use in incense and medicines.

The oil obtained by steam distillation of sandalwood is of pale yellow colour, optically clear with a characteristic odour conforming to pharmacopoeial standards. The oil is widely used in perfumes and finds a place in the manufacture of soaps, face creams, toiletry powders and attar.

Fine powdered sandalwood, is used for cosmetics. Sandalwood, one of the finest wood for carving and turning is said to be the next to ivory for intricate workmanship and a large number of idols and other utility articles are made out of it. Heart wood of sandal is described as astringent, antipyretic, cooling exhilarating. The paste of the sandalwood is also used in religious functions.

Other Uses

Heartwood of sandal tree has been found the finest wood for carving and turning next only ivory for intricate workmanship. Utility articles like jewellery cases, cabinet panels, chess boards, penholders knives, caskets etc., are made out of it. However, India's production (mainly from Karnataka and Tamil Nadu) has declined and craftsmen doing intricate work of art since ages, are now shifting from sandalwood to other raw materials that are easy to produce (Anonymous, 2008).

Botanical Description

Santalum album is a small, evergreen tree attaining a height of 12 to 13 m and 1 to 2.4 m wide with slender, drooping as well as erect branching. It is a partial root parasite and found in association with other tree species. Leaves are opposite and decussate or rarely with whorled arrangements. Flowers are purplish brown.

The trees flower at an early age to 2 to 3 years. Most of the trees flower twice a year (March to May and September to December) whereas, others once in a year. Over lapping of flowering and fruiting seasons usually results in continuous supply of flower and fruits. Fruit is a drupe, purplish brown when fully ripe and single seeded.

Sandal is a slow growing tree under forest conditions (1cm girth per year). However, it can grow at the rate of 5 cm girth/year or even more under favourable soil moisture conditions. In Himachal Pradesh under sub-tropical climatic conditions, heart wood formation starts at about 8 years of age (Bhardwaj, 2014).

Cultural Practices

Climatic requirements: Tropical and sub-tropical to sub-temperate climate conditions are the most suitable for the cultivation of sandal trees. Rainfall of 600 to 1600 mm is found desirable for their growth and the trees occur at an altitude from sea level to 1800 m.

Soil requirements: Sandal trees are found in sandy loam, clay loam, red laterite and even black cotton soils. Trees growing in stony or gravelly soils are known to have highly scented wood.

Nursery raising: The seed beds of size 10 m x 1 m are prepared with sand and soil in the ratio of 2:1. Seed @ 4 kg per bed is spread uniformly over the bed. These are then covered with 2 cm of soil, watered and mulched with straw. The straw is removed when leaves begin to appear. The beds are watered at least once a day depending upon the climatic conditions. To follow the nursery schedule, seed sowing should be carried out in the month of December- January.

The above seedlings are transplanted at 2-4 leaf stage in containers (root trainers/polybags). Root trainers with potting media, consisting of sand, soil and compost in the ratio of 40 : 10 : 50 can be used for production of sandalwood seedlings in about six months.

Nutritional enrichment of potting media by adding neem cake and application of slow release fertilizers (neem coated urea/sulphur coated urea) boost the growth of seedlings. To further improve the growth of seedlings and better survival, application of bio-fertilizers (*Glomus fasciculatum*) is recommended.

Field planting: Healthy sandalwood seedlings having height of about 25- 30 cm and collar diameter 3.0 mm are planted in pits of 50 cm^3 size with a spacing of 3m x3m to 5m x 5m at the onset of monsoon.

Intercultural operations: Soil working to a radius of 50 cm is done once in six months. Pruning of the host plant is done periodically to allow maximum sunlight available to sandal. To facilitate maximum heartwood formation, side branches on the lower half of the main stem are pruned at regular intervals.

The above operation *i.e.* soil working, weeding and pruning are required to be followed in the second and third years. To improve initial growth, farm yard manure mixed with leaf mulch can also be applied.

Harvesting and yield: Dead and dry sandal trees including roots up to 2.5 cm diameter are extracted from the forests (roots contain oil). After uprooting, the tree top and branches that have no heartwood, are chopped off. The trunk branches and roots are cleaned by chipping off the bark and portion of sapwood. Rough cleaned wood is weighed and is given for final cleaning. Final clean wood is separated as i) Billets ii) Roots and iii) Chips and stored in sandal depots.

A sandalwood tree having a girth of 80 cm at breast height can yield about 75-80 kg heartwood under favourable conditions which can fetch a price of Rs. 1500 or more kg^{-1}.

References

Anonymous. 2008. Sandwood availability in Mysore declining. *The Tribune*. 128(326): 14 (Monday, 24 November)

Bangalore, M.(2007). Sandal Tree. Gram Vikas Jyoti. January- March Issue, pp 59-61.

Bhardwaj, S.D. 2014. *Santalum album*-its occurrence, phenology, reproduction, utilization and economic importance. (Personnel communication) Solan, HP.

23

Soapnut Tree (*Sapindus mukorossi* Gaertn.)

Common name: Ritha or Reetha

Family: Sapindaceae

Introduction

Ritha is commonly grown in China, Florida, Hawaii and India. In India, it is mostly found in Himalayan states like Uttarakhand, Himachal Pradesh and Jammu and Kashmir and North Eastern Hill states, particularly Assam. Apart from serving its leaves as fodder and a fruit as good substitute for washing soaps especially woolens and silk fabrics, fruits of the tree have also medicinal uses.

Medicinal Values

Fruits of the Ritha tree, which are globose fleshy and saponaceous in nature, are regarded as expectorant, useful, in salivation, chlorosis and epilepsy. They are also used as fish poison. Sprinkling of fruit soaked water acts as repellent for snakes and lizards. For this purpose fruits are required to be soaked about a week in the water and thereafter the water is sprinkled after rubbing the outer surface of fruit shell a number of times.

Botanical Description

Ritha, a very handsome tree green in summers and deciduous in winter grows generally 6 to 9 m in height, but may grow up to 18 m or more in height

and about 1-8 m in girth. Its bark is dark greenish or grey or brown in colour and is fairly smooth. Leaves are alternate, paripinnate, 30-50 cm long, leaflets 5-10 pairs, glabrous, lanceolate, acuminate. Flowers are small, polygamous and mostly bisexual in terminal thysses or in compound, cymose panicles. Fruit is globose and fleshy. Seeds are enclosed in black, smooth and hard endocarp (Chauhan, 2006).

Cultural Practices

Climatic requirements: It is a tropical to sub-tropical plant and shows its best growth in warm climate, having annual rainfall between 150-200 cm. the tree prefers full sun light or partial shade.

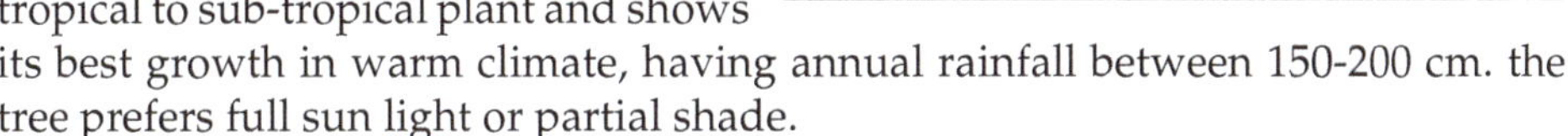

Soil requirements: Soap nut tree grows best in deep loam soils with more clay content.

Method of planting: The tree can easily be propagated through seeds. Soap berry seeds take 1-3 months to germinate fully in summer. Soak the seeds overnight in warm water to scarify seed shell. Seeds are sown in a well prepared fields at a soil depth of 2.5 cm. Keep the seeds well moist and warm under bright light.

Flowering and fruiting: May to June is time for the flowering of soap nut tree whereas the fruits begin to appear during October to January.

Yield: Approximately 45 kg of fruit will yield 13.6 to 16.0 kg of seeds.

References

Chauhan, N.S. 2006. Medicinal and Aromatic Plants of Himachal Pradesh, Second Edition. Indus Publishing Company PS-5, Tagore Garden, New Delhi-110027.

24

Yellow Myrobalan (*Terminalia chebula* Retz.)

Common name: Harad or harar, reer, reed

Family: Combretaceae

Introduction

Harar is a well known species of Sub-Himalayan or Siwalik region and grows upto an elevation of 1500 m above mean sea level. It also abounds in the deciduous forest of Peninsular India. Vernacular names of Harad have been mentioned in Table 24.1.

Table 24.1: Vernacular Names of Harad

Sl.No.	*Language*	*Name*
1.	English	Chebulic Myrobolan
2.	Hindi	Harar
3.	Himachali	Harad
4.	Dogri	Harar, Reer or Reed
5.	Marathi	Hirda
6.	Kannad	Anale
7.	Tamil	Kadakai
8.	Telugu	Karaka
9.	Burmese	Panga

It is an important associate in dry deciduous forests of Siwaliks, *Kandi* belt of Jammu (J&K State), Himachal Pradesh and Uttrakhand as well as Punjab and Haryana. In Jammu region, it occurs in areaas of Jammu, Udhampur, Kathua and Rajouri districts upto 1000 m. In Himachal Pradesh, it grows gregariously on poor rocky sites upto 1050 m either pure or with *Pinus roxburghii* in Kangra and Hamirpur districts, whereas, in sporadic form, "harar" is found in Bilaspur, Nahan and Nalagarh areas.

Harar is known to have laxative or purgative properties. Besides this, "harar" possesses anti-amoebic, anti-tumour and cardiotonic properties (Anonymous, 2002). It has proved useful in diabetes and arrests the activity of HIV virus. "Harar" is a component of famous Ayurvedic preparation – *"Triphala"*, and fruits are also used for drying cotton, wool and leather, and for tanning purpose.

Medicinal Values

The fruit is well known for its medicinal value as laxative, astringent as treatment for chronic stomach ulcers. It is main constituents of *"Triphala"* which is known for stomach disorders. Finally, powdered fruit is used as denitrifice in toothache and considered useful in curing gum bleeding and ulceration. It is also useful in stomatitis as gargle. The fruit paste is used in burns and fruit extract as wash in watering eyes and also in asthama.

The main quality is to regulate the functioning of the body. The dried "harar" fruits are exported in the form of preservative (*murbha*), preserved in honey or sugar syrup. Since, it regulates body functioning, it is also found useful in diabetics especially to pregnant and nursing mothers. It arrests the activities of HIV virus and is used as one of the ingredient in formulations against AIDS.

Table 24.2: Traditional Methods of Use for Controlling Various Disorders/Diseases

Disorder/Disease	*Method*
Stomach trouble/ Stomach ache	The grounded "harar" and "bahera" mixed with salt is a good remedy for stomach trouble. If taken regularly at least once a week, can check stomach aches.
To clear bowels	The dried "harar" is traditionally been exported in the form of "harar" preserve (*muraba*) for use as purgative.
Vomiting and hiccups	"Harar" powder mixed with honey can be taken in case of vomiting. For hiccups its powder should be taken with water.
Itching and minor skin troubles	In case of itching and skin troubles like boils, fruit is fried in oil, massage which will give instant relief.
Spleen enlargement	For spleen enlargement, its powder should be taken with jaggery or salt with a glass of hot milk.
Chronic constipation	The powder taken daily with jaggery cures chronic constipation.
Dysentery and diarrhoea	For this purpose, its powder should be taken with pipli powder in warm water.
Piles	Cooked in cow's milk, it will help in piles troubles if taken with jaggery.

Other Uses

Fruit of the "harar" is the best source of tannin for tanning leather, and for dying cotton and wool. The fruits are extensively used as auxilliaries with other dye stuff. In combination with iron salts it gives black dyes and with alum yellow dyes. About 30 per cent requirements of tanning industry is met from the fruit of this species. It is remarkable to note that collection and processing of "harar" fruit is a thriving business and has developed into a cottage industry in many areas of Jammu Siwaliks.

Botanical Description

"Harar" is a moderate to large sized deciduous tree, attaining a height of 15 to 24 m and a girth of 1.5 to 2.5 m. Zadoo (1996-97), however, recorded the girth of a tree named *'Raja Reed'* near a village Mathwar in Jammu as 3.44 m at breast height. It has a cylindrical bole of 4 to 9 m with round crown, spreading branches and a short trunk. Its bark is dark brown, often longitudinally fissured, exfoliating woody scales. Leaves are elliptic or ovate in shape, 8.16 cm in length and 4.8 cm wide. They become silky and shining and almost glabrous when mature.

Flowers are bisexual yellowish white in colour with terminal panicle spikes. Fruits drupe, ovoid narrow at base, usually 2.5 to 3.5 cm long, 5-ribbed on drying. Wood of the tree is very hard, fairly durable and is used for agricultural implements and other purposes. The tree is, however, important mainly on account of its fruit, which is the best of commercial myroblans used for tanning in the Indian sub-continent.

Cultural Practices

Climatic requirements: The tree shows fairly a wide adaptation to climate. Under natural conditions, it grows in localities with maximum and minimum temperatures from 37 to 48 degree and 1 to 15 degree Celsius (Anonymous, 2002) respectively.

"Harar" is an important species of the forests of Jammu, Himachal Pradesh, Punjab, Haryana and Uttranchal Siwaliks. It can withstand dry conditions prevailing in *Kandi* belts, where it is contributing a lot to the economic well being for the people by providing an additional income to the poor villagers, ensures livelihood to the poor villagers.

Soil requirements: "Harar" can be found on a variety of geological and soil formations. In Siwaliks, including *Kandi* belt, it grows on boulders and conglomerates having a thin soil layer. It grows on clay and sandy soils also. However, the best growth of the "harar" can be found on loamy sand, sandy loam and loam soils.

Methods of planting: Plantation can be established through nursery raised seedlings during July-August. Generally, 1-2 years old plants with a ball of earth are planted for their better survival. Seed sowing in the nursery is done in the months of June or July. The fruit stones are sown after removing the hardened fleshy covering. Removal of hardened fleshy covering can be done by soaking the fruit stones in cold water for 24 hours followed by stratification in cow dung for about 5 weeks. Optimum spacing in nursery is 15 15 cm. Germination starts within 15 days after sowing the seeds and takes about 90 days to complete, with 60 per cent germination. The seedlings become ready for out planting after 1 or 2 years. The seedlings are raised in nursery with clay-loam or sandy loam soils.

Grafting: The bud grafting technique has been proved successful, especially in "Raja Harar" species. For this purpose, bud material can be had from the superior tree grown in the nursery for further experimentation and multiplication. Vegetative propagation is the only alternative to achieve the twin objectives of quality yield and germplasm conservation. The plus trees of *Terminalia chebula* can be multiplied through vegetative propagation and seed orchards can be established for producing genetically improved seed.

Harvesting: Harvesting of fruits is done when they become ripe *i.e.*, when their colour turns yellowish or reddish. Generally, the harvesting starts from middle of October to middle of November and continues up to end of December depending upon the climate. In cooler climate, the ripening of the "harar" fruits can be prolonged upto March and they begin to fall soon after ripening. After harvesting, fruits are spread for sun drying in open.

Post harvest management: Following steps are involved in processing of the fruits:

1. **Sun drying:** The harvested fruit are sun dried for 3 to 4 days to reduce the moisture content and to dry them.
2. **Grading:** The fruits are graded according to their size, *viz.*, small, medium and large.
3. **Roasting:** The fruits are roasted in an iron pan (*Karahi*) having sand. The pan is kept over a *"bhati"*, heated by burning firewood or brushwood. A wooden spatula is used to stir the fruits constantly in the pan while roasting process is going on. A sieve is used to separate the roasted fruits from the hot sand once they are sufficiently roasted. This stage is determined by change in colour of the fruit to golden brown. It usually takes 15 to 20 minutes to get roasted to the desired extent.
4. **Pressing with a gunny bag:** The hot roasted fruits are pressed by hand with gunny bags. This helps in filling any holes on the surface of fruit and to provide a typical luster and strength to the fruit.
5. **Spreading under shade inside the house and storage:** The roasted and processed fruits are spread on the floor in the rooms within the houses for 5 to 7 days, and, thereafter stored in gunny bags, are ready for despatch to market at Amritsar.

References

Anonymous. 2002. Package of practices for agroforestry species. Directorate of Extension Education, Y.S. Parmar University of Horticulture and Forestry Nauni, Solan, Himachal Pradesh (India).

Rajagopalan, U. 2006. Harad and Baheda – Excellent curative herbs. (Personnel communication).

Zadoo, B.L. 1996-1997. "Raja" Harar (*Terminalia chebula*) Retz – Its importance and need for conservation in Jammu Siwaliks. *Forest News Letter*. 3 (November, 1996 - January, 1997 Issue), 32-38.

Fruit Trees

1

Almond (*Prunus amygdalus* Baill)

Common name: Badam

Family: Rosaceae

Introduction

Almond, is one of the fruit crop which is grown not only in India, but is also in China, Greece, Iran, Italy, Morocco, Spain and Pakistan (Trivedi and Verma, 2014). Global production of almonds is around 1.7 million tonnes (Mt). In India, domestic production of almond is miniscule at 1200 tonnes. Its cultivation is restricted to selected hilly areas of Jammu and Kashmir, Himachal Pradesh and Uttrakhand. India imports 98 per cent of its almond requirement. The exporting countries for India are USA, Australia, Iran and Afghanistan. India ranks 3rd in the world in import and consumption of almond. Consumption of almond in India is increasing annually by 20 per cent due to increase in disposable income and awareness about its medicinal and nutritional values (Trivedi and Verma, 2014). Thus promotion and popularization of suitable varieties among the farmers for cultivation in Himachal Pradesh, Uttrakhand and Jammu and Kashmir would fetch good returns to the farmers.

Medicinal Values

Almond is believed to improve memory. If almond and milk is consumed, it strengthens our memory power (Singh, 2010). Almond milk is prepared by crushing the almond without the outer cover and adding water and sugar to it.

Almond possesses high nutritional profile and are valued for their ability to strengthen the body and smoothen the mucus membranes. Powdered almonds have been historically made into cakes, breads and pudding for people suffering from diabetes. An emulsion or a drink smoothens the intestines and can be used in lung ailments and coughs (Trivedi and Verma, 2014). In Ayurveda and Unani systems of medicines, it is a primary tonic to promote vitality, used to treat anaemia (due to its high iron content and vitamins), mental fatigue, constipation, impotency, disorders relating to respiratory system, and keeps body warm during winter season.

Almonds are a good source of fat which is mostly in the form of mono-saturated fatty acids and as such decreases unwanted cholesterol of body. Almonds contain omega-6 fatty acid which is good for the heart. Almonds have enough amounts of phosphorus and calcium which are required for bone growth. Magnesium and manganese are also available in almond, which are used in carbohydrate metabolism as co-factor of enzymes. Presence of vitamin B_6 in the almonds helps in protein metabolism and restricts the level of homocysteine which is otherwise detrimental to heart. Thus almonds are friendly to the heart like oats and soybean and do not cause increase in weight. Almonds are rich in phenolic compounds which act as strong antioxidants (Trivedi and Verma, 2014). Vitamin E and dietary fibers help in preventing cancer. Vitamin E also contains alpha tocopherol which nourishes the skin and gives a radiant glow (Kaushal, 2014).

Eating almonds on regular basis helps to increase the levels of gut bacteria in abdomen. These bacteria then help in healthy digestion giving clear and healthy skin. Almonds have anti-aging properties. Massaging pure almond oil on face, especially under eyes, helps in delaying wrinkles.

Botanical Description

The tree is a small, deciduous, fully mature tree attains the height of 4-10m with trunk diameter about 30-40 cm. The first young branch is green in colour, which after exposure to sunlight becomes purple and appears grey after the second year of growth. Leaves are about 7.5-12.5 cm long with petiole of around 2.0 to 3.0 cm in length (Trivedi and Verma, 2014). Flowers appear single or in groups during spring season, white or pale pink in colour with 5 petals and diameter 3.0-5.0 cm. The flower is bisexual in nature. Tree starts bearing in the third year after planting and reaches full bearing after six years. Fruit is called drupe containing nut with only one seed but occasionally two.

Cultural Practices

Climatic requirements: Almond require sub-temperate to temperate climate which is found in hilly areas having an altitude of 1000-1500m or more. Almonds require less chilling environment than other temperate fruits like apple, cherry, peach, plum etc. Almonds require only 250-400 chilling hours. Average rainfall required for almonds is 300 to 900m.

Soil requirements: Sandy loam and loamy sand are the best soil for the growth of almonds. Soil pH should be in the range of 6.5 to 7.5 though it can be grown in soils having pH 6.0. Soil rich in organic matter provides proper moisture and aeration for roots. Soils must be well drained.

Propagation: Almonds are cultivated by several methods like root stock propagation, hard wood cutting, grafting and micro-propagation. Micro-propagation is an advance technique used to produce plants from tissues but it requires sophisticated infrastructure and skilled persons. Pre-treatment called stratification is required for breaking the seed dormancy. Seeds should be stored at 1 to 5° Celsius for one month to break dormancy. After this treatment sow the seed at 5-8cm deep in the field at a spacing of 10-15 cm x 1-2 m.

Other technique used for propagation is hard wood cutting. This method is comparatively easy and very common. Trees developed by this method are genetically identical to source plant. The wood cuttings should be 1 to 3 cm in diameter and 30 to 40 cm long. For enhancing rooting, these wood cuttings are dipped in indole butyric acid (IBA) and naphthalene acetic acid (NAA) which are subsequently planted in field about 15 cm deep and 15 cm apart. Once the root stock is available, any method of grafting like whip grafting, cleft grafting, bark grafting may be used during December or January.

Varieties: Usually varieties have been selected on account of earliness or lateness and qualitative attributes. Only few varieties recommended for commercial cultivation include 'California Paper Shell', 'Drake, 'IXL', 'Merced', 'Ne Plus', 'Pranyaj', 'Waris', 'Primorski' and 'Shalimar'.

Irrigation: Although the almonds can tolerate hot climate yet for better production, irrigation is essential. Almond plants are able to tolerate limited rainfall in summer. Because of this, almonds can be easily cultivated in areas where irrigation is problem.

Harvesting and yield: Almond trees start producing fruit at the age of 3 to 4 years, but yield maximum production at the age of 6-7 years and can continue to bear fruit for more than 50 years. The fruit matures in the autumn, 7–8 months after flowering. Average yield in almond is 1068 kg acre^{-1} or 2670 kg^{-1} ha.

References

Kaushal, K. 2014.6 Amazing reasons to eat more almonds *Ayurveda*.1(24): 16.

Singh, C. 2010. Simple ways to tone up your brain, *Ayurveda*. May-June issue, pp.20.

Trivedi, A.K. and Verma, S.K.2014. Almonds - Rich source of nutrients and antioxidants. *Indian Farmers Digest*. 47(12): 22.

2

Aonla (*Emblica officinalis* Gaertn.) Syn. *Phyllanthus emblica*

Common name: Amla

Family: Euphorbiaceae

Introduction

Aonla or Indian goose berry or amla (*Emblica officinalis* Gaertn. syn. *Phyllanthus emblica*) being rich source of vitamin C, is essential for curing scurvy disease. In India, it is known by many other names apart from *amla* or aonla as described in Table 2.1. Traditionally Indian gooseberry is a forest or the household tree in the northern part of the country, especially in the Siwaliks. However, because of the increasing demand of aonla for Ayurvedic System of Medicine, it is now being grown commercially all over the country particularly in Uttar Pradesh, Maharashtra, Gujarat, Andhra Pradesh, Karnataka and Tamil Nadu (Wali, 2011).

The Siwaliks including Kandi belt of Jammu as well as those in Punjab, Haryana, HP and Uttarakhand, were hitherto very famous for native fruits. They include mango, mulberry, jamun, phalsa, girna, karonda, kakoa, aonla, harar and behra.

The people of Siwaliks locally call amla tree as *"Amli"* which find less space with the farmers by virtue of its smaller fruit size.

Table 2.1: Common Names of Aonla in India

Sl.No.	*Language*	*Name*
1.	Hindi	*Aonla*
2.	Assamese	*Chukna Amlaki*
3.	Bengali	*Amloki*
4.	Gujarati	*Amran*
5.	Malayalam	*Amlakam Nellikai*
6.	Marathi	*Amla*
7.	Tamil	*Neelikkai*
8.	Oriya	*Aora*
9.	Sanskrit	*Amilika*
10.	Dogri	*Amla, Amli*
11.	Himachali	*Amla, Amli*
13.	Punjabi and Dogri	*Amla*

Source : Jain, 1991.

Amla grows abundantly in sub-tropical forests in Jammu and Kathua districts, locally known as *Rakhs*. But due to denudation of forests, the number of these trees has gone down. However, in view of its being nutritive and having therapeutic properties and other medicinal values, its cultivation has been again realized. Moreover, it has less water requirement in comparison to other traditional crops *i.e.* maize, pearl millet and wheat in Kandi area, It can also be grown successfully as an agroforestry crop, giving dual income to the farmers as is being done in Tamil Nadu, Uttar Pradesh and Madhya Pradesh. The food and medicinal value *vis-a-vis* technical knowhow of aonla cultivation is described as under:

Food Value

A single full grown aonla fruit possesses about 20 times more vitamin C than that of an orange (Singh *et al.*, 2015). Besides, it is a good source of carbohydrates (14 per cent), mineral matter particularly calcium (0.05 per cent), phosphorus (0.02 per cent) and iron (1.21 per cent). Aonla also contains vitamin B_1 to an extent of 10 mg 100 g^{-1}. Its calorific value is 59 100 g^{-1}. Chemical composition of aonla fruits has been shown in Table 2.2.

Table 2.2: Chemical Composition of Aonla Fruits

Sl.No.	Component	Percent/mg 100 g^{-1}
1.	Moisture	81.2 per cent
2.	Protein	0.5 per cent
3.	Fat (ether extractable)	0.1 per cent
4.	Mineral matter	0.7 per cent
5.	Fibre	3.4 per cent
6.	Carbohydrates	14.0 per cent
7.	Calcium	0.05 per cent
8.	Phosphorus	0.02 per cent
9.	Iron	1.2 per cent
10.	Calorific value	59 mg 100 g^{-1}
11.	Nicotinic acid	0.2 mg 100 g^{-1}
12.	Vitamin C	500 mg 100 g^{-1}
13.	Vitamin B_1	10 mg 100 g^{-1}

Medicinal Values

Indian gooseberry is believed to have medicinal and rejuvenating properties (Anonymous, 2011). It is a natural source of vitamin C and is said to be equivalent to two oranges. Its regular consumption helps to prevent number of health disorders which are as follows:

(i) Fermented liquor prepared from aonla fruits is useful in indigestion, anaemia, jaundice, and certain heart complaints, bad cold and promoting urination.

(ii) Fruits of aonla are one of the three constituents of the well known preparation called *Triphala* in Ayurvedic System, the other two being the harad (*Terminalia chebula*) and bahera (*T. belerica*). *Triphala* is used as laxative and in treatment of enlarged liver, piles, stomach complaints and eye pain.

(iii) Experiments on patients suffering from pulmonary tuberculosis showed that vitamin C of aonla fruits is more quickly absorbed and assimilated in human system than synthetic vitamin C, yet the factors responsible for this are still unknown.

(iv) Raw fruits have cooling effect and are mild laxative in nature. Dried fruits are found useful in diarrhoea and dysentry. Even pickled fruits are prescribed in Indian System Medicine. Seeds of the fruit are reported to cure asthma and stomach disorders.

(v) Fruits are also used for preparing products like Chyavanprash, Amritkalash and Kayakalpam. Kayakalpam is actually a paste which gives strength and complexion to our body.

(vi) Aonla, when taken with honey, helps to balance the body.

(vii) Its juice is good antidote for snake and other insect bites.

(viii) Girls eating aonla, attain their puberty at proper time and at right age.

(ix) In Sidda Medicine System, it has a vital role in curing diseases.

(x) Aonla used as a main ingredient in preparing hair tonic and shampoos. Even its dried fruits are used for washing the hair, a traditional practice.

(xi) A table spoonful each of fresh aonla juice and honey mixed together forms a valuable medicine for the vitiated conditions of tridosha, diabetes, cough, asthma etc. (Verma and Verma, 2009).

Medicinal values of aonla against different diseases have been shown in Table 2.3.

Table 2.3: Medicinal Values of Aonla

Disease	*Parts Used*
Liver disorders	Fruits of *Emblica* are good liver tonic and removes its various disorders.
Indigestion, anaemia, jaundice, cold in nose and certain heart complaints	A fermented liquor made from *Emblica* fruits are considered the best for curing the said diseases.
Diseases caused by deficiency of Vitamin C like scurvy, pulmonary tuberculosis	*Emblica* fruits are rich source of Vitamin C, which is known to be more quickly assimilated in human system than synthetic Vitamin C.
Diarrhoea, dysentery, haemorrhages and bronchitis	Dried fruits are used in preventing the said diseases. Even pickled fruits are prescribed in Indian System of Medicines.
Asthma and stomach disorders	Seeds of the fruit have been found useful in curing these diseases.
Enlarged liver, piles, stomach complaints, pain in eyes	The fruits are one of three constituents of the well known Indian preparation known as *Triphala* (the other two constituents being *Bahera* and *Harar*). *Triphala* has been found as laxative and in treatment of the mentioned diseases.
Diuretic, constipation	Aonla is valued for the said disorders.

Source : Jain, 1991.

Other Uses

1. Many entrepreneurs in the small scale sector are coming out with new products such as aonla juice, aonla squash, aonla preserve, aonla candies and jam, which have increased the demand for aonla in the market.
2. Fruits are also used in inks and dyes. As aonla contains tannin, so fruits along with twigs and bark are used for tanning.
3. Wood is also used as timber and leaves as manure in cardamom plantation; also help in amelioration of alkali soils.

Distribution

Leading aonla growing states in India are - Uttar Pradesh, Madhya Pradesh, Tamil Nadu, Kerala, Andhra Pradesh, Gujarat, Maharashtra, Punjab, Himachal

Pradesh and Jammu and Kashmir. Pratapgarh in Uttar Pradesh contributes nearly 80 per cent of production in the country.

Botanical Description

Aonla is a small or medium sized deciduous tree. It has small leaves, 10-13 mm long, 2-3 mm broad, very closely set in pinnate fashion, branches appear rather feathery in appearance.

Male and female flowers are borne on same tree. They are usually pale green in colour, generally exist in small dense clusters below the leaves. Male flowers are small but numerous on short stalks, while the female flowers are fewer in number but small akin to male flowers.

Fruits are 1.5-2.5 cm in diameter, fleshy, round and indistinctly marked into six lobes. They are yellowish or pale green in colour. Fruit size in the cultivated varieties is bigger as compared to wild one.

Cultural Practices

Climatic requirements: Aonla is a sub-tropical fruit but its cultivation in tropical climate is quite successful. It requires relatively warmer climate and grows throughout tropical and sub-tropical conditions in India. The tree is not affected much by hot wind. However, it is damaged by frost to some extent. The young plants must be protected from hot winds during May to June and from frost during winter upto 3 to 4 years in North India. The mature tree can tolerate freezing temperature as well as high temperature upto 46°C. The warm season appears conducive for initiation of floral buds. It can generally be grown in semi-arid sub-tropical region with annual rainfall of 350 - 500 mm.

Soil requirements: Aonla can be grown in varied type of soils ranging from sandy loam, silty loam to loam. In fact, it can be grown in light as well as heavy soils; the only exception is very sandy soils. But well drained fertile loamy soil is the best. In Siwaliks, it is found to grow even on marginal and wastelands. The aonla trees can also be grown in moderately alkaline soils (pH 7.5 - 8.0). However, high salinity causes excessive accumulation of chlorine in leaves, which is harmful to the plants. Sodicity results in accumulation of sodium in toxic amounts in all cultivars.

Propagation: The plants raised from the seeds are not true to type and produce small sized fruits of inferior quality. Of various methods of vegetative propagation, shield budding is the commercial method. Older trees of inferior type can be rejuvenated by resorting to top working as is done in case of *ber* plants.

Layout of orchard: The orchard should be laid out before the plants are put in pits of 1 m^3. Pits are required to be dug at least one month prior to planting.

Planting time: At the time of digging upper soil is kept separate. After 15-20 days exposure to the sun, these pits are filled with the upper soil first and sub-surface at the top. Prior to filling the pit both kinds of soils are mixed with well rotten farm yard manure or compost (3 - 4 kg each). To each pit (1 m^3), 30 g of lindane, 5 per cent Aldrin dust mixed in 2 kg soil should be added to control termites. Grafted or budded plants are planted during rainy season at a distance of 9 m. The newly planted trees usually shed their leaves a few days after planting despite good care.

In rainfed areas including *Kandi* belt of Siwaliks, seedling root stocks can be raised *in situ* for budding with superior varieties. In such areas, the planting should be taken-up during early part of the monsoon *i.e.* in the month of July. After planting, limited irrigation should be given upto 2 years.

Varieties: The most popular varieties being planted are Krishna, Chakaiya and Kanchan developed by Narendra Deo University of Agriculture and Technology, Kumarganj, Faizabad (UP); Anand-1 and Anand-2 from Gujarat Agricultural University; BSR-1 from Tamil Nadu Agricultural University. Besides these varieties, Banarasi amla and Neelam-7 are famous for growing not only in U.P. but also in the Siwaliks of Himachal Pradesh, Punjab and Jammu and Kashmir. Banarsi amla, the best variety, is fairly large and shiny and yellow in colour. Chakaiya fruit is oval-shaped with soft pulp and nice flavour but the quality is poor. Red tinged aonla fruit is medium to large-sized with red flesh near the base. Green-tinged amla, is a variety with large fruit and green in colour. White streaked amla has medium fruit with white streaks. Katha variety possesses small fruits with greenish white streaks.

The varieties like Banarsi, Chakaiya and Neelam-7 have been found successful for cultivation in *Kandi* areas of Jammu, Jammu and Kashmir State. The yield potential of suitable varieties for commercial cultivation in North India has been mentioned in Table 2.4. The data shown in the Table 2.4 indicate that the highest amount of vitamin C is present in Narendra-7 (Neelam) variety, followed by Chakaiya and Narendra-6 (Amrit).

Table 2.4: Varieties of Aonla Suitable for Commercial Cultivation in North India

Sl.No.	*Variety*	*Amount of Vitamin C per 100 g Pulp (mg)*	*Yield (kg) at the Age of*	
			5 yrs	*10 yrs*
1.	Narendra-6 (Amrit)	640	36	70
2.	Narendra-7 (Neelam)	733	42	80
3.	Narendra-10	625	30	80
4.	Kanchan	600	48	85
5.	Chakaiya	655	25	100
6.	Banarsi	580	50	70

Irrigation: Although it is an arid plant and its initial water requirement is less, yet after one month of their planting, irrigation induces greater growth. Young plants, therefore, require irrigation at an interval of 10 - 15 days especially during summer months (May and June). Drip irrigation has proved very useful in aonla orchards of Tamil Nadu and Kerala to yield 30 kg of fruit tree^{-1} or even more in the third year.

It is point to mention that after fruit setting the aonla tree goes to the dormancy stage. As such it does not require much water for its growth and development, especially local cultivars. However, in the high yielding varieties of aonla, if irrigations as per the schedule (Table 2.5), are provided yield increases considerably. The interval for irrigating the young aonla plants is 10-15 days, while for full grown trees is 15-25 days.

Table 2.5: Irrigation Management for Aonla Fruit from January to June

Month	*No. of Irrigation*	*Time of Irrigation*
January	1	15th to 30th January
February	1	15th to 28th February
March	1	1st to 15th March
April	1	15th to 30th April
May	1	1st to 15th May
June	1	15th to 30th June

Manure and fertilizers: If 15 - 20 kg well decomposed/well rotten farm yard manure or compost (brown or black in colour and amorphous in nature) is applied to young plants and 20 - 30 kg to the well established plants during September to October every year, there is no need to apply inorganic fertilizers.

Pesticides: Aonla being relatively insect and disease-free does not require the use of pesticides.

Other cultural practices: Where the lands are sloppy, water conservation management practices should be followed. Staggered or continuous V-shaped trenches made at suitable distances, are very useful. These conserve water more in the root zone of the plant and are also for checking the soil erosion. The vegetative barriers of *Khas* (*Vetiver*) grass or napier in the lower side of the trenches also help to reduce soil erosion.

Harvesting and yield: Local or *desi* aonla bear fruits quite late, usually 5 - 6 years after planting and fruits ripen during October to March. However, with the introduction of high yielding varieties of aonla, it starts fruiting by the end of second year after planting. An average yield during this period is found to be 10 kg tree^{-1}. The maximum yield of 100-150 kg or more tree^{-1} can be achieved at the age of 5-6 years.

The farmers of *kandi* belt in Jammu get good returns from local amla varieties. They are earning Rs. 2500 to Rs. 3000 per tree. They sell them at the rate of Rs. 8–10 kg^{-1}.

Forms of Aonla Consumption

(i) **Pulp :** For extracting pulp from aonla, its fruits after complete washing are boiled (20-30 minutes) till become soft and then seeds are separated. The pulp is, thereafter, extracted by adding double the water to the weight of segments.

If pulp has to be preserved for longer, it should be heated to about 70°C and cooled to about 40°C. Potassim metabisulphite (1 g kg^{-1} of pulp) should be mixed. The pulp then gets ready and should be filled into clean sterilized glass or food grade plastic containers and sealed with wax.

(ii) **Jam :** For jam making 1 kg pulp, 0.75 kg sugar and 2 g citric acid is required. These are cooked to desired consistency (68 per cent sugar). Essence of flavour particularly of pine-apple can be added. Prepared jam is filled in clean and sterilized bottles and sealed.

(iii) **Chutney :** Ingredients : aonla fruit – 1.25 kg, sugar – 1 kg, salt – 50 g, ginger – 15 g, garlic – 15 g, spices – 25 g, red chillies – 10 g and glacial acetic acid – 10 ml.

Steam the aonla fruits (after complete washing) for 15 minutes in pressure cooker. Remove the seeds and pass them through a mixer to obtain fine pulp. Add sugar and salt to the pulp and water (half of the pulp). Add chopped garlic and ginger to the pulp, and cook the mixture. Tie the spices loosely in a muslin bag and immerse into the mixture. Cooking is done to a desired consistency. Add the glacial acetic acid. Fill this chutney into glass jars, allow cooling and sealing.

(iv) **Pickle:** Different kinds of pickles can be prepared from aonla fruits. Ingredients: Aonla segments – 1 kg, salt – 150 g, turmeric – 10 g, red chillies – 15 g, black pepper – 10 g, methi powder – 30 g and cooking coil - 500 ml.

Fruits are washed and boiled for 15 minutes. Segments are separated and seed is removed. It is kept in sun in polythene bag or glass jar to remove the moisture. The oil is heated to a smoking point and fruit segments are then fried in oil for a short while, salt and spices are mixed, 5 ml glacial acetic acid is added. Cool it and pack in suitable containers.

(v) **Preserve (*Murabha*):** Select large blemish free fruits of *aonla* and after thorough washing with water keep them in 2 per cent common salt solution overnight. Repeat this treatment for two more successive nights. On the fourth morning, remove the fruits from salt solution; wash them thoroughly to remove the salt.

Prick the fruits at many points with a stainless steel fork. Tie these fruits in a piece of muslin cloth and then boil them in 1 litre of water (in which 20 g of alum has been added) till they become sufficiently soft. Weigh the material in the proportion - Amla = 1 kg; Sugar = 1.5 kg.

Take half of the sugar and add it in 1 litre of water. Dissolve, boil and filter through a thick piece of muslin cloth. Add boiled aonla to it and leave them in sugar solution for night. Drain away syrup and add remaining 750 g of sugar to it. Boil the above solution and add fruits, and boil the material till the syrup becomes fairly thick. Put the product in cleaned wide mouthed glass jars or bottles and allow them to cool. Cap them and store in a cool dry place

References

Annonymous, 2011. Berry blast. *The Kashmir Times*. 68 (157): 12 (June 7, Thursday)

Gupta, R.D. 2004. Aonla for health. *The Daily Excelsior*. 40 (206) : 4 [*Sunday Magazine*].

Gupta, R.D. 2005. Peasants of Siwaliks can grow amla-I: (*Phyllanthus emblica*) for health and wealth. *The Kashmir Times*. 62 (263) : 13.

Gupta, R.D. 2005. Peasants of Siwaliks can grow amla-II : (*Phyllanthus emblica*) for health and wealth. *The Kashmir Times*. 62 (284) : 13.

Jain, S.K. 1991. Bitter sweet portions – Nature cure. *Indian Express – Magazine*, March 24, 1991.

Naidu, S. 2003. Vitamin See. *Economic Times*, 7th July, pp 12.

Singh, R., Kumar, A. and Sahu, R. P. 2015. Fruits and Vegetables rich source of antioxidants. *Indian Farmers Digest*. 48 (2): 20-21.

Verma, K.C. and Verma, S.K. 2009. Aonla-ayurvedic wonder plant. *Indian Farmers Digest*. 42(10) : 22-24.

Wali, P. 2011. Indian goose berry. *The Daily Excelsior*. 47 (37): 3 *Sunday Mag.*

3

Apple
(*Malus domestica* Mill.)

Common name: Seb

Family: Rosaeceae

Introduction

Apple is a sub-acid and one of the most valuable fruits. It is a fleshy with tough skin, ranging in colour from greenish-yellow to red and usually 8-10 cm in diameter, and has pinkish or yellow-white flesh. The apple is considered "protective" and most wholesome food. Apart from the energy value, it plays an important role in the normal development of metabolic changes *i.e.* both physical and chemical that occurs within the body and enables continued growth and functioning.

The apple tree is a native of Eastern Europe and Western Asia and has been known to be cultivated since time immemorial. Its mention has been made both in the annals of old China and Egypt, and the Bible. The Scandinavians have called it the "Food of Gods" and believed that apple fruit had lot of medicinal values to regenerate both mind and body of the human beings.

In the Indian sub-continent apple is mostly grown in the Kashmir valley (Jammu and Kashmir State) as well as in some parts of Jammu like Bhaderwah and Bani areas; in Himachal Pradesh, Kulu, Kotkhai of Shimla *vis-a-vis* parts of Solan and Sirmour district and Uttarakhand.

In light of the huge importance of the apple fruit as mentioned above, it becomes necessary to know about the food and medicinal values in details as below:

Food Value

The apple fruit is a highly nutritive and carbohydrate rich food. It is rich source of minerals and vitamins. Its sugar content is about 9.5 per cent. Of this, fruit sugar constitutes about 60 per cent, glucose nearly 25 per cent and cane sugar or sucrose about 15 per cent. The acid content generally increases with sugar content, mostly dominated by malic acid which is completely used in the body.

The skin of the apple fruit should not be discarded. When taken raw with the skin and the flesh it contains more vitamin C than the inner flesh. The skin also contains vitamin A five time more than the flesh. However, these days it is apprehended that the apple skin is toxic due to the presence of pesticides and chemicals, which are invariably sprayed to save the fruits from various diseases and insects.

Medicinal Values

Apple is invaluable for the maintenance of good health and in the treatment of many ailments. The active principle of apple is pectin, natural therapeutic ingredient found in the upper part and the pulp. Pectin aids in detoxification by supplying the "glacturonic acid" needed for the elimination of certain harmful substances. It also helps to prevent decomposition of protein matter in the alimentary canal. The malic acid content in the apple is beneficial to liver, bowel and brain.

The particular ailments in human beings which can be cured through the use of apple (Gupta and Kher, 2009 and Gupta 2015) are as under:

(i) **Anaemia:** Apple being rich in iron, has proved beneficial in the treatment of anaemia. It will be particularly useful in the form of freshly prepared apple juice. It may be taken up to 0.5 kg to 1 kg daily for effective results.

(ii) **Constipation and diarrhoea:** Apple is beneficial in the treatment of both constipation and diarrhea. For preventing constipation, two medium sized apples should be taken every day for clearing bowels properly. Cooked or baked apples are good for controlling diarrhoea.

(iii) **Dysentery:** The use of apple has also been found to be useful in preventing chronic dysentery especially among children. Ripe and sweet apples should be crushed into pulp and given to child frequently in a day ranging from one to four table spoons according to age.

(iv) **Stomach disorders:** For the upset or disordered stomach, whole apple is prepared by slicing and pounded it till it becomes slightly mashed. And after sprinkling honey or cinnamon should be chewed thoroughly prior to swallowing. This should be taken several times between meals for effective results. In all, natural medicinal substance, pectin in the mashed apple serves to create a protective coating by virtue of its qualities as an absorbant and demulscant.

(v) **Headache:** Apple can be used in treating all types of headache. A ripe apple after removing its rind, the inner hard portion taken with little salt every morning during empty stomach provides relief from headache.

(vi) **Heart disease:** Apples are of special value to the heart patients. It is because apples are being low in sodium content but rich in K and P. From ancient times, apples are considered effective remedy for functional disorders of heart. Recently Dr Elizabeth B Conen of California University (USA) has revealed that people who consume plenty of K through food items are likely to escape from heart trouble. Apples being a good source of K help in alleviation of heart disease.

(vii) **Hypertension:** Apples are invaluable due to its diuretic effect causing, thereby, increased secretion of urine and ultimately bringing the blood pressure to normal.

(viii) **Kidney stones:** Use of apples by the patients suffering from kidney stones has been found helpful for their removal. The ripe fresh fruit gives more beneficial results.

(ix) **Eye disorders:** The apple peel water is excellent medicine for the inflamed eyes both as beverage and eyewash.

(x) **Rheumatic afflictions:** Apples are found excellent medicines for gout, arthritis and rheumatism caused due to uric acid. Presence of malic acid in apples is believed to neutralize the uric acid and offer relief to the patients.

(xi) **Dry cough:** Sweet apples are valuable in dry cough if about 250 g of apple taken daily upto a week.

(xii) **Dental disorders:** Tooth decay can be prevented by regular consumption of apples as they possess mouth cleansing properties. It is because malic acid in apple besides neutralizing uric acid also exerts an antiseptic influence upon the germs present in the mouth and teeth when it is chewed thoroughly.

(xiii) **Promotes vigour and vitality:** The apple is the best fruit to tone up the weak conditions of the human body. It removes the deficiency of the vital organs and renders the body strong and stout. It gives strength to brain also as it contains lot of P as compared to fruits and vegetables.

Precautions

As apple has cooling effect so its direct consumption may cause gastric trouble. However, its effect can be pacified when its use is made after sprinkling small amount of black pepper powder and black salt.

Prior to consumption the apples, must be washed thoroughly in order to remove the chemicals (residues) which are sprayed to prevent insects and diseases.

Botanical Description

The seeds are small in size but tastes hot *i.e.*, garam taseer, and bitter. Apple tree like pear, peach, plum and apricot is slightly shade and direct sun tolerant and grows well in light intensity ranging from 2000-8000 foot candles (Sharma and Mir, 2001). In apple, as the shading increases, the percentage of fruit drop also increases progressively and remaining fruits have less weight. Moreover, in shaded apple, starch peak is reached later and fruit maturity gets delayed. The chilling requirement of apple is more as compared to peach. The minimum temperature required for peak flowering of apple is 5-6°C and for fruit ripening varies from 15-19°C. High temperature reduces aroma and colour development in fruit (Sharma and Mir, 2001).

Cultural Practices

Climatic requirements: Apple is a typical temperate fruit crop requiring 1000-1500 hours of rest period below 7°C. The average temperature during growing season should be 21°C to 24°C. Sunshine is essential for colour development of the fruit. In general, a cool climate with low winter temperature and little rainfall in summer is most suitable for apple cultivation.

Soil requirements: Apple is grown on a wide range of soils. However, a well drained slightly acidic (about 6.5 -6.7 pH), loam soil adequately deep (> 45 cm) is ideal. Apple cannot withstand water logging which causes collar rot and death of trees.

Varieties/Cultivars: Some of the important cultivars suitable for J&K state are:

- ☆ ***Early season cultivars***: Mollies Delicious, Starkrimson, Red June, Irish Peach, Benoni and Tydeman's Early.
- ☆ ***Mid season cultivars***: American Mother, Razakwar, Jonathan, Cox's Orange Pippin, Red Gold, Queen's Apple, Rome Beauty and Scarlet Siberian.
- ☆ ***Late season cultivars:*** American Apirouge, Ambri, King of Pippins, Lal Ambri, Sun-hari, Chamure, Red Delicious, Golden Delicious, Baldwin, Yellow Newton and White Dotted Red.

Propagation: The apple plantations raised on seedling rootstock by grafting on dwarf rootstocks will result in early bearing and increase productivity per unit area of land. The common method of propagation of clonal rootstocks is by mound layering (stooling). The desired cultivars can be grafted or budded on seedling or clonal rootstocks and ideally through tongue grafting.

Planting: The normal system of planting is square at a distance of 6 m - 7.5 m during the month of March. In high density plantation on clonal rootstocks, the number of plants per hectare can vary from 500 and 1666 plants per hectare with a planting distance of 5 m x 4 m and 3 m x 2 m, respectively. During planting, the prime consideration should be the planting of sufficient pollinizers to ensure effective pollination.

Training: Training in apple trees is done to give desired shape to a tree during early years. The apple is trained to a modified leader system. The plant is pruned to 90 cm above immediately after planting. In the subsequent years, 3-4 scaffold limbs around the main trunk are developed in different directions.

Pruning: In non-bearing trees, very light and corrective pruning is done. Thinning is done to space laterals, narrow and criss-crossed branches are removed. Pruning in bearing trees should be done to maintain balance between vegetative growth and spur development and for this, light to moderate pruning should be done in bearing trees. Over-crowding branches need be removed. Large cuts should be covered with Bordeaux paste.

Manures and fertilizers: Apply good quantity of well rotton FYM in the basins and application of fertilizers need to be followed depending on the age of plantation as given below:

Table 3.1

Age of Tree (Year)	*Fertilizer Requirement (g/tree)*		
	Urea	*Diammonium Phosphate*	*Muriate of Potash*
1	35	20	50
2	70	45	100
3	105	65	150
4	140	85	200
5	175	110	250
6	245	150	350
7	315	195	450
8	385	235	530
9	455	280	630
10	525	325	730
11	630	385	900
12	735	450	1050
13	840	515	1190
14	940	580	1345
15th year and onwards	1050	645	1500

These fertilizer doses have been recommended for apple fruit trees grown in Jammu and Kashmir and may vary for apple trees grown in other states depending on soil conditions.

Irrigation: In case there is a prolonged dry weather, 2-3 irrigations should be given in June, July and August. In rainfed areas, mulching with FYM, straw or cut grass can conserve moisture and improve soil conditions.

Harvesting and yield: The apple should be harvested at proper picking maturity. The commonly followed maturity indices are: (i) when apples attain proper size peculiar to variety (ii) the fruit should attain intense skin colour and (iii) when in most of apple cultivars seed colour turn brown. The yield of apple varies with rootstocks, cultivars, soil fertility, weather conditions and orchard management practices. However, on an average, an apple tree yields 30 to 50 kg fruit per year.

References

Annonymous, 2013. Package of practices for fruit crops. Directorate of Extension, SKUAST- Jammu (in Press).

Gupta, R. D. and Kher, R. 2009. Apple a day keeps the doctor away. Davis ayurveda for Holistic Health. 1(12): 16-17.

Gupta, R.D. 2015. Why apple a day. The Daily Excelsior. 51 (45): 8 (Sunday Magazine)

Sharma, J.P and Mir, A.P. 2001. Dynamics of cold Arid Agriculture. Kalayni Publishers, Ludhiana - New Delhi-Noida (UP) –Hyderabad-Chennai-Calcutta (Now Kolkata)-Cuttack.

4

Banana
(*Musa paradisiaca* L.)

Common name: Kela

Family: Musaceae

Introduction

Banana has been recognized as the fourth most important global food commodity after rice, wheat and milk. Banana is a multipurpose tree like plant, which is mainly cultivated in Tamil Naidu for its leaf, green fruit (plantain), tender pith (pseudo stem) and fruit (Anuratha and Ganesaraja, 2013). Fruit pulp of banana is rich source of easily digestible carbohydrates apart from number of vitamins like riboflavin, niacin and ascorbic acid, and minerals such as calcium, magnesium, potassium, phosphorus, sodium, iron, manganese and zinc vis-à-vis several medicinal properties. Hence, it is considered as fruit for all ages of people from infants to old ones.

Nutritional Value

As already mentioned banana is an important source of energy and cheapest source of nutrients for the bulk of our population. Banana, in fact, provides more balanced food than many other food crops. In terms of energy, banana is even better as compared to potato. The approximate composition of fruit is given in Table 4.1.

Medicinal Values

As bananas are a great source of potassium (about 400 mg in 100 g of fruit) which helps to lessen one's risk of high blood pressure and stroke and plays a key role in muscles function (Singh, 2013). Bananas are delicious and sweet to eat, making them a good sugar substitute and natural energy source. The fibre in banana helps restore normal bowl action. A medium sized banana has around 108 calories.

Table 4.1: Nutritional Contents of Banana and Plantain in 100g of Fruit

Nutrients	*Banana (Raw)*	*Plantain (Raw)*	*Vitamins*	*Banana (Raw)*	*Plantain (Raw)*
Carbohydrate	23g	32 g	Vitamin A	64 IU	112 IU
Protein	1g	1 g	Vitamin C	8.7 mg	18.4 mg
Dietary fibre	3g	2 g	Vitamin E	0.1 mg	0.14 mg
Energy	89 calories	122 calories	Vitamin K	0.5 mcg	0.7 mcg
Calcium	5 mg	3 mg	Thiamin	0.031 mg	0.052 mg
Magnesium	27 mg	37 mg	Riboflavin	0.073 mg	0.054 mg
Phosphorus	22 mg	34 mg	Niacin	0.0665 mg	0.686 mg
Potassium	358 mg	499 mg	Vitamin B6	0.367 mg	0.299 mg
Sodium	1 mg	4 mg	Folate	205.6 mcg	225.6 mcg
Iron	0.26 mg	0.6 mg	Pantothenic acid	0.334 mcg	0.260 mcg
Manganese	0.27 mg	Nil	—	—	—
Zinc	0.15 mg	0.14 mg	—	—	—

Source: Balamohan *et al.*, 2012.

Magnesium present in banana serves several important metabolic functions in the human body. It plays a role in the production and transport of energy. It is also important for the contraction and relaxation of muscles (Prasad, 2012). Magnesium is involved in the functioning of certain enzymes in the body. Ascorbic acid (vitamin C) in banana helps to cure scurvy disease; characterized by bleeding gums. Unripe fruit is cooked as vegetable, especially for those who suffer from diabetes and haemoptysis. In the dried or preserved state, it is anti-scorbutic and also useful in diarrhoea.

Chapattis made of the dried and powdered fruit is highly effective in dyspepsia, flatulence and acidity. Milk shake of banana is helpful in gastritis. Fermented liquor made from banana is good in dyspepsia. Anemic persons are benefited with meals of banana as it contains iron. Roots are helpful in anemia. Young tender leaves make an excellent eyeshade in eye diseases.

Flower juice mixed in curd is effective in dysmenorrhea and menorrhhagia. extracts of various parts of banana are mixed in many other medicines (Singh, 2005).

Other Uses

In certain parts of Tamil Nadu, bananas are exclusively cultivated for their leaves. Banana fruits are processed to powder flour, chips, jam, jelly, wine etc. Attractive handicraft items like mats, carry bags, etc., can also be made from the banana fibre. The fibres can also be used for making paper pulp.

Botanical Description

The banana is a tropical, world's largest herbaceous perennial herb. The plant is tall often reaching 2-9 m, tree-like with a sturdy main pseudostem (clustered, cylindrical aggregation of rolled leaf stalk bases). As new growth occurs in the centre of the pseudostem, the edges are forced apart. The leaves arranged spirally at the top have large blades with a pronounced central midrib and obvious veins. Leaves grow up to 2.7 m (8.9 ft) in length and up to 0.6 m (2.0 ft) in width. Each pseudostem produces a group of flowers which may also be called the 'banana heart' which bear fruits in a hanging cluster. The perennial portion of the plant is the rhizome, often called a corm. Root development may be extensive in loose soil up to 9 m laterally. The inflorescence is a spike and grows through the centre of the pseudostem bearing male and female flowers. The female flowers appear in rows further up the stem (closer to the leaves) from the rows of male flowers. A red bud at the end of the spike contains the male flowers which die quickly. The male flowers have functional stamens but aborted ovaries. The banana fruits called "leathery berry" comprise of a protective outer layer, or skin, with numerous long, thin strings that run between the skin and the edible inner portion. Each bunch of fruits ("hands") along a central stem consisting of two transverse rows of fruits called "fingers". The seeds are tiny black specks running through the center of the fruit which do not germinate. Fruits mature in about 60 - 90 days after flowers first appear. In commercial plantations, the parent banana plant dies after harvest and is replaced with daughter plant. However, plantation can grow for 25 years or more if managed properly.

Cultural Practices

Climatic requirements: Bananas perform very well under humid tropical climatic conditions but can be cultivated successfully even under humid or semi-humid sub-tropical conditions also. The growth and yield of most of the varieties are better at a temperature range of 25 to 35° Celsius though they can be cultivated at a temperature range of 10 to 40° Celsius (Balamohan *et.al*, 2012). However, under cooler climates, the crop duration is extended than the normal climate.

Soil requirements: Bananas can be grown on a variety of soils such as light sandy soils of coastal plains, loamy alluvial soils of river banks and clayey black cotton soils etc. However, it grows vigorously on well drained sandy loam soils having at least a depth of 1m. The most suitable pH should be in the range of 6.5 to 8.0. Soils with a pH more than 8.0 are not very suitable for its cultivation. At low pH levels (acid soils), the incidence of *Fusarium* will be severe.

Propagation: Bananas are mostly propagated through suckers. There are two kinds of suckers *viz.*, sword and water sucker. The former have a well developed corms attached to their mother plant and a pointed tip with narrow sword shaped leaves during their early stages. The water suckers are broad leaved and detached from mother plants.

Sword suckers having 1.5 to 2.0 kg weight, free from any disease/nematodes, should be selected. They are prepared by trimming off the roots and decayed portion of the corm and cutting the pseudo stem leaving 20cm from the corm (Balamohan *et.al*, 2012).

Planting time and spacing: Banana is generally planted before monsoon rains when the temperature is neither hot nor cold. Mostly, planting is done during June to July. Under wetland system, however varieties like 'Poovan', 'Monthan' and 'Rasthali' are planted during February to April, whereas, 'Nendran' and 'Robustas' are planted during April-May.

Irrigation: First irrigation should be given after 3-4 days of planting. Subsequent irrigations are to be given once in a week for irrigated garden lands and once in 10-15 days for wetlands.

Adaptation of drip irrigation in horticultural crops *viz.*, banana, grapevine, citrus and some vegetable crops etc., has not only resulted in increased yield but also saving water compared to conventional surface flood/furrow/border/basin irrigation method (Chaudhary, 2005). The yield of banana by conventional method of irrigation was recorded 5759 per ha, whereas, its yield was found 8759 per ha following drip irrigation (Chaudhary, 2005).

Manures and fertilizers: Banana requires high amount of NPK for its proper growth and development. The recommended doses of NPK per ha are 1,111kg, 500kg and 1,106kg, respectively (Balamohan *et al.*, 2012). It would be better to use NPK on the basis of soil test. However, the results of various trials in different localities revealed varying NPK requirement in 'Robusta banana' as 200, 30 and 300kg per ha.

Weed management: Yield losses caused by weeds in banana plantation are reported to be 40-70 percent depending upon cultivars and weed intensity. Therefore, regular weeding is done manually. Use of weedicides should be discouraged due to their ill effects on health.

Pests and diseases: For control of pests like corm weevil, pseudostem weevil and banana aphids, organic pesticides should be used instead of inorganic ones. Similarly organic chemicals are required to be applied for controlling diseases like sigatoka leaf spot and panama wilt as well as bunchy top. Herbal extract of neem

(*Azadirachta indica*) can be successfully used in plant protection. In disease control, cultural techniques combined with microbial activity have contributed a lot.

Harvesting and yield: Utmost care in harvesting of bunches should be taken in order to retain fruit quality. Avoid injury or damage to the fingers as they lead to black patches and rotting after the fruit ripens. It is recommended to cut the bunch leaving 30 cm of peduncle above the first hand. Do not stop irrigating the plants at least one week before harvest as it reduces fruit quality and shelf-life.

The average yield of various varieties varies from 30 to 80 tonnes per ha per year. It is 40-50 tonnes per year for 'Robusta' and 'Dwarf Cavendish', and for 'Monthan', it is 30-40 tonnes per ha per year.

References

Anuratha, A. and Ganesaraja, V. 2013. Banana Special–a boon for banana growers. *Agrobios Newsletter*. 11(12): 42.

Balamohan, T.N, Karunakaran, G., Kumar, M. and Manoranjitham, S.K. 2012. Advances in production technology of Banana. *Indian Farmers Digest*. 45(6): 33-39.

Choudhary, D.R. 2005. Drip irrigation system for sustainable of irrigation. *Farmers Forum*. 5(6): 17 to 18 and 25.

Prasad, M.2012. Magnesium for you. *DAV Ayurveda*. 1(21): 18-19.

Singh, C.2013. Top 10 healthiest fruits. *DAV Ayurveda*. 1(22): 16-17.

Singh, S.S. 2005. Fate of the Medicinal Plant Resources of Jammu and Kashmir. Directorate of Environment and Remote sensing, Jammu.

5

Ber (*Zizyphus mauritiana* Lam.)

Common name: Ber

Family: Rhamnaceae

Introduction

Ziziphus mauritiana is also known as Chinese Apple. This species is believed to have originated in Indo-Malaysian region of South-East Asia. It is now widely naturalized throughout the world tropics from Southern Africa through the Middle East to the Indian sub-continent and China, Indo-Malaya and into Australasia and Pacific Islands. In Northern Australia and Fiji, it has now become a weed. Apart from having medicinal values, *Ziziphus mauritiana* can also be used for a number of other purposes like timber, fire wood and as frame work in house construction and for making charcoal etc.

Nutritional Values

Ziziphus mauritiana is quite nutritious and rich in vitamin C. Apart from vitamin C, it is also contains thiamin (vitamin B_1), vitamin B_3 as well as minerals like calcium, iron and phosphorus. It is a rich source of carbohydrates (17 per cent).

Medicinal Values

Fruit is acidic and astringent. It is laxative and given in liver complaints (Singh, 2005). Fruit improves digestion and act as blood purifier. Bark is given in diarrhoea in decoction and powder form. Powdered bark is a good dressing for old wounds and ulcers. Bark decoction is administered in fever and delirium.

Root bark juice is a strong purgative and is also applied externally in gout and rheumatism. Tender leaves make a good paste for application on boils, abscesses, carbuncles. The leaves are also useful in bilious problem, burning of skin and for purifying blood. Leaf drug is also given in scorpion bite.

The seeds are sedative and are taken sometimes with buttermilk to halt nausea, vomiting and abdominal pains in pregnancy. They check diarrhoea and poulticed on wounds. Mixed with oil they are rubbed on rheumatic areas. Leaves besides helping in liver troubles, are also useful for curing asthma. The dried ripe fruit is a mild laxative.

Other Uses

Ber timber is hard, strong, fine grain textured, tough, durable and reddish in colour. It has been used to make legs for boat ribs, frame work in houses etc.

Botanical Description

Ziziphus mauritiana is a small, spiny or thorny tree or a large evergreen shrub up to 3-12 m in height with trunk 40 cm or more in diameter having a medium life span. Bark dark grey or dull black, irregularly fissured (Orwa *et al.*, 2009). It has spreading crown, stipular spines and many drooping branches, zigzag branchlets. Leaves are alternate, in two rows, oblong-elliptic, variable in size with tip rounded or slightly notched base. Inflorescence is axillary cymes, 1-2 cm long, with 7-20 tiny flowers, greenish-yellow, faintly fragrant. The fruit, a drupe is of variable shape and size. It can be oval, obonate, oblong or round. The fruits skin is smooth, glossy, thin and bright in colour. Seed is round to ovoid, irregularly furrowed stone, containing 1-2 elliptic brown kernels each 6 mm long.

Cultural Practices

Climatic requirements: *Ziziphus mauritiana* is a hardy tree that copes with extreme high temperatures; and thrives under xerophytic conditions with an annual rainfall ranging from 470-1200 mm.

Soil requirements: It can be grown on a wide variety of soil types, including cracking clays, sodic soils and deep alluvial soils. It can even grow on laterite soils and flourishes well in alkaline soils with neutral pH or as high as 9.2. However, deep, sandy loam to loam textured soil with neutral or slightly alkaline pH is considered best for its growth.

Propagation: Propagation is most commonly through seeds. The seeds normally germinate within 6 weeks, whereas, seeds free from pulp only need 1-3 weeks to germinate. As seedlings do not tolerate transplanting, therefore, the best alternative is to sow the seeds directly in the field or use polythene bags in the nursery bed. Seedlings are ready for budding in 3-4 months. The seedlings should be grown in full light.

Harvesting and yield: Fruits mature differentially depending upon the time of fruit set. Fruit matures in February-March or March-April.

Generally, the yield is in the range of 80- 200 kg of fresh fruit tree^{-1} year^{-1} when the trees are in their prime bearing age of 10-20 years.

References

Orwa C, A. Mutua, Kindt, R., Jamnadass, R. and Anthony, S. 2009. Agroforestry Tree Data base: a tree reference and selection guide version 4.0.

http: //www.worldagroforestry.org/sites/treedbs/treedatabases.asp

Singh, S.S. 2005. Fate of the medicinal plant resources of Jammu and Kashmir. Directorate of Environment and Remote sensing, Jammu, J&K State.

6

Sweet Cherry (*Prunus avium* L.)

Common name: Cherry, glass

Family: Rosaceae

Introduction

Cherry appears to have been first domesticated in the Greece about 300 BC. Most of the European countries cultivated cherries, with Germany as the centre of cultivation. Most of the present day cherry cultivars are either of European or have originated from America. Cherries accomplish their earlier reproductive phase during severe winter and mature in late spring or summer. In Europe, the cherries have red, black or yellow varieties. They are drupes, meaning a fleshy exterior surrounding a bit or stone.

In India, the cultivation of cherry is mainly confined to the Kashmir valley in the state of Jammu and Kashmir, parts of Uttrakhand and Himachal Pradesh. In Kashmir valley, cherry is the earliest fruit crop which comes into the market and fetches good price. In Jammu and Kashmir, cherry has been grown since long and is very popular among fruit growers. The cherries are used as dessert as well as for juice and jam making. In Europe, a wine called "Krisheasser" is distilled from the fermented pulp of cherries (Mir *et al.*, 2009).

Cherry is not only a tasty fruit but also increases the red blood cells as has been investigated by the experts of London-United Kingdom (Anonymous, 2011). Besides, the cherry possesses a number of medicinal values.

Medicinal Values

Cherry consumption increases number of red blood cells. It is because the cherry has a red pigment known as anthocyanin which assists in increasing red blood cells. It also cleanses the kidneys and liver. Cherry has been found useful for patients suffering from arthritis and high blood pressure.

In fact, the presence of anthocyanins in the cherries serves as powerful antioxidants. The power of this antioxidant is the key to variety of health benefits. Studies have suggested that these disease fighting pigments possess anti-inflammatory, antiaging and anti-carcinogenic properties.

Recent studies have indicated that cherries are very high in iron and disease fighting flavonoids (Singh, 2013). They also have potassium, magnesium, vitamins C and E, folate an heart protective carotenoids. Cherries significantly reduce inflammation, arthritic pain, bad cholesterol and cancer risk.

Cherry is a good source of vitamin C, which helps in curing scurvy disease (Gum disease). Nutritive value per 100g of cherries is carbohydrates 11.8 per cent, protein 1.1 per cent and minerals 0.8 per cent. It is further added that cherry provides 6.4 Kcal of energy per 100 g and is very low in fat. Cherry also contains other important nutrients such as beta carotene, vitamins, potassium, magnesium, iron, fibre and folate.

Botanical Description

"The term cherry red is not just used to describe luscious lips, it is equally indicative of sound health". And not without reason, the reddening of the cherries is due to the powerful antioxidant called anthocyanin. The cherry or sweet cherry is a large sized fruit tree. Locally cherries are called glass. Generally tree bears fruits on spurs which remain productive for about 12 years.

Cultural Practices

Climatic requirements: Sweet cherries are grown on elevated land. Mostly sour cherry production occurs normally when the ambient temperature is about 15° Celsius for a spell of 2-3 months during its productive stage. Hence, the temperature must be greater than 15° Celsius, generally in the range of 16-25° Celsius. Severe rains at fruiting time develop cracks and lead to fungal disease. During hot days when the atmospheric humidity is low, paddy straw or dried grass should be used as mulch material to save the leaves suffering from water scarcity and shrivelling of the fruits.

Soil requirements: Cherries thrive best on well drained sandy loam soils. The soils should be rich in organic matter to hold or retain enough soil moisture during the period of high water use. However, under no situations they withstand water logging.

Propagation and planting: Cherry seeds of late maturing varieties viz; 'Misri' are sown in the nurseries and seedlings are raised. Seedlings should be maintained in the nursery till they attain the pencil thickness (1-2 years). Then they are budded by T- budding method in the 3rd week of July to August when bark splits easily. Grafting can be done in February - March when both scion and stock are in dormant condition.

As most of the orchardists prefer buded plants than those of grafted, so cherry suckers are mostly used as root stocks. In recent years, however, many of the progressive orchardists are using 'Mazzard' and 'Mahalab' root stock (Mir *et al.*, 2009).

For planting pits of 1 m x 1 m size already dugout in the month of November and then refilled mainly by top soil and well decomposed FYM in the ratio of 1:1. The planting is done in the same month before frost *i.e.* between 15th February to ending March at a spacing of 5.5 m x 5.5 m. Care is to be taken that bud union is kept above the ground.

Training and pruning: Like other fruit trees, training and pruning is also considered essential for cherry. This is practised at different stages of tree starting from the time of planting till the meaningful productivity of the tree.

Tree is head back to a height of 1.2 m from the ground level to modified leader system. Criss-cross branches should be removed and maintain 5-6 branches of scaffold limbs at 15-20 cm apart in different directions. Cherry bear fruits generally on spurs which remain productive for about 12 years, so less removal of wood is required than any other deciduous fruit tree.

Manures and Fertilizers: Fertilizer management involves proper placement and timing of fertilizers. For better growth, development and quality of fruits, the application of manures and fertilizer is important. Manure should be applied in the month of December, whereas full dose of P and K and half dose of N should be applied before flowering *i.e.* around March and rest dose of N be applied three weeks after fruit set.

Irrigation: The water requirement is relatively more during spring and early summer while the fruits are under the process of development. Thereafter, the requirement of water decreases.

Harvesting: Cherries are harvested in late spring or summer after the full maturity of fruit is attained and fruits start ripening. Since, picking of cherry involves maximum care to maintain quality and it becomes labour intensive. Over ripening should be avoided. After harvest, fruits are packed in cardboard boxes having different sizes viz; 1, 2 and 5 kg capacity respectively and sent to the nearby city or town on the same day. Average yield in India varies from 15-20 kg per tree.

References

Anonymous, 2011. Cherry fulfills the scarcity of blood. *The Hind Samachar*. 81(4): 10

Mir, M.M, Askary, T.H. and Hasan, Wajid. 2009. Cultivation of cherry in Kashmir valley. *Indian Farmers Digest*. 42(6): 34-36.

Singh, C. 2013. Top 10 healthiest fruits. *Ayurveda*.1(22): 16-17.

7

Citrus Fruits (*Citrus* spp. L.)

Common name: Nimbu, santra

Family: Rutaceae

Introduction

The most important commercial citrus cultivars in India are the mandarin (*Citrus reticulata* Blanco) which is also known as loose skin orange, followed by sweet orange (*Citrus sinensis*) and acid lime (*Citrus aurantifolia* Swingle). Commercially, kinnow mandarin is grown in states like Punjab, Haryana, Himachal Pradesh, western part of Rajasthan and Uttar Pradesh. Apart from this, mandarin is also grown in north-eastern region and Darjeeling areas by the names as Khasi Mandarin and Darjeeling mandarin, respectively.

In Khera district of Gujarat, Akola in Maharashtra, Ferriyakulam in Tamil Nadu acid lime is mostly grown, whereas, in Vidarbha region of Maharashtra and adjoining areas of Madhya Pradesh Nagpur mandarin is grown. Contrary to this, mosambi is grown in Marathawada region of Maharashtra, Sathgudi in Andhra Pradesh and Coorg mandarin in Coorg area of Karnataka (Singh, 1999). Although citrus cultivars of mandarin like Coorg, Nagpur, Khasi and Darjeeling have ecological differences yet are quite akin to each other.

Despite producing more than four million tonnes of orange (worth rupees 1, 16,000 million) during 2000 on 0.2 million ha, India is only third in the list of orange growers, the other two being Brazil (17 m tonnes), USA (9 m tonnes) out of the annual world production of 64 m tonnes of oranges.

All citrus species yield not only valuable fruits but their every part is medicinal, *viz.* roots, barks, leaves, flowers and fruits. They are suitable as home remedies as well as in Ayurvedic and Siddha systems of medicines. Like most of the citrus fruits, oranges are rich in vitamin C and are good for skin's collagen (Mohindra, 2014).

Medicinal Values

Acid lime (*Citrus aurantifolia*) also called Leumu or Nimbu and Kazi Nimbu has large number of advantages. Its use in water quenches the thirst. It is useful for fever, cures indigestion and other stomach diseases. Improves light of the eyes, increases appetite and kills the bacteria of teeth when it rubs on them. It provides strength to skin and helps in dissolving fat. It also removes the pimples and spots of skin.

Acid lime is a very rich source of vitamins, minerals and salt, and therefore, it is a highly nutritious food (Singh, 2005). The fruit is aperients, astringent, antipyretic and sedative. Fruit juice in water is effective in lowering hypertension. The rind of fruit is a storehouse of valuable medicinal ingredients. The rind is rubbed on effected part of the skin for cure or just rubbed for cosmetic purpose. Seed oil gives valuable tincture.

Warm lemon water serves as the perfect good morning drink, as it aids the digestive system and makes easier the process of eliminating the waste products from the body. It prevents the problem of constipation and diarrhoea by ensuring smooth bowel functions. One of the major health benefits of drinking lemon water is that it paves away for losing weight faster, thus acting as a great weight loss remedy. Lemon water flushes out body toxins and thus is extremely beneficial for the body. Lemon, a fruit popular for its therapeutic properties helps promote one's immune system and, thus, protects from the clutches of most types of infections. It also plays the role of blood purifier. Lemon is a fabulous antiseptic bestowed on us by the Mother Nature. Lime water juice works wonders for the people having heart problems, owing to its high potassium content. So one must make it as a part of daily routine to drink a glass of warm lemon water in the morning and then open his gateway to enjoy its health benefits (Anonymous, 2007).

Recent studies indicate that among the antioxidants present in *Citrus lemon* (Bara Nimbu or Nibu), particularly flavonoids and flavonol glycosides have antibiotic properties which protect us against ailments like leukemia and cancer. They also protect us against infectious diseases like cholera. Lemon is also known to aid healing process like blood clotting (Gupta, 2013). Apart from vitamin C, lemon is also found to contain limonin, a liminoid which help to reduce the total cholesterol levels in those who have high cholesterol levels.

High level of vitamin C in lemon juice helps to reduce stress and depression. Individuals with a weak immune system are prone to lose vitamin C from their bodies whenever they are faced with stress. Lemon juice has a very calming effect on our nerves, making it ideal drink whenever faced with high level stress.

Gargling with lemon juice has proven to help in curing of throat related infections such as sore throat, oral ulcers and tonsillitis. Drinking lime juice daily has also proven to be quite beneficial for those suffering from peptic ulcers. A lemon also helps in reducing the cellulite content of the body, thereby, detoxifying the body and reduce obesity and effective against osteoarthritis. In skin, it improves collagen levels and repairs tissue damage, helps to prevent sunburn, besides wrinkles and fights free radicals which can cause blemishes and acne.

Sour orange (*Citrus aurantium*) is extremely rich in vitamin A and B_1. Preparations from various parts are given in febrile inflammation, scorbutic. Rind powder is given in ringworms infection. Its juice is diaphoretic, purgative and stomachic. Fruit peel is effective in nutritional deficiencies and in digestive problems. The essential oil from leaves is also a tonic. It is stimulant and is given in fever, inflammation, nervous disorder, scurvy and hysteria. It is a lotion for eyes, skin, hands and all other parts. It is also given in dysentery, diarrhoea, rheumatism and animal bite poisons. Decoction of leaves is given in cachexia, enlarged spleen and is emenagogue. Dry rind is effective in dyspepsia and general debility. Fresh rind is rubbed on face in acne and eczema.

Although orange is considered as one of the most popular and widely consumed fruit having a global appeal yet it is orange juice that is a star the world over. Besides making an excellent cocktails and mocktails, orange juice can be used straight as a great refreshing drink. Its health benefits are immense. The abundance of vitamin C in it helps fight flu and other common viruses. It protects from prostate and breast cancer and improves circulation of blood.

Researchers at the Brigham Young University have studied the orange in detail (Anonymous, 2010) to understand its role for maintaining good health. This popular fruit is well known for its vitamin C and blood protecting antioxidants. Orange has been found better than taking a vitamin C capsules for meeting body's requirement of vitamin C. It has been stated that every time we take carbohydrates, it increases the amount of free radicals in our blood. Over the time, this increases our chance of hardened arteries and heart disease. But eating orange fruit safeguards us from that effect for few hours after every meal. Consume more of fresh whole fruit such as orange to cut diarrhoea as it serves diarrhoea deterrent. Lime juice and sherbet made at home must be preferred for curing diarrhoea.

Oranges contain the substance inositol which regulates serotonin and insulin level resulting in fewer mood swings and relief from depression. It can also help in bringing down fats and reduction of blood cholesterol.

Grape fruit is rich in folic acid which is crucial for oxygen delivery to the brain. Grape fruit is also a key in serotonin production. A study by Tufts University in Boston found that blood levels of folate were lower in people with depression. Try grape fruit with brown sugar and angostura bitters.

Researchers have noted that supplement companies often mix high concentration of extracts from blueberry and blackberry and orange and throw them all together and hope to get good results. They have further identified several combinations of antioxidants that were most synergistic – the compounds hesperidin and naringenin, in particular, appeared to contribute the most punch in the combination.

Citron fruit is not only the cheap source of vitamin C all over the world after Amlaki and Guava but is also effective for correction of anorexia due its sour taste. It also increases the salivation and secretion of other gastric juices related to digestive mechanism. Fresh juice of citron is important in cardio-vascular disorders (Chaudhuri, 2006).

Fruits of all citrus plants are toning astringent and antiseptic. Being acidic, they are capable of combating oiliness (Anand, 2006).

Products made from Oranges

Orange juice, sweet orange oil, the orange blossom, orange blossom honey and marmalade are the main products which are made from oranges.

(i) Orange juice, one of the commodities traded on the New York Board of Trade, Brazil is the largest producer of orange juice in the world.

(ii) Sweet orange oil, a byproduct of the juice industry produced by pressing the peel, is used as flavouring of food and drink and for its fragrance in perfume and aroma therapy. Sweet orange oil consists of 90 per cent d-Limonene, a solvent used in various household chemicals such as to condition wooden furniture and as hand cleaning agent.

(iii) The orange blossom, which is the state flower of Florida, is traditionally associated with good fortune, and was popular in bridal bouquets and head wreaths for weddings for some time. In Spain, orange blossom after drying is used to make tea.

(iv) All parts of the orange are used to prepare marmalade. The pith and pips are separated, and typically placed in muslin bags where they are boiled in the juice to extract their pectin, aiding in setting process.

The zest of an orange peel is often the basis of orange, clementine and honey, and tangerine tea. It may be unpleasant to eat in its natural state but when dried and infused with boiling water, the peel produces a strong aromatic flavor and is known for its immune system boosting properties.

Botanical Description

Citrus is most important fruit crop in the world after banana. There are number of citrus species which are evergreen, thorny, aromatic shrubs and rarely small trees of warm regions. They are also known as orange group of plants and belong to the Rutaceae family.

The orange is a hybrid of ancient cultivated origin, possibly between pomelo (*Citrus maxima*) and tangerine (*Citrus reticulata*). It is a small tree, growing 10 m tall with thorny shoots and evergreen leaves 4-6 cm long. The word orange ultimately comes from Sanskrit Narang. Oranges originated in South-East Asia, either in India, Pakistan, Vietnam or southern China (Anonymous, 2006).

Although there are 600 varieties of oranges yet main varieties being cultivated in commercial scale in India are Blood Red, Mosambi and Satguri. Blood Red is the most important variety in Maharashtra, Haryana, Punjab and Rajasthan, whereas Satguri is extensively grown in Andhra Pradesh. Although Blood Red variety used to grow in Jammu region (J&K state), but now has almost got extinct.

Cultural Practices

Climatic requirements: Citrus is one of the important fruit crops and is grown commercially in tropical and sub-tropical climatic regions of our country. The foothills of Punjab where citrus plants are grown represent semi-arid or sub-moist type (Gupta and Banerjee, 1991). In sub-mountainous area of Jammu locally known as *Kandi* belt, citrus is also grown *vis-à-vis* that of Himachal Pradesh and Uttarakhand, where despite good rainfall (1000 mm annually), scarcity of water is prevalent.

Soil requirements: Citrus fruits can be grown on wide variety of soils. As for example, in north-west India, where citrus is grown, the soils are mostly alluvial in nature with soil texture ranging from sandy loam to loam and medium in fertility. On the other hand, the soils of north-east India are though alluvial in nature, yet vary in soil texture from sandy loam to clay loam and low to medium in fertility. Citrus growing areas of central India are mostly clay in nature (Srivastava *et al.*, 1999) with medium to high fertility status. Citrus growing areas of south India are red and laterite type with texture from sandy loam to loam and low fertility status.

An ideal soil pH range for citrus growing is considered to be between 6.5 and 7.5 because in this range, soils are more fertile and least troublesome. However, citrus can be grown even in soils having pH as low as 4.5 or as high as 8.5 or 8.9 provided some suitable practices like application of liming in acidic soils and liberal use of organic manures in soils having high pH (8.5 to 8.9).

Requirements for planting: To get the high economic return from the planting citrus orchards, the first and the foremost requirement is to use certified virus free material. For this purpose, the best cultivar for a particular region, a virus free material should be selected based on its true to type nature, superior quality, high yield and pomological description using standard nomenclature. Such kinds of citrus plant material can be obtained from well known citrus plant growing nurseries, preferably from the Horticulture Divisions under Horticulture/Agriculture Universities, Indian Council of Agriculture Research Institutes, and/or Departments of Horticulture and Agriculture.

Time and method of planting: Although citrus plants being ever green in nature can be planted during February to March and July to September (rainy season), yet they should be planted preferably during rainy season. The selected site for planting citrus species should be well ploughed twice or thrice and properly leveled after deep ploughing. Prior to planting, the field should be properly laidout. One meter deep pits with one meter diameter should be dug at least one month before the monsoon at a distance of 5 m 5 m which results in 400 plants per hectare. Rows are required be planted with top soil mixed with 30-40 kg of well decomposed FYM. The level of soil should be kept slightly higher than the pit so that after irrigation it settles to the ground level. Pits should be treated with chloropyriphos (75 g pit^{-1}) to protect it from infestation of white ants (termites). After defining the exact position of the planting pit with the planting board, the soil should be scooped to the centre of the pit and the soil around, the plant pressed down firmly. Thereafter, a small basin should be made around the plant for watering. Immediately after planting, the plants should be properly watered. It is desirable to provide a stake for the plant to avoid damaging, at the grafting joint and to keep it erect.

Water requirement: Generally water requirement of citrus trees is higher than most of the other sub-tropical fruits, it is because of their evergreen nature and active growth and development. The water requirement is dependent upon edaphic (soil, its type, structure, texture, aeration and available depth), climate (temperature, radiation, wind speed, evaporation, rainfall, sunshine hours, vapour pressure) and crop factors (age of the plant, stage of the crop and canopy). As for example, under Dehradun valley condition, water requirement of mature kinnow mandarin has been recorded 60-90 liters per plant at 7-10 days interval depending upon season (Shirgure *et al.,* 1999).Under Maharashtra condition, water requirement of young, middle and mature Nagpur mandarin varied from 4.7 to 11.2 liter per day, 35.0 to 101.9 liter per day and from 57.0 to 166.0 liter per day, respectively.

Irrigation: In Kinnow mandarin, irrigation at an interval of 3-4 days during summer and weekly in winter is sufficient under Punjab soil conditions. Under Vidarbha soil condition in Nagpur mandarin, 28-30 irrigations are sufficient to mandarin for proper growth and productivity in surface method of irrigation. On the whole, in maintaining the health and production of citrus plants at higher level in winter the interval varies from 10-15 days and in summer it is much shorter.

Manures and fertilizers: Before planting, apply 30-40 kg of well decomposed organic manure (FYM, compost) with one kg of superphosphate in each pit and mix it thoroughly in the soils. After planting, follow the fertilizer schedule according to the age of plant as per location specific standard package of practices adopted by State Agricultural Universities. A fully grown fruit bearing plants should be given 60 kg of FYM, 1 kg of urea, 2.5 kg of superphosphate and 1.5 kg of muriate of potash.

Time of application: Optimum manuring at the right time results in better growth and increase in production. Generally, young plants (1 to 5 years) are manured once in a year, while full bearing trees (6 to 12 years of age or more) are manured twice in a year. Total amount of phosphatic and potassic fertilizers are applied at one time and nitrogenous fertilizer is applied in two or three split doses. In India, manuring is generally done twice a year, once during June and another after harvesting during December-January.

How to apply: As roots of most of the citrus plants are confined to the radius of 30 cm and up to the depth of 60 cm. Hence for efficient use of manure and fertilizer, it is exhorted to apply them within this radius. Field (basin) should be given hoeing and manures/fertilizers should be applied in the form of a ring, *i.e.* a basin, around the tree trunk in a moist condition. Manures or fertilizers should be well mixed with the soil followed by a light irrigation.

Harvesting and yield: In India, citrus fruits are harvested at one time leading to mixing of good and poor quality fruits. In acid limes and mandarins, pulling off the fruit is the common practice. Kinnows are harvested by conventional stick beating or by garden secateurs. The picked fruit has to be put into tube which then carries to the ground.

Conventionally, Nagpur mandarins are harvested by snap method using ladders and bags causing buttonhole injury to 8-9 per cent in fruits. Clippers developed by modifying secateurs reduce such injuries. However, harvesting is faster in snap method than in clipping.

The productivity trend under citrus in different citrus growing states as reported by Singh (1999) reveals entirely different scenario. Highest productivity was obtained in Tamil Nadu (17.6 tonnes ha^{-1}), followed by Madhya Pradesh and Gujarat (16.0 tonnes ha^{-1}), Andhra Pradesh (15 tonnes ha^{-1}) and Bihar (10.0 tonnes ha^{-1}). The productivity of citrus in Punjab and Haryana ranges from 8.1 to 8.2 tonnes ha^{-1}, while in Himachal Pradesh the productivity of citrus is very low as compared to Punjab and Haryana. It is about 1 tonnes ha^{-1}. Owing to very low productivity of citrus in Himachal Pradesh, many farmers belonging to Indora block in Nurpur tehsil of Kangra district are abandoning orchards. It is attributed to change of climate.

Orchardists narrated though the climate change had been taking a toll on their citrus and manage orchards for the past seven years, *i.e.* with effect from 2005 but the year 2012 had been bad in particular. The farmers had suffered heavy losses and due to continued loss, a number of farmers are now uprooting their citrus orchards and going for cultivating their traditional rice and wheat crops (Lalit Mohan, 2012).

Although the yield of citrus crop in Punjab is low as compared to south and central India, yet to give an unprecedented boost to the production and productivity of citrus fruits, the state of Punjab has established five citrus estates, namely Abohar, Tahliwala Jattan, Badal, Hoshiarpur and Bunga (Anonymous, 2010). In all the above five estates, all facilities related to "citrus production" would be provided under one roof.

References

Anand, T.K. 2006. Food for your health and beauty. *Ayurveda*. December Issue, p. 16.

Anonymous, 2006. Orange. *Ayurveda*. December Issue, pp. 36-37.

Anonymous, 2007. Warm lemon water benefits. *The State Times*. 12 (271): 6 [September 30].

Anonymous, 2010. Good reasons to have a whole orange. *The Kashmir Times*. 67 (355): 12 [December 23].

Anonymous, 2010. Horticulture - taking Punjab to greater heights. *The Tribune*. 130 (32): 15 [February 2, Tuesday].

Chaudhuri, S. 2006. Cheaper C's. *Ayurveda*. April Issue, p. 13.

Gupta, R. 2013. Lemon - A health beneficial fruit. *Personnel Communication*. [The author is Plant Physiologist in the Office of MDO, Udhampur].

Gupta, R. D. and Banerjee, S. K.1991. Problems and Management of Soil and Forest Resources of North-West Himalayas. Mahajan Book Centre (Now Sahil Book Depot) Last Morh, Gandhi Nagar, Jammu, J&K, India.

Lalit Mohan, 2012. Indora farmers abandoning orchards. *The Tribune*. 132 (120): 10 [May 1, Tuesday].

Mohindra, V.2014. 10 amazing fruits that can make you look beautiful. *Ayurveda* 1 (24) : 10-11.

Sharma, R.R., and Faza Ahmad Md. 1999. Nutritional management in citrus. *Intensive Agriculture*. 37 (12): 13-15.

Shirgure, P.S., Srivastava, A.K. and Singh, S. 1991. *Intensive Agriculture*. 37 (1-2): 12-14.

Singh, S. 1999. Citrus industry of India current status and future strategies. *Intensive Agriculture*. 37 (1): 1-6.

Sood, S. 2010. 10 herbs for tears. *Ayurveda*. 1 (18): 4-5.

Srivastava, A.K., Singh, S. and Shirgure, P.S. 1999. Citrus soils - their nature and properties. *Intensive Agriculture*. 37 (1-2): 28-31.

8

Coconut (*Cocos nucifera* L.)

Common name: Nariyal, khopa, khopra

Family: Araceae

Introduction

Coconut, commonly referred to as "Tree of Life" as well as "*Kalpa Vriksha*" and "King of Palms", is a versatile tree, which translates as a tree that gives all that is necessary for life. It provides livelihood to billions of people across the world. Coconut palm serves as variable source of food and drink to millions of people, especially those dwelling in the tropics. India ranks second in the world so far as coconut production is concerned (Mathew, 2008). The country, in fact, has been bestowed with agro-climatic conditions very much suitable to coconut plantation. Out of 29 States and 3 Union Territories in the country, coconut is cultivated in 18 States and 3 Union Territories. However, nearly 80-90 per cent of the total areas under coconut lies in the four southern states of India *viz.*, Kerala, Tamil Nadu, Andhra Pradesh and Karnataka. India has obtained the highest productivity of coconut among the other major coconut growing countries of the world. The crop's contribution to the GDP (gross domestic product) is Rs. 8,300 crores (Rs. 83,000 millions) and it provides livelihood to more than 10 million population.

Coconut is considered to be the most sacred and is used for a number of religious customs/rituals. One of the most common offerings in a Hindu temple is a coconut, especially in Sri Vaishno Devi Shrine. It is also offered on occasions like

wedding, use of new vehicle, construction of new bridge, house and shop, etc. and a number of festivals.

The fibre covering of the dried coconut is removed except for a tuft on the top. The marks on the coconut make it look like the human being. The coconut is broken, symbolizing the breaking of the ego. The juice within, representing the inner tendencies (*vaasnas*) is offered along with the white kernel -the mind to the Lord.

The coconut also symbolises selflessness. Every part of the tree, the trunk, leaves, fruit, coir, etc. is used in innumerable ways like thatches, mats, tasty dishes, oil, soap, etc. Coconut takes in salty water from the soil and converts it into sweet nutritive water that is beneficial particularly to the sick persons. It has lot of medicinal uses and is used in preparation of many medicines in Ayurveda as well as other systems of medicines.

Medicinal Values

Coconut's water being cooling in effect cures fever, quenches the thirst and prevents cholera. When coconut water is dabbed on face and left over for 30 minutes, it cures acne (Anand, 2006). It gives relief to the patients suffering from cough. But after taking coconut water, one should not drink water for a while. Coconut water is proved beneficial for small pox suffering patients and diabetic patients.

As coconut oil consists of 66 per cent of medium chain triglycerides (MCTS) so it is cholesterol free. Thus, when this oil is consumed, its MCTS are processed by the body in different ways. They require less energy and fewer enzymes to break them down for digestion. They are broken down immediately and help in the process of excess calorie burning and weight loss. Coconut oil also contains lauric acid (50-55 per cent), which helps in lowering the blood cholesterol and also in the formation of anti-agent components. These substances also assist preventing obesity, heart diseases, cancer and other chronic degenerative diseases (Sahni, 2011).

Coconut oil lowers the cholesterol as a direct result of its ability to stimulate the thyroid (Anonymous, 2011). When sufficient thyroid hormones are present in the body, cholesterol is converted into steroids. These steroids include progesterone, dehydro-epiandrosterone and pregnenolone. These are essentially for preventing heart disease and other degenerative diseases. Damage to the body from free radicals which cause age spots, wrinkles and sagging skin is also effectively countered by the antioxidant properties of coconut oil.

Coconut oil has been found excellent for deep frying and baking, thanks to its higher smoking point. Generally, oils have a tendency to release certain byproducts when overheated. Mostly these byproducts are toxic and are found to be a cause for several diseases including cancers (Anonymous, 2011).

Very little amount of coconut oil gets absorbed into the food stuffs, while deep frying compared to other cooking oils which result in comparatively low consumption of oil. It is not damaging to the health even if one uses it on a regular basis. Another plus point of coconut oil is that it possesses anti-bacterial properties.

It is concluded that coconut water is the natural supplement of electrolyte solution and administered in various diseases caused by increased body temperature. Ripe fruit is good for urinary diseases and muscular growth. Besides, nourishing hair as hair oil, it is used for cooking which is also helpful for obese to curtail the extra fats. *Narikela Lavana* is one of the good formulated medicine for indigestion, hyperacidity, abdominal colic pain, etc.

Botanical Description

Coconut is a tall, perennial, oil yielding, unbranched, erect, cylindrical, pillar-like stem reaching up to 25-30 m mainly in coastal areas. The stem is without cambium, develops from the single terminal bud called the 'cabbage' which is the palm's only vegetative growing point. Stem growth is fastest at early stages, which can be as much as 1.5 m per year. The palm has adventitious roots produced from the basal portion, no taproot or root hairs but lot of primary roots.

The first leaves of a coconut seedling have the pinnae fused together and appear as entire leaves, subsequent leaves tend to split into leaflets forming a rosette. A mature leaf is 3-4 m long and has 200-250 leaflets and remains on the tree for about 3 years before shedding. Leaves are fully opened with strong petioles attached to the stem. The age of an adult palm is correlated with the number of leaf scars. The number of scars on the stem, divided by 13, gives the approximate age of the palm in years (Mahindapala and Pinto, 1991).

The inflorescence known as 'spadix' enclosed in double sheath or spathe, borne singly in the leaf axil. Flowers are monoecious and male flowers outnumber female flowers. The coconut fruit is a large, fibrous dry drupe ovoid in shape, up to 40 cm long and 30 cm wide. Exocarp is smooth, green to red brown or even ivory in colour. Mesocarp is white and endocarp is a nut with hard shell. Shell and the kernel enclose thin brown seed coat called testa. Its fruit is very hard on outer surface and soft kernel inside with tasty and nutritionally rich water. It is available both in raw and ripe forms. Fruit takes about one year to mature.

Cultural Practices

Climatic requirements: Coconut can be grown in humid tropical climatic conditions. The coastal ecosystem is the most suitable for its cultivation. Nearly 90 per cent of the total areas under coconut lies in the four southern States of India, namely Kerala, Tamil Nadu, Karnataka and Andhra Pradesh (now Andhra Pradesh and Telengana).

Soil requirements: Deep (not less than 1.5 m depth) and well drained soils are the most suitable for the cultivation of coconut. Shallow soils with underlying hard rocks, low lying areas which are subjected to water stagnation and heavy soils should be avoided.

Propagation

1. **Selection of mother palms for nursery:** The mother palms from which nuts are to be selected for nursery sowing should bear the following characteristics:
   (a) Regular bearing habit and producing not less than 80 nuts per annum.
   (b) Age 20 years or more. If the mother palm is the progeny of elite planting material and yields consistently higher yields for a period of not less than 6 years, seed nuts can be collected from such palms. There is no need for insisting 20 years in such conditions.
   (c) More than 30 fully opened leaves with short, strong petioles and wide leaf base should be firmly attached to the stem.
   (d) Bearing at least 12 bunches of nuts are required with strong bunch stalks. These nuts should be of medium size and oblong-shaped.
   (e) Husked nuts should have weight not less than 600 g and mean copra content should be 150 g per nut or more.
2. **Collection and storage of seed nuts:** The selected nuts for propagation should be matured ones (at least 11 months old). These should be stored in shade prior to sowing in the nursery.
3. **Planting of seed nuts:** The fields where the seed nuts have to be planted must be well-drained with light soil texture. Beds of 1.5 m width and of convenient length with 75 cm space between beds are recommended. The seed nut should be planted in pits 25-30 cm deep and covered with soil so that the top portion of the husk alone is visible (Selvakumari *et al.*, 2013).
4. **Nursery production and plantation:** After germination, the seed nuts are transferred to the main nursery that can be either in field or polybag. The germinated nuts are placed 30 cm apart in prepared furrows. Two-thirds of the nut should be covered with soil. A narrow bed of 20 to 25 cm, measuring 150 to 180 cm is needed using well textured and friable soil. The optimum period of field nursery is 6 to 8 months.

Spacing trials have shown that plant density of 175 palms (as grown in the nursery stated above) per hectare with a spacing of 7.5 7.5 m is optimum under rainfed conditions (Thamban, 2000). Usually the seedlings are planted before rains in May-June.

Manures and fertilizers: Regular manuring from the first year of coconut planting is essential to ensure good vegetative growth, early flowering and bearing of high yields. The first dose of fertilizer (50 g N + 40 g P_2O_5 + 135 g K_2O_5) should be applied in the third month after planting when the southwest monsoon ends on the west coast, if seedlings are planted before the rains in May-June. In the second year, 160 g N + 120 g P_2O_5 and 405 g K_2O should be applied in two splits during May and September-October. This dose may be doubled during the third year. From the fourth year onwards fertilizers should be applied @ 500 g N + 320 g P_2O_5 and 1200 g K_2O (Srinivasa Reddy and Upadhyay, 2001).

Recently, a low cost technology for the large scale production of vermicompost from the coconut plantation wastes has been standardized with an epigeic local earthworm. Studies at Central Plantation Crops Research Institute have shown that nitrogen fixed by symbiotic cover legume, *Mimosa invisa* – Rhizobium association substituted 50 per cent of the nitrogen requirement of coconut palm, when grown as green manures in the basins.

Long-term studies also showed that highest yield of about 90 nuts per palm per year was recorded in plot receiving both organic and inorganic fertilizers, followed by 80 nuts per palm per year under rainfed conditions. Application of organic manure alone contributed for a yield increase of 10 nuts per palm.

Pests and diseases: Among the various diseases, the root (wilt) disease caused by phytoplasma is a major one. Being a non-lethal, debilitating disease, an integrated approach is to be followed for management of root wilt disease (Ramjegathesh and Valarmathi, 2012). For this purpose, application of *Bacillus subtilis* and *Trichoderma viride* in coconut palms to manage root (wilt) should be made.

Follow integrated nutrient management; apply organic manure @ 50 kg per palm per year. Organic manure must be well decomposed. Apply balanced use of chemical fertilizers. Coconut ripens in 12-13 months after opening of the inflorescence. Harvested nuts are stored in heap under shade for few days. Then nuts are ready for marketing. To get optimum yield fully mature nuts should be harvested.

Harvesting and yield: Yield is usually estimated in terms of the number of nuts produced per palm or unit area and weight of equivalent copra. The national productivity per palm is only 43 nuts, whereas, it is capable of yielding 60-80 nuts per palm per year under rainfed conditions and 90 to 120 nuts per palm per year under irrigations.

References

Anand, T.K. 2006. Food for your health and beauty. *Ayurveda*, December issue, pp. 16.

Anonymous 2011. Coconut oil busts bad fats with flavour. *The Kashmir Times*, 68 999): 7 (The Sunday Times).

Kumar, S.D. and Zechariah, J. 2011. Medicinal and aromatic plants in India. *Agrobios Newsletter*, 10 (5): 45.

Mahindapala, R. and Pinto, J.L.J.G. 1991. Coconut cultivation. Coconut Research Institute, Bandirippuwa Estate, Lunuwila, Sri Lanka. 156 p. Mathew, M. 2008. In search of hidden treasury. *Agriculture Today*, 11 (4): 18-19.

Ramjegathesh, R. and Valarmathi, P. 2012. Coconut root (wilt) disease – over view. *Agrobios Newsletter*, 10 (11): 46-47.

Sahni, N. 2011. Fitness-mantra. *The Tribune*, 131 (49): 4.

Selvakumari, P., Ponnuswami, V., Deepa Devi, N. and Rohini, N. 2013. Nursery propagation of coconut. *Agrobios Newsletter*, 11 (12): 43-44.

Srinivasa Reddy, D.V. and Upadhyay, A.K. 2001. Nutritional requirement of coconut palm. *Intensive Agriculture*, 39 (5-6): 22-23.

Thamban, C. 2000. Central Plantation Crops Research Institute. *Swarna Sedyam*, 4 (2): 40-42.

9

Custard Apple
(*Annona reticulata* L.)

Common name: Sittaphal

Family: Annonaceae

Introduction

Custard apple is believed to be a native of the West Indies, which was carried away in early times to Central America to Southern Mexico as well as in Peru and Brazil. Later on it was spread to south Africa, Malaya, Philippines and south-east Asian countries such as Pakistan and India.

In India, custard apple is grown in parts of West Bengal, especially around Kolkata and runs wild in many areas (Jamwal, 2014). Although the custard apple tree is not so attractive yet its fruit has lot of food value besides variety of medicinal uses.

Food Value

Food value per 100g of edible portion of the custard apple varies widely. Its general composition stands given in Table 9.1.

Medicinal Values

The leaf decoction is given as a vermifuge. Crushed leaves or a paste of the flesh may be poultice on boils, abscesses and ulcers.

The unripe fruit is rich in tannins which after drying and pulversing is employed against diarrhoea and dysentery. In severe cases, the leaves, bark and green fruits are boiled together for 5 minutes in 1 litre of water to make an exceedingly potent decoction.

Table 9.1: General Composition of Custard Apple (Approximate analysis)

Constituent	*Percent*
Edible portion	28-55
Moisture	73-74
Protein	1.60
Fat	0.30
Mineral matter	0.70
Carbohydrates	23.90
Calcium	0.20
Phosphorus	0.04
Iron	1.00
Sugar	12.4 – 18.15
Acidity	0.26 – 0.65
Caloric value	105.00

Source: Prasad *et al.*, 1995.

The fruits of the custard apple are the best source of blood tonics. The roots have strong purgative values and as such are used to remove constipation. The leaves have supportive characteristics. Since the dried seed kernel contains about 27 per cent oil, it is used in soap and paint industry. The insecticidal properties of oil is well documented and is reported to be used by Vietnamese farmers to protect the rice crop from insect pests. Fragments of the root bark are packed around the gums to relieve tooth ache. The root decoction is taken as febrifuge.

Other Uses

The leaves of the custard apple have been employed in tanning and they yield a blue or black dye. A fibre derived from the young twigs is superior soft, fibrous and durable, moderately close grained with a specific gravity of 0.650 and as such is used to make yokes for oxen. Use of the fruit in ice-cream, confectionery and certain other milk products is also proved.

Botanical Description

The custard apple tree is erect, small, deciduous or semi-evergreen tree with a rounded or spreading crown and trunk 3.0 to 3.2m in thickness. The height of the tree ranges from 4.5 to 10.0m. The ill smelling leaves are deciduous, alternate, oblong or narrow, lanceolate 10-20 cm long, 2.5 cm wide with conspicuous veins. Flowers in drooping clusters are fragrant, slender with a outer fleshy, narrow petals 2.5 cm long, light red or purple with spot on the inside at the base.

The compound fruit, 8 to 16 cm in diameter may be systematically heart shaped lopsided, or irregular or nearly round or oblong with a deep or shallow depression at the base. The skin, thin but tough, may be yellow or brownish when ripe with a pink or brown colour.

Cultural Practices

Climatic requirements: Custard apple is generally classified as semi wild fruit due to its spontaneous spread in forests, wastelands and other un-cultivated places. It can be grown under wide climatic limits. Being a tropical tree, it has a tremendous potential to adopt dry climate and can withstand any degree of drought.

In India, it does well from the plains up to an elevation of 1200m. The critical climatic requirement is that it suffers from hot and dry climate during flowering.

Soil requirements: As the custard apple is very shallow rooted tree and as such it can be grown to varied types of soils viz; sandy to clayey having good drainage. However, better growth and reasonably good yields are obtained in sandy loam to loam soils with well drained conditions.

It is pertinent to mention that custard apple is so versatile that even it grows in crevices, joints, faults and fissures of rocks. Further, its natural plantation can be seen commonly on the tops or slopes of the gravelly and rocky undulating terrain.

Propagation: Although seed is the usual means of propagation yet the custard apple tree can be multiplied by budding or grafting on to its own seedling or on to soursop, sugar apple or pond apple.

Pests and diseases: Custard apple is so hardy that it has no serious pests and diseases. However, sometimes it is heavily attacked by chalcid fly. In India, the ripening fruit must be covered with bags or nests to avoid damage from bats.

Harvesting and yield: The custard apple has the advantage of cropping in late winter and spring when the preferred members of the genus are out of season.

Fruits get ripened after four days of their harvesting. Fruits can be safely stored at room temperature with a shelf life of four days when they are treated with calcium carbide. The best way to get the fruits ripened is to wrap them in perforated polythene bags and can be ripened in wheat grains stores. The ripe fruits being soft require careful handling in marketing and are mostly sold in local market. They should be picked up when have all green colour.

The custard apple tree is naturally a fair to heavy bearer. With adequate care, a mature tree will produce 34 to 45 kg of fruit per year.

References

Jamwal, J.2014. Some more useful not very common fruits (Rural Round up). *Personnel Communication.*

Prasad, J., Chary, G. R. and Gajbihye, K. S. 1995 Custard Apple- A boon to Vidarbha. *Intensive Agriculture*. 33(5-6): 17 – 20.

10

Grapes (*Vitis vinifera* Linn.)

Common name: Angur

Family: Vitaceae

Introduction

The grape vine cultivation is prevailing since the time of antiquity. It was introduced into North-India from Iran and Afghanistan during 1300 AD by the Muslim invaders and in South India in 1863 by the Christian missionaries from France. The discovery of cultivar 'Anab-e-Shahi' was made by the Shankar Pillay in 1930 (Bhatt *et al.*2004). It was really a boost to viticulture in India and South India in particular. Now, grapes are being cultivated on large scale both in southern parts (Karnataka, Maharashtra, Andhra Pradesh etc) and north Indian region (Haryana, Punjab, Uttar Pradesh, Himachal Pradesh and Jammu and Kashmir, Uttrakhand) as well. It is also being cultivated in many of the valley areas of North-Eastern hills, especially Arunachal Pradesh.

Now promising varieties of grapes have been developed and its cultivation has also picked up owing to its medicinal uses.

Medicinal Values

Grape fruit (fresh and dried) is given to those suffering from anaemia. It is also for those who are afflicted with wasting diseases. Grape fruits are prescribed in bilious dyspepsia, hemorrhages, dysuria and other urinary troubles. Also prescribed in chronic bronchitis and heart diseases and gout. Fresh juice is given to children in teething and for removing constipation. Recommended for treatment of epilepsy and in colds and fevers. Juice of soar grapes has been found good in bruises and sprains. Dried fruit (resin) is given in cough, fever, catarrh, jaundice, in acute conditions of enlarged liver and spleen.

Ash of the wood is known to prevent stone in bladder, in piles and swelling of testes. Juice of unripe fruit is very effective in throat troubles. Leaf juice is given in diarrhoea and powdered seed is recommended in chronic dysentery (Singh, 2005).

Grapes are rich source of flavonoids which are good for skin and health. Grapes are abundantly rich with antioxidants and anti-inflammatory properties. Apart from its raw form, its juice also aids in preventing pimples and fading the acne scars from the skin. Grapes contain hydroxyl acids which help to keep skin smooth and wrinkle free (Mohindra, 2014). Red grapes contain iron, potassium, fibre and powerful antioxidants. Although red wine gets most of the publicity, dark coloured grapes are the original source of the flavonoids, anthocynins and resveratrol which have shown to prevent heart disease and cancer. One cup of red or purple grapes has 60 calories (Singh, 2013). Fresh grapes are laxative in nature. Wine can also be prepared from grape juice.

Botanical Description

Grape is a vine belonging to the succulent or fleshy fruits group having berries. In tropical climate vines remain evergreen. The plant has characteristic shredding bark, forked tendrils, lobed leaves, elongated clusters, seeded or seedless berries attached to the pedicels.

Fruit is a type of berry which grows in clusters of 15- 300. Bunches are usually compact, prominently shouldered, conical and big in size. Berries are green to amber in colour, medium in size and round in shape. Purple colour of fruit is due the presence of anthocyanins. Grapes are typically an ellipsoid in shape.

Cultural Practices

Climatic requirements: Grapes are suitable for areas where summers prevails during April to September with mean daily temperature from 18 to 27° Celsius while the coldest months are December to February with temperature from 3 to 15° Celsius. The average annual rainfall varies from 600 to 800 mm (Sannigrahi, 1997) with relative humidity ranging from 40 to 80 per cent. Regions having heavy rainfall or hail storms are not suited for grape cultivation.

Soil requirements: Grapes can be grown on well drained, sandy loam to loam soils, slightly acidic to neutral in soil reaction. The soils must be rich in organic matter, potassium, calcium and magnesium with adequate amount of phosphorus and sulphur. The availability of iron, manganese, zinc and copper should be medium to high.

Varieties: 'Pusa seedless', 'Thomson seedless', 'Perlette' and 'Beauty seedless' are the main varieties which can be grown depending upon their performance.

Nursery raising: The grapes are mainly propagated by branch cutting, budding, layering and grafting. Cuttings are collected during pruning in December and January. About 30 to 40 cm long healthy (7-8 mm diameter) matured vines with 4 or 5 buds are selected. They are then put in polythene bags filled with 6.5 kg humus rich soil. Polythene technique in the nursery helps in easy transportation and transplanting of saplings without damaging roots. Sufficient moisture is required to be maintained in the polybags to help better root and shoot growth.

Land preparation: The pits of size 60cm x 60cm x 60cm are dug at a distance of 3m x 3m during October (immediately after the monsoon) and filled with top soil and farm yard manure (1:1 mixture). The layout is usually the square system to facilitate good wire netting system. Rows are perfectly made in east-west direction.

Transplanting: In the month of February and March when atmospheric temperature increases, each pit soil is mixed with 10 kg well decomposed FYM or compost. One year old disease free, healthy saplings raised in polybags are transplanted after removing polythene cover.

Irrigation: Grape is a drought resistant crop. During summer, one irrigation in every month is essential in the hilly areas but during winter, vines remain in dormant condition and irrigation is not essential. One irrigation should be given in February when buds start growing with the rise in temperature.

Training and pruning: Bower system of training has been found very effective. In this system vines are allowed to grow upto 2m height. Where it reaches to criss cross network supported by angle iron pillers or wooden posts, the top portion is pinched off to facilitate side shoot production. Two vigorous shoots in opposite direction are selected at the core level for training as primary arms. The secondary and tertiary shoots are selected accordingly to get a good spread over.

Pruning, the judicial removal of plant parts is usually done in the month of December-January to get quality fruits and good yield. All the overcrowded diseased and damaged shoots are removed completely.

Weeding: During rainy season, there is large growth of weeds. To remove them, time to time weeding in pit sites and grass cutting in vineyard are useful for getting good growth of grape vines and eventually high yield.

Manures and fertilizers: Usually for grape vines in the age group of 3-5 years, 0.5 kg each of urea and potash and 1.0 kg of single superphosphate are applied along with 20 kg of FYM, while for vines above 5 years, 0.5 kg urea, 1.5 kg single superphosphate and 0.75 kg muriate of polash should be added along with 25 kg of FYM. Half of the fertilizers and full dose of FYM are added to each pit and mixed with soil over root zone thoroughly while other half is added just after setting up of berries.

Pest and diseases: Pest infestation on grape vine is very minimal in the hilly areas of North-Eastern region and those of Siwaliks. Sometimes flea beetle and mealy bugs are noticed which can be controlled by spraying any neem based pesticide. Similarly for controlling powdery mildew and anthracnose, any organic pesticide should be used. However, owing to its non-availability, spraying of 0.2 per cent Bavistin just at leaf emergence and twice afterwards at monthly interval controls anthracnose effectively. Powdery mildew can be checked by spraying wettable sulphur (0.2 per cent) or by dusting it at 5-7 days interval during infection.

Harvesting and yield: Grape vines ripen in June onwards depending upon the temperature and start production after 3-4 years. About 15 to 20 kg grape yield is recorded per vine. The berries are good and uniform in size. The cluster length and weight varies from 30 to 40 cm and 300 to 400 g respectively.

References

Bhatt, L, Mishra, D.S and Mishra, N.K. 2000. Grape cultivation in Uttranchal: problems and Prospects. *Indian Farmers Digest*. 37(5-6): 9-10.

Mohindra, V. 2014.10 Amazing fruits that can make you look beautiful. *Ayurveda*.1(24): 10-11

Sannigrahi, A.K.1997. Prospects and problems of Grape cultivation in Arunachal Pradesh. *Farmers and Parliament*. 32 (5): 17-18.

Singh, C.2013. Top 10 healthiest fruits. *Ayurveda*.1(22): 16-17.

Singh, S.S. 2005. Fate of the Medicinal Plant resources of Jammu and Kashmir. Directorate of Environment and Remote sensing, Jammu Government of Jammu and Kashmir.

11

Guava
(*Psidium guajava* L.)

Common name: Amrood

Family: Myrtaceae

Introduction

Guava (derived from 'Arawak' via Spanish, 'Guayaba') is a genus comprising of 100 species of tropical shrubs and small trees. It is native to Mexico, the Caribbean, Central America and Northern-South America. Guava is cultivated in many tropical and sub-tropical countries for the tasty fruit and several species are grown commercially. Guava is the fifth most important sub-tropical fruit crop of India after mango, banana, citrus and apple. The fruit is commonly eaten as whole, but is often prepared in variety of ways as a desert. In Asia, fresh raw guava is often dipped in preserved prune powder or salt. Boiled guava is also used extensively to make candies, jellies, preserves, jams, marmalades and juices. In Egypt, guava juice is very popular. The whole fruit is edible, from seeds to rind but many people choose to cut out the seeds and the middle portion of the guava. Besides these, guava is well known for its medicinal properties.

Medicinal Values

Leaves are utilized for medicinal purpose, as a remedy to diarrhoea and for their microbial properties. In Cuba, the leaves are used in barbecues providing a nice smoky flavour and scent to the meat. Recent studies have shown that guava can help the diabetics to lower sugar level. The claims to assist combat diabetes are still at the testing level and to be concluded (Jamwal, 2007).

Bark is rich in tannin, resin and calcium oxalate, carbohydrates and salts. Leaves and roots contain mineral salts, chlorophyll, tannin and volatile oils.

Ripe fruit is aperients. Fresh fruit is a rich source of potassium even more than banana which helps in controlling heart rate and blood pressure. It is also recommended in gout. Water in which cut fruit is soaked overnight is very effective in diabetes. Root bark is used to cure chronic dirrhoea in children when given in decoction. It is given in cholera to arrest vomiting and diarrhoeaic symptoms. Red variety of guava is considered superior for this purpose, decoction of leaves is applied locally to bad ulcers, scurvy. The decoction also makes an effective gargle for curing swollen gums and ulceration of mouth. An excellent poultice is made from ground leaves (Singh, 2005).

Guava is one of the richest sources of Vitamin C and therefore accelerates the healing of wounds after surgical operation. Fruit is a moderate source of B-complex vitamins such as pantothenic acid, niacin, vitamin-B_6 (pyridoxine), vitamins E and K, as well as minerals like magnesium, copper, and manganese. Twigs make good tooth brush (*Datoon*).

Botanical Description

The plant is an evergreen shrub or a small tree reaching upto 6-7 meters with branched trunk at the base. Mature guava trees are not frost sensitive and can survive the temperature as low as 5° Celsius for short periods of time but younger plants will be damaged. Thus the plants are resistant to drought but highly susceptible to frost. They can be grown up to an altitude of 1200 m above mean sea level.

The leaves are opposite, simple, elliptic to ovate, 5-15 cm long. The flowers are white in colour with five petals and numerous stamens. The fruit is round to pear shaped, ranging from 3 to 10 cm in diameter, and up to 2 cm in some selected cultivars. Fruits have typical fragrance. It has a thin delicate rind, pale green to yellow at maturity in some species, but pink to red in others, a creamy white or orange salmon flesh with many small hard seeds and a strong characteristic aroma.

Cultural Practices

Climatic requirements: Plants require tropical to sub-tropical climate. Under sub-tropical conditions it mainly occurs on the slopes of Shivalik hills.

Soil requirements: Guava mostly grows on sandy loam or loamy sand soils. But can grow in coarse textured soils also viz; gravelly sandy loam or gravelly silty loam. Plants can grow in soils with pH range of 4.5-7.0 and in alkalinity up to 8.5.

Variteis/cultivars: In Guava, a number of selections namely 'Lalit', 'CISH-G-1', 'CISH-G-2', 'Allahabad Safeda', 'Lucknow-49' and 'Red fleshed' have been developed for domestic and export markets.

Planting and management: Proper management and care is must for successful cultivation and production of fruit crop. The grafted seedlings are cheap, easy to establish, hardy and come earlier in flowering and fruiting than the seedlings raised by seeds. Survival and establishment of seedlings is better if they are raised in polythene bags or earthen pots and transplanted to field when they are about 1 year old or 30-45 cm tall. The seedlings are planted in pits of 90x90x90 cm size at a distance of 5-6 m (Prasad and Rathore, 1988).

For the vegetative propagation, the rootstock is raised from the seed. Seed is sown in raised beds in August. For air layering, a ring of bark 3 cm long on previous years shoots having 1 cm diameter is removed and covered with moss and wrapped with polythene film. It takes 30-40 days for rooting in monsoon season. The rooted air layered shoots are detached and planted separately as individual seedling. For getting good yield, regular manuring, pruning and weeding is very essential. Pruning is done twice in a year. Pruning of established trees should be carried out to retain a manageable height and to open out the canopy. The lateral branches should be allowed to grow 60 to 90 cm and the tips are cut to encourage further branching.

Irrigation: To obtain optimum yield under low rainfall conditions, runoff water can be utilized judiciously after collecting in ponds. During summer, the irrigation interval should be 7 days and during winter it is between 15-20 days.

Conservation of soil moisture with organic mulches, bentonite clay, barriers and anti-transparent material can help in successful cultivation of guava.

Shelterbelt plantation: The wind velocity during summer is around 20 km hour^{-1} in Rajasthan, therefore, make it difficult for successful guava cultivation. As shelterbelt plantation minimizes the adverse effect of scorching wind during summer, it is advisable to plant 3-4 rows of shelter belt trees. Recommended shelter belt tree species are *Acacia* species (*A.nilolica, A. catechu*), *Prosopus juliflora, Cassia fistula, Zizyphus nummularia* etc.

Harvesting and yield: Fruits are harvested when fruit colour changes from dark to light green. Fruit should be harvested every 2-3 days to prevent over ripening. Yield depends upon the age of the crop, cropping pattern followed and weeding as well as pruning. Fruits of 'Lalit' are of medium size weighing about 150 g each.

On an average under optimum-conditions, the yield of guava tree is 3.5 q in grafted plant and 0.99 q in seedling tree.

References

Jamwal, J.2007. The new horticulture plantations. The Kashmir Times. 64(16): 13 (April 27, Friday).

Meena, V.K. and Bhangarva, G.S.2006. Improved technology in Guava plantation. *Agricultural Extension Review* 18(1): 19-21.

Prasad, R.N. and Rathore., S.S. 1988. Fruit trees for agro-forestry in Arid Region. *Farmers Parliament* 23(10): 1-2 and 24-25.

Singh, S.S. 2005. Fate of the Medicinal Plant resources of Jammu and Kashmir. Directorate of Environment and Remote Sensing, Jammu, Government of J&K.

12

Kiwi Fruit (*Actinidia deliciosa*) Syn. *A. chinensis deliciosa* C.F. Liang and A.R. Ferguson

Common name: Kiwi

Family: Actinidiaceae

Introduction

Kiwi fruit was identified as a promising fruit in India during the year, 1984 and was first introduced in hill regions of India. As a result, its cultivation was successful in Himachal Pradesh, Sikkim and Arunachal Pradesh. The farmers earn good income from the sale of Kiwi fruits. However, its full potential is still to be exploited, especially in the hilly regions of Uttrakhand and Jammu and Kashmir. Kiwi fruit was first planted in Lal Bagh garden at Bangalore as an ornamental tree (Mishra *et al.*, 2006; Kaim and Singh, 2011). Over the period, it was extended in the mid hills of Sikkim, Meghalaya, Arunachal Pradesh and other hilly states.

Besides the specific flavour, kiwi fruit is very nutritious particularly vitamin C, and also possesses lot of medicinal uses.

Medicinal Values

Kiwi fruit contains more than twice the vitamin C of an orange. Thus, its use prevents scurvy disease of the gums. Kiwi fruit is also an excellent source of phosphorus, potassium, calcium and magnesium. Fruit is rich in vitamin A and vitamin E. Kiwis have shown to boost the immune system and reduce respiratory diseases. One medium sized Kiwi fruit contains 47 calories and 3 g of fibre. The composition of kiwi fruit at maturity is given in Table 12.1.

Other Uses

In addition to being used as fruit, Kiwi fruit can be preserved in the form of jelly. It can be combined with other fruits in salads and desserts. It is a delicacy for breakfast and makes delicious pies and preserves and can be used in ice creams and converted into a variety of juices.

Table 12.1: Composition of Kiwi Fruit at Maturity

Sl.No.	*Component*	*Value*
1.	Edible portion (per cent)	90-95
2.	Energy value (Kcl/100g)	49-66
3.	Water (per cent)	80-88
4.	pH of juice	3.3-3.8
5.	Total solids	15-18 (g per 100g fresh weight)
6.	Organic acids (as citric acid)	1.0-1.6
7.	Total sugars	7.5-13.0
8.	Reducing sugars	6.0-12.0
9.	Proteins	0.5-1.2
10.	Lipids	0.3-0.9
11.	Pectin	0.3-1.1
12.	Crude fiber	1.1-2.9
13.	Minerals	
	☆ Ash	0.6-11.0 (mg/100g fresh weight)
	☆ Potassium	185-576
	☆ Calcium	25-60
	☆ Magnesium	14-60
	☆ Phosphorus	20-40
	☆ Sodium	3-10
	☆ Sulphur	25
14.	Ascorbic acid	80-300

Source: Kaim and Singh, 2011.

Botanical Description

Botanically kiwi plant is a vine up to a height of 8m. It is dioecious plant. Stem is woody at the base and twining as the top and tendrils provide for support. Leaves are oval, alternate, long petioled. Flowers are white initially turning yellow. Fruits appear in clusters, with brown skin, covered with hair and are oval shaped. Flesh is green containing black seeds arranged in circular pattern in the centre

Cultural Practices

Climatic requirements: Kiwi fruit is very climate specific and requires 1200-1500 hours below 7° Celsius during winter months which is met at 1200-1500 m elevation (Joshi *et al.*, 2007). It is, thus, a temperate crop and need to be planted only in the regions having snowfall. However, the amount of snow fall is immaterial because it grows both in low and high snowfall conditions. Another important aspect is that the maximum temperature should not exceed 30° Celsius. According to Mishra *et al.* (2006), Kiwi is a temperate fruit which requires about 700-800 chilling hours below 7° Celsius to break its rest period in the winter. The plant is a climber which cannot survive in temperature higher than 35°C. Rainfall to the extent of 150 cm per year is sufficient which must be well distributed throughout the growing period.

Soil requirements: The deep yellow, brown and well drained soil having loamy texture and good fertility is considered to be the best for the production of Kiwi fruits (Joshi *et al.*, 2007). It can even be grown on well drained sandy loam soils. The soil should be neutral in reaction (Mishra *et al.*, 2006).

Propagation: Sufficient planting material is required to be arranged for planting. Fortunately Kiwi fruit can be easily propagated by vegetative techniques. However, the commercial methods of propagation are grafting, cutting and layering (Mishra *et al.*, 2006). The propagation through cutting is, however, the most rapid and suitable method of multiplication. The cuttings 0.5-1.0 cm thick with relatively short internodes and 15-30 cm long are the ideal ones.

The cuttings should be planted 6-8 cm apart in rows spaced at 12-15 m. The nursery bed should be thatched or shaded with shade net. Adequate irrigation and drainage should be provided in the nursery bed. The plant will be ready for planting in one year.

The Kiwi plants are also propagated through grafting. Though it takes about 2 years to develop a nursery plant through grafting, yet this method is the easiest and most economical. Layering can be done on one year old branches in the month of July.

Planting time: The ideal time for Kiwi planting is winter season. The planting distance should be kept 5x5m with size of pits 1x1x1m.

Manures and fertilizers: Once the vines of Kiwi grown in orchards begin cropping these are fertilized during spring or early summer. The recommended doses of N, P and K ha^{-1} are 200 Kg, 55 kg and 100-150 kg respectively. A one year old Kiwi fruit requires 4-5 kg of well decomposed FYM.

Irrigation: Since Kiwi fruit has large leaves, so it transpires large amount of water. In the initial stage, the orchard requires regular irrigation. But once it is established, it requires irrigation only at flowering and fruiting stages.

Training and pruning: Kiwi fruits require attention on training and pruning. It requires a permanent structure made of wire and iron angles. The main aim of training is to establish and maintain a good framework of main branches and fruiting arms. It also facilitates soil management, spraying and harvesting properly. No branch should be left on main stem up to 1.5m height during 1st year. Next year, branches should be trained in the opposite direction.

Pruning is essential in Kiwi to get proper fruiting every year. It can be done twice in a year during winter and summer. In Kiwi, only 3-6 buds of the current season's growth are productive. Thus, pruning should be carried out in such a way that the fruiting areas are available every year requiring the wood to be young. Dormant pruning must be completed by mid-February every year.

Harvesting and yield: It takes about 4-5 years for a Kiwi vine to start perfect bearing and give commercial production after about 10 years. Harvesting period varies from region to region but generally harvesting is done during October-December. 4-5 kg fruits per plant is the yield during early years of fruiting.

References

Joshi, K., Pande, H and Pandey, K.K. 2007. Promotion of Kiwi fruit in Uttarakhand. *Indian Farmers Digest*. 40(3): 21-23.

Kaim, J.C. and Singh, M.2011. Production Technology of Kiwi fruit. *Indian Framers Digest*. 44(3): 32-35.

Mishra, Y.D., Singh, R.K. and Tiwari, H.C. 2006. Cultivation of Kiwi fruit in Uttaranchal hills. *Indian Farmers Digest* 39 (11): 22-24.

13

Malay Apple (*Syzygium samarangense* Merr. and Perry)

Common name: Samarang rose apple, jumrool, jamral

Family: Myrtaceae

Introduction

Malay apple is native to Indo-Malayan region throughout Southeast Asian region mainly India, Indonesia, Malaysia, Phillipines, Srilanka and Thailand. It is widely cultivated in the coastal forests (Pusdekar and Thul, 2013). It was introduced in Philippines in pre-historic times and now it is widely grown throughout the islands of Philippines. It is also grown in Thailand, Cambodia, Laos, Vietnam and Taiwan. The tree is frequently cultivated in India mainly in West Bengal, Goa and South India and is of much economic value. Its fruits are eaten raw or may be cooked as sauce. Its wood is used for constructing huts besides having medicinal values.

Medicinal Uses

The flowers are astringent and used to treat fever and to stop diarrhoea. It has been investigated that their main component is tannin. Besides tannin, they also possess oleic acid, B-sitosterol, desmethoxymatteucinol etc. Bark with salt applied to treat cuts and wounds. The roots are found effective to treat itching and provoke menstruation. Locally healers traditionally make plant remedies to get relief from cough, diabetes, headache, high blood pressure, inflammation, constipation and catarrh. High antioxidants and polyphenols present in fruits help to prevent inflammation. They show weak antibiotic action against *Staphylococcus aureus*, *Mycobacterium smegmatis*, and *Candida albicans*.

Other Uses

Wood is reddish, soft to hard but inclined to warp. Although it is difficult to work, yet is employed for construction, railway ties etc. in some countries.

Botanical Description

It is tropical tree which grows to about 12-15 m tall with evergreen leaves. The opposite leaves are nearly sessile, elliptic-oblong, rounded or slightly cordate at the base, yellowish to dark bluish green, 10-25 cm long and 5-12 cm wide, very aromatic when crushed. The flowers are fragrant and yellowish-white in colour, 2.5 cm in diameter with few petals and a number of stamens.

The waxy fruit is bell shaped or oblong, edible berry with colours ranging from white, pale green, purple or crimson, or even black (4-6 cm long) in wild trees. There may be 1or 2 almost rounded seeds. The fruits in some cases are entirely seedless.

Cultural Practices

Climatic requirements: The tree requires extra tropical climate, growing only at lower altitudes up to 1200 m in India. It does best in parts of the country having a long dry season.

Soil requirements: The tree grows vigorously on a wide range of soils from sandy to clayey. The soil must be well drained and fertile, otherwise the crop will be small with poor quality of the fruits. In India, it grows best on the banks of ponds, lakes and streams with optimum drainage and no ponded water.

Propagation: The trees grow spontaneously from seed but superior types are reproduced by layering, budding on to their own rootstocks or on seedlings. If planted in orchards, the trees are spaced 8-10 m apart. Seed germinate in 2-4 weeks and 8 months old seedlings are ready for transplanting, whereas, through cuttings after 6 weeks after rooting.

Season: In India, the tree blooms during March and April and fruit ripen in May and June. In Srilanka, fruits get ripened from March to May, whereas, in Java the flowering occurs from April to June and fruiting from June to August.

Yield: The Malay Apple is a heavy bearer on fertile soil. When the tree is about 5 years old, it may yield a crop of about 700 fruits with 20-70 kg per tree.

References

Pusdekar, M. G. and Thul, A.V. 2013. *Syzygium samarangense*-An under-utilized fruit crop of Tropics. *Agrobios Newsletter*.12(6): 59-60.

14

Mango (*Mangifera indica* Linn.)

Common name: Aam

Family: Anacardiaceae

Introduction

Mango is one of the most ancient tree of India, which is most popular and familiar to common man and probably the commonest fruit tree in the country. It had been receiving royal patronage from remote past to the Mughal Rulers, especially Akbar the great, who is accredited with the largest number of varieties developed during his regime. Cuckoo's joy/Mango, is available throughout India especially in summer and warm places which are very popular due to its delicious taste of the fruit.

"When cuckoo begins singing on the branches of mango tree the whole of the environment gets filled with joy and becomes provocative to men, women, boys and girls. This evinces that mango fruit season is drawing near".

Mango fruit is so well known that it hardly requires any addition. In the jungle among various varieties of mango, the local or desi mango still holds its supremacy over new varieties in quality and quantity. But its stocks have declined fast. Its trees in old orchards and farms are being sold to wood based industries as a result of ignorance and greed of the owners (Singh, 2005). At this rate of destruction, the last tree of 'Desi mango' may disappear soon.

Presently, India covers an area of about 22.05 lac ha and production of 137.92 lac tonnes is closely linked with the history literature and life of Indians (Bhalarao, 2011). It is a fruit which is allured by everybody young or old, and rich and poor due to its easy availability. Amongst various fruit crops, mango occupies a prominent place and is acknowledged as the king of the fruits in the country. This fruit is consumed fresh all over the world more than any other fruit. Mango is the only fruit which possesses flavours of different flowers like rose, saffron, mogra, khas etc. and various fruits like guava, jack fruit, lemon apple and many others.

As a raw fruit, it is utilized for preparing pickle, mango preserve (Murabba) and mango leather (*Amb pappra*). Not only this, the mango fruit is well known to possess food value vis-à-vis number of medicinal values which are presented here under.

Food Value

We are very well acquaint with the importance of fibre in our diet. If one eats or takes mango regularly during the season his bowls and stomach become cleared every morning. Thus, it is a comfortable food for those who suffer from acute constipation.

An average size mango can meet upto 40 per cent of daily fibre requirement. Besides being delicious and rich in vitamins, minerals and antioxidants mango contains enzymes with stomach soothing properties. Presence of maltase enzyme is one of the glaring examples. Mango is an excellent source of vitamin A and vitamin C (Ascorbic acid) as well as good source of phosphorus, potassium and calcium, iron and beta carotene.

Medicinal Values

Very little is known about the medicinal values of mango. It is one of the important trees offering values for common household remedies. Ripe fruit is a rich source of vitamin A and C efficacious for eyes. Fruit juice is restorative tonic in run down health. It is the most invigorating. Juice is fed in apoplexy with amazing results.

In Java and the Philippines tender leaves are eaten as salad in various ailments. Tender leaves are also masticated in gum troubles. Ash of burnt leaves is a house hold remedy for burns, scalds and is rubbed over affected parts. Fumes from burning leaves are inhaled in throat trouble and hiccups. Leaf juice is given in diphtheria, rheumatisms, arthritis. It produces a tonic action on mucous membrane.

Dried mango flowers are astringent and their powdered form is given to cure diarrhoea, dysentery, catarrh of gall bladder. Gum from stem bark is an effective dressing for cracked feet. It is a good substitute for gum arabica of commerce. The gum is also antisphlitic. Extract from all parts of mango possess antibacterial and antifungal properties. Powdered kernel is given in bleeding piles. It is also anthelmintic. Mango kernel is used as a therapeutic agent in diarrhoea and other gastro-intestinal dysfunctions.

In cholera peels of unripe mango fruits are given after grinding them with curd. Powdered mango peels and inner bark of mango trees are used to expel intestinal worms, in uterine bleeding, leucorrhoea, bleeding piles, bleeding ulcers, various urinary infections and diabetes.

In the Ayurvedic System of Medicine, various medicinal properties of the mango tree are attributed to its different parts (Vatsyayan, 2009). Its leaves, roots and bark are astringent, acrid, haemostatic, antiemetic and constipating in nature, whereas, seed kernel is refrigerant, killer of abdominal worms and is also uretine tonic. In high fever due to heat waves, administration of mango mixture made from boiled unripe fruit is highly effective. Powder or decoction of fruit pulp is used in abnormal genital discharges like over leucorrhoe, metrorrhagia, polymenorrhea etc., both in oral and vaginal routes (Chaudhuri, 2006).

The tender leaves have proved beneficial for diabetes. An infusion is prepared by soaking 15 g of fresh leaves in 250 ml water over night and squeezing them well in the water in the morning. About 100 ml of the filtrate should be taken every morning to control early diabetes (Gupta and Mahajan, 2009). Alternatively, about 49g dry powder of leaves is taken once daily in empty stomach. Recently, it has been found that the mango fruit provides various health and skin benefits (Mohindra, 2014). Rich in vitamin A and antioxidants, mangoes aid in making the skin soft and even toned.

Botanical Description

Mango has fascinated the Indian from time immemorial. The luster of mango leaves can be seen in many paintings and sculptures of Ajanta and Elora. The unique fragrance of its blossom and unrivalled sweet taste of its fruit have been immortalized in the ancient Sanskrit literature includes the *Sri Valmiki Ramayana* and voluminous work of Sri Kalidasa. Ayurvedic texts also did not lag behind and have abundantly eulogized the medicinal qualities of mango. It is known as 'Amra and Rasala' in Sanskrit.

Mango is a medium to large evergreen tree with rounded canopy (low to dense and open). Bark is usually dark grey-brown to black, figured irregularly. Tree has a long and unbranched tap root. Leaves are simple alternatively arranged 15-45 cm in length, with prominent secondary veins and appear in flushes. Leaves contain 'mangiferin' (xanthone). Flowering inflorescence is pseudo-terminal, narrow to broad, conical upto 45 cm. Flowers upto 70 per cent are bisexual and remaining are male. Fruit is a fleshy drupe varying in size, colour, flavour and taste depending on the variety and climate. Important feature of mango fruit is the formation of conical projection at the proximal end called beak. Skin is smooth or rough, gland dotted and of varying colours.

Status of Mango in J&K

In Jammu region of Jammu and Kashmir State, hitherto there were large number of mango groves particularly in Jammu district, covering the tehsils of Akhnoor, Jammu and Samba including Kandi area. Many of the people from far flung areas used to visit Jammu city for relishing the delicious sucking mangoes during the season. Lahore and Sialkote cities which are now in Pakistan, are some of the places where from the well to do families used to come to Jammu city before, 1947 to taste various kinds of sucking mangoes as well as to relieve themselves from the sweltering heat from cold water of Ranbir canal. Similarly, a large number of villagers of Jammu and other districts of the Jammu region used to move to Akhnoor town with their relatives for rejoicing the mango season. In fact, the Akhnoor town was previously used to be considered as an abode of mango fruit during the season and the people of the town along with their relatives used to sit on the banks of the river Chenab with mango baskets to rejoice.

But presently, many of the mango groves have become extinct and as such now very less quantity of sucking mango fruit comes to the market for sale. Moreover, three to four decades ago, the ripe sucking mango fruit was so much flooded into the market and people used to prepare mango leather of various kinds *i.e.*, sweet, sour and salty, locally known as *amb pappar*. But due to lack of mango fruit, this vocation has almost vanished now.

Cultural Practices

Climatic requirements: Mango can grow from sea level to an altitude of about 1400 m provided there is no high humidity, rain or frost during the flowering period. Therefore, it would be better to limit the cultivation upto 600 m elevation (Tiwari *et al.*, 2007). It can do well in areas having an average rainfall as low as 25 cm with irrigation facility to as high as 250 cm. However, mango cannot do well in areas which experience frequent rains or high humidity (over 50 per cent).

Soil requirements: Mango can be grown on a wide variety of soils ranging from alluvial to lateritic provided it is deep (2.0 m to 2.5m) and well drained. The suitable soil pH for mango is 5.5 to 7.0 or up to 7.5.

Propagation: A number of vegetative methods have been standardized of which inarching, veneer grafting and stone grafting are the most common. Generally mango is propagated on seedling mango root stocks.

Layout and planting: After selecting the places or locations for growing the plants, pits of 1x1x1x m are usually dugout during summer months. The pits are filled with soil mixed with 20-25 kg well decomposed FYM, 2.5 kg superphosphate, 1 kg muriate of potash per pit.

July to September is the suitable time for planting. A spacing of 10m x 10m is considered sufficient. "Chausa', however, should be planted at a spacing of 12m x 12 m whereas, the cultivar 'Amarpalli' could be planted at a spacing of 2.5 x 2.5 m with a density of 1600 plants ha^{-1}.

Manures and fertilizers: Mango trees having 10 years of age and above are given 100kg FYM year^{-1} tree^{-1} along with 2.2 kg urea, 4.7 kg superphosphate and 1.7kg muriate of potash as recommended by Pantnagar University. These doses may vary for other states depending upon soil conditions.

Farm yard manure, phosphatic and potassic fertilizers should be applied during the month of December and urea should be given before flowering in the month of January or February. Manure and inorganic fertilizers should be applied in a trench of 30 cm wide and 20 cm deep, 1.5 to 2.0 m away from the tree trunk.

Irrigation: In case of bearing trees irrigation at regular intervals (10-15 days) is necessary during the fruit development period, starting from fruit set to full development stage. This helps in improving fruit size and in reducing fruit drop. However, one should stop irrigation at least 2-3 months before the flowering. Furrow or drip irrigation methods can be followed.

Harvesting and yield: The mango fruit ripens in about 90 days after flowering. All fruits on a tree are generally harvested when few fruits ripen naturally and fall from the trees. Another criterion for harvesting is a slight change of colour in some of the fruits. Fruits should be harvested individually along with pedicel (8-10mm) during morning or evening hours and collected gently in baskets to minimize damage.

After harvesting, mangoes are graded according to their size. Based on fruit weight, mangoes are classified into A, B and C class in range of 200-350g, 351-550g and greater than 550g, respectively.

To maintain the quality, proper packing is must which is done either in wooden boxes or bamboo boxes. However, these days perforated cardboard boxes are used. In these boxes, either fruits are individually wrapped with tissue paper packing or paper shavings are used for cushioning. Generally, green but mature fruits store better than those harvested ripe from the trees. A temperature range of 12-13 degree Celsius along with 90 per cent relative humidity is considered to be the optimum for storage of most of the mango cultivars.

The average yield of mangoes irrespective of varieties in our country ranges from about 8-9 tonnes ha^{-1}.

New Approach in Mango Production

According to researchers, for mango yield, one can cultivate "dwarf mango trees" of 6-7 feet (1.8-2.1 m) height to have access to the fruit directly and also increase yield of the tree. Ultra High Density Plantation (UHDP) which is currently being adapted and further researched by Jain Irrigation system Limited, Jalgaon in their field laboratory at Ellaymuthir in Udumalpet, Tamil Nadu.

The main important aspect regarding UHDP in mango is "Change in cultivation pattern, clubbed with proper drip irrigation and fertilizer would help India further consolidate its position as the world leader in mango production and export".

Methods of Planting: Mango grafts of commercial varieties are planted at very close spacing of 3m x 1m or 3m x2 m which allow attaining good growth during initial years and orchard attains commercial production from 3rd year onwards. Special techniques of training and pruning are used to develop sufficient number of branches and shoots. Special care for nutrition management and pest control is followed.

Jain Irrigation System Limited (JISL): This system of irrigation started large scale mango plantation during 1996 by not adopting conventional planting distance of 10 mx10 m, instead they adopted 4.5 m x 4.5 m which is considered as medium density plantation (MDP). Results of MDP were remarkable and the orchard could start commercial production in the 4th year itself and later on yields were better than those from the conventional besides several advantages.

During 2005, JISL expanded its mango plantation in Ellaymuthur village of Udumalpet (Taluka), District Coimbatore and replicated the same planting method. Although JISL, realized commercial yield from 4th year onwards yet the high yields were not enough to recover investments on land development and orchard establishment. Hence, they were further in search of better technology which can give higher yield. Accordingly, three scientists from JISL participated in a conference on mango in South Africa during 2006 and visited UHDP of Tommy Atkin variety of mango on large scale at a farmer cum scientist's field. It gave the impression, that Indian varieties could also be grown successfully under this planting system subsequently JISL determined to begin a semi commercial scale trial in Ellaymuthur farm and planted about 2 ha of UHDP of cultivar 'Alphonso' and 'Ratna'. Encouraged by the initial success of UHDP, they have extended UHDP to about 20 ha by including other commercial varieties like Totapuri (Bangalora) Imam Pasand, Baneshan etc.

Pruning and reduction in gestation period: Under the first UHDP for mangoes in the country, special pruning and canopy management techniques were adopted. Mango has a long gestation period. That's why to reduce it from nine years to five in medium density planting and further down to four in UHDP. The yield ha^{-1} for high volume varieties under traditional planting methods is 10 tonnes which rose to 25 tonnes ha^{-1} under UHDP.

Soils Suitable for adopting UHDP: Ultra high density mango planting can be done at all those places where traditional mango is grown. Specifically it is resorted in well drained soils with pH range from 6.5 to 7.5 having no hard pan or calcium carbonate layer upto 1.5m depth.

Benefits to orchardists: Very low gestation period permits orchardists to reduce cost burden and generation of early income. This technology is very suitable even for small and marginal orchardists (farmers). Yield of the mango is very high and brings more income for the orchardists. It renders sensible bankable project to financial support due to assured high early returns.

References

Bhalarao, P.P. 2011. Ultra high density plantation- A new approach in mango production. *Agrobios News letter* 9(10): 19-20.

Chaudhuri, S. 2006. Cheaper C's. *Ayurveda*. April Issue, pp 3-5.

Gupta, R. D and Mahajan, N. 2009. 10-Herbs for Diabetes. *Ayurveda*. 1(13): 8-9.

Mohindra, V. 2014.10 Amazing fruits that can make you look beautiful. *Ayurveda*.1 (24): 10-11.

Singh, S.S. 2005. Fate of the Medicinal Plant resources of Jammu and Kashmir, Directorate of Environment and Remote Sensing, Jammu, J&K State.

Tiwari, J.P., Mishra, D.S., Chand, S. and Kumar, R. 2007. Mango cultivation: A commercial venture. *Indian Farmers Digest.* 40 (12): 27-31.

Vatsyayan, V.P. 2009. Summer's delight king of fruits. *Ayurveda* 1(13): 29.

15

Olive
(*Olea europaea* L. sub sp. *O. cuspidata*)

Common name: Jaitoon, kao, kau

Family: Oleaceae

Introduction

The olive, native to Mediterranean, Asian and parts of African countries, is evergreen slow growing tree measuring from about 8 m to 15 m in height. The olives have been cultivated since ancient times as a source of olive oil which is more popularly known as "Zatoon Oil". Hence, it is mainly cultivated for olive oil *vis-à-vis* fine wood from the tree and as fruit.

The olive is mostly grown in countries like Spain, Italy, Greece, Turkey, Syria, Tunisia and many other countries including India. Spain, however, has attained monopoly with an area of nearly 2.4 million ha with production of more than 6 million tonnes. In India, parts of Himachal Pradesh and Jammu and Kashmir are the major states where olive is grown. In Jammu and Kashmir, olive can be cultivated in certain pockets of Rajouri, Poonch, Doda, Ramban, Udhampur and Reasi districts and upper reaches of Kathua district. Locally, olive (*Olea europea/cuspidata*) is known as *Kao*. It is evident from the production of 250 kg olive oil produced from olive trees, grown in Udhampur, Ramban and Doda districts (Anonymous, 2008). Olive oil can be sold at high rates ranging from Rs. 1000 to Rs. 1200 L^{-1}. Thus, the production of oil from olive trees is an asset for the farming community. In European countries, olive oil is the main cooking medium. It is further added that along the river Chenab from Reasi to Thathri high quality timber from olive trees is obtained. The wood is also used for making tool handles, agricultural implements and walking sticks. The leaves of the *Kau* are used as fodder due to richness of Ca (3.2 per cent), Mg (0.8 per cent) and K (0.6 per cent) and moderate content of N (0.8 per cent) and P (0.1 per cent) as reported by Gupta and Kher (2009).

Medicinal Values

Olive oil has ideal fat composition, and as such is preferred as edible oil by the Europeans for salad dressing and making dishes (Sharma, 2008). But unfortunately, oil has a very low smoke point and, therefore, it is not advised as sole cooking medium in the Indian context.

The oil from olive fruits is rich in poly-unsaturated fatty acids and free from cholesterol and, hence, has proved useful for heart patients. Besides, it is used in cosmetics, as massage oil and also in soap industry. Recently a vital component of olive oil has been identified that gives protection from heart attack and stroke (Anonymous, 2009).

The mono-unsaturated fats found in olive oil result in low rates of cancer in addition to heart diseases. Olive oil also contains antioxidants like polyphenols which play a pivotal role in warding off some of age-related disorders. Olive leaves appear to selectively block entire virus specific system in the infected host, have the ability to penetrate infected cells and stop viral replication (Dragomirescu and Sood, 2009). As an antioxidant it protects blood vessels from damage and heart from coronary occlusion and infarcet. Olive leaves are also found affective against *Plasmodium falciparum, P. vivax* thereby, preventing malaria. Reports about the benefits of the olive leaf extract in the treatment of malaria are present in literature since 1827. Even one report in 1906 indicated that olive leaves were superior to quinine for malarial infections.

Today Indians are switching to olive oil as cooking medium, on account of reasons of health and wellness. And the biggest challenge is to educate Indian consumers on the benefits of olive oil as a cooking medium. Today, the domestic olive oil consumption is rising 25 per cent annually (Chaudhary and Grover, 2013).

Other Uses

The fruit is pickled and also used as salad, besides being medicinal by important (Bhushan and Jasrotia, 2013). However, olive fruits are mainly utilized for extraction of oil and on a limited scale for table purposes. The oil is not only tasty and nutritious but is also free from cholesterol and is rich in poly unsaturated fatty acids.

What makes olive oil different from other oils must be understood. It is derived from a fruit rather than a seed with one of the highest concentrations of polyphenols, mono-unsaturated fatty acids and vitamin E, compared to other oils and its method of extraction. These special features of olive oil are responsible for several health benefits, which consist of high antioxidant level through polyphenols, vitamins E and A which lower bad cholesterol, blood pressure and cancer (Khosla, 2013).

Botanical Description

Olea cuspidata (Kao or kahu) is a wild species of olive which grows at an altitude of between 1000 to 1300m. This wild species when grafted with *Olea europaea* provides good quality olives for producing quality olive oil (Gupta, 2011). There are seven olive trees in Israel that are over 3,000 years old. But all are healthy and continue to produce fruits, while all others are 1500 to 2000 years old and have proven that the olive tree can live thousands of years (Dragomirescu and Sood, 2009).

It is small, evergreen tree, growing up to 8-15 m with thorny branches. Leaves are opposite, dark greyish-green on upper surface and whitish below. Flowers appearing in spring are white in racemes, produced on old growth, generally in the leaf axils. Fruit is a drupe, 1-2 cm long, fleshy, turning purple-black at maturity, containing one seed only.

Cultural Practices

Climatic requirements: Quite akin to apple trees, olive trees also require chilling hours though varying from variety to variety. Areas where olives are cultivated must be free from frost and hailstorms. The olive tree has a wide range of adaptability and as such can be cultivated at an altitude varying from 1000 m to 2000 m or even more. However, the best growth of olive has been seen in between 750 m to 1200 m especially in Jammu region of Jammu and Kashmir. For optimum productivity, the mean annual temperature should be between 15° and 30° Celsius with an annual average rainfall of 1000 mm.

Soil requirements: The olives can be grown on a wide range of soils ranging from clay/clay loam to loamy sand/sandy loam/sandy. However, the clayey soils must have sufficient amount of organic matter and well drained. Similarly, the sandy type of soils must be rich in organic matter and/or must be supplied through addition of organic manures to make such soils retain moisture and plant nutrients.

The most suitable soil pH for olive cultivation is 6.5 to 8.0. Although the olive can be grown even on calcareous soils yet presence of excessive free $CaCO_3$ is deterimental to its growth. It is because more content of $CaCO_3$ in soils render them alkaline which, restricts the uptake of plant nutrients especially Zn, Fe, B etc.

Although olive tolerates soils which are high in boron and calcium contents but on such soils it records poor growth (Badiyala *et al.*, 1997).

Propagation: The most economical method of olive propagation is semi-hard wood or terminal cuttings. The olive is also propagated from seedlings of *Olea cuspidata*. Recommended planting distance is 8 m 8 m for irrigated areas; 6 m 6 m for rainfed areas. The olive trees are generally trained by modified leader system giving them a globular, semi-spherical shape. Bearing olive trees require minimum pruning. Dead and interfering branches should be thinned out to promote vigorous growth.

Harvesting: As all fruits of olive trees do not ripe at a time so harvesting has to be done frequently. Harvesting of fruits may be done by hand picking or by shaking the limb of the tree with a bamboo pole over a cloth or polythene sheet already spread under the tree canopy.

References

Anonymous. 2008. *The Kashmir Times*. 65 (16) : 6 (January 16).

Anonymous. 2009. Olive oil protects against heart attack. *The Shadow*. 3 (334) : 2 (Friday, April 3).

Badiyala, S. D., Bhagat G. R. and Gupta, R. D. 1997. Soil requirement of different fruit crops. *Farmers and parliament* (Farmer, Parliamentary Forum, New Delhi). 32 (5) : 11-12, 24-25.

Bhushan, B. and Jasrotia, A. 2013. Olive-a viable opportunity. Personal Communication.

Choudhary, M. and Grover, K. 2013. Olive oil-an emerging cooking medium in Indian households. *Agriculture Today*. 17 (3): 52-59

Dragomirescu, A. and Sood, S. 2009. Lesser known miracles of olive leaves. *Ayurveda*. 1 (13) : 4-5.

Gupta, R. D. 2011. Wild fruit cultivation in Kandi areas of Jammu. *The State Times*. 16(95): 6 (April 5, Tuesday).

Gupta, R. D. and Kher, Ravi. 2009. Olive cultivation in Jammu. *The Daily Excelsior*. 46 (70): 6 (12th March, Friday)

Khosla, I. 2013. A heart friendly oil. *The Sunday Tribune Spectrum*. 133 (346): 3 (15 December)

Sharma, A. R. 2008. The truth about edible oils. The Tribune 128 (217) : 14 (Jalandhar, Wednesday August 6).

16

Papaya
(*Carica papaya* L.)

Common name: Papita

Family: Caricaceae

Introduction

Papaya is an important fruit crop of tropical and sub-tropical regions of the country. The important papaya growing states of the country are Andhra Pradesh, Tamil Nadu, Maharashtra, Bihar and North-Eastern states preferably Assam and Tripura. Because of shorter life of the plantation, the area under cultivation varies appreciably in year to year.

Papaya is an excellent whole some fruit, delicious and nutritious. Unripe fruit is cooked as vegetable. Apart from this, in USA un-ripe fruit is used in making soft drinks and is also candied. Above all papaya, is having high medicinal values.

Medicinal Values

All parts of papaya plants possess medicinal values. Papaya leaves are rich in vitamin C and vitamin E. Latex contains digestive enzymes. Ripe fruit is stomachic digestive, carminative and diuretic. Syrups and wines made from papaya are expectorant, sedative and tonic. Milky juice of unripe fruit is anthelmintic. Latex is used as cosmetic and for curing many skin diseases. Green fruit is laxative and diuretic.

Seeds of papaya are vermifuge. Roots are given to the patients suffering from piles. They also act as a generative tonic. Leaves make excellent poultice for nerves, pains and elephantoid growth. The enzyme *papain* is protein digestive. It is also used in food processing and brewing industry world over. *Papain* has milk coagulating properties also besides those of digestive.

Use of papaya in home is beneficial in constipation. To remove pimples rub internal papaya skin after grinding. Use of ripe papaya is also useful to remove the disorders of spleen and liver.

To kill the worms of children, give the boiled seeds of papaya. In addition to improving digestion, its use is also made as the purgatives. In this regard, raw papaya is boiled and then given. It is essential for high blood pressure. Eat papaya before or after a meal, is helpful not only in relieving constipation but also helps one to look young (Bagde, 2009). Papaya is a rich source of vitamins A, B E and as such taking of papaya helps to rejuvenate skin and improve the complexion.

Papaya is an excellent source of potassium. In addition to being a food that is potassium rich, the papaya also helps to prevent cancer and easing ulcer pain. The *papain* in papaya helps cure all sorts of digestive problems (Prasad, 2010).

Papaya prevents rheumatism and reduces urine acidity (Chavan, 2012). Apart from this, papaya corrects menstrual cycle and is found effective in case of delayed periods. However, papaya must not be used in pregnancy. It is because it causes utretic contractions and may leads to abortion.

Papaya helps to lose weight; fibres in papaya help in making stomach feel fuller so one does not have a strong urge to eat more. This can make one to eat lesser resulting into lowering of fat.

Papaya is very good for skin. As it is abundant in antioxidants due to concentration of vitamin A, vitamin C and vitamin E and pepsin (*papain*) as well as alpha hydroxy acids so it is helpful in removing dead skin. Besides mashed papaya applied on skin will exfoliate it in a gentle and natural way. Papaya, in fact has both the restorative as well as preventive properties when it is used in the field of skin care. The fruit enzymes of papaya are widely known to provide and enhance the skin disorders, healing affects like the very common acne problem. Also, fruit enzymes in yellow fruit prove to be great for skin exfoliation (Chavan, 2012).

Botanical Description

It is an excellent herbaceous tree of many varieties. Some are dwarf; others grow to a height of 8-10 m. Some are short lived (2-3 years), others 15-25 years.

A special character of the papaya is the occurrence of *papain* in it. *Papain*, the sap of the tree is the dried latex obtained by making incisions in the papaya fruit, just before it matures.

A hydrolyzing enzyme which breaks down protein in aqueous solution, *Papain* has often been called vegetative pepsin because it contains enzymes somewhat similar to pepsin. However, unlike pepsin it acts in acidic, neutral or alkaline media. Today, 95 per cent of *papain* is used in food and beverage industries.

Cultural Practices

Climatic requirements: Preferablly humid tropical climate is considered the best for cultivation of papaya. It can be cultivated upto an elevation of 1000 m.

Soil requirements: A well drained sandy loam soil is the most suitable for papaya cultivation.

Propagation: Papaya is commonly propagated by seed. About 250 to 500 g of seed is sufficient to raise seedlings for one hectare of land depending upon of seeds 2 or 3; (Biswas and Kerimenla, 2012; Pathak and Singh, 2000). Since it is cross pollinated crop, plants raised from the seed shows different levels of performance. Hence, it becomes utmost necessary for the farmers to collect the seeds from good and reliable sources.

Sowing of seeds is done in the April-May or May- June so that the seedlings get ready for transplanting before the onset of monsoon *i.e.* July-August.

Nursery management: Seeds are sown in the nursery beds, prepared after ploughing at least twice and planking, at a depth of 1 cm in rows spaced 15 cm and between two seeds 2.5 cm. But most commonly seeds are sown in the polythene bags. Four to five seeds are sown in each bag filled with sand, soil and farm yard manure in the 1:2:1 ratio *i.e.* in equal proportions. After about two months of the sowing, seedlings will be ready for transplanting.

Transplanting: Pits of 50 x 50 x 50 cm size are to be dug at required distance. Pits should be filled with top soil along with 20 kg well decomposed FYM and 1 kg neem cake. Two seedlings (dioecious var.) should be planted in each pit and single seedling planting for gynodioecious line (var.) followed by light irrigation.

Planting density: The planting density depends on the number of factors like cultivars to be grown, soil type and rainfall pattern of the area. Generally spacing of 2m x 2 m for tall variety with 2500 plants ha^{-1} and 1.5 m x1.5 m for dwarf variety with 4400 plants ha^{-1} is recommended.

Irrigation: The papaya plants respond well to irrigation. Adequate irrigation helps in rapid fruit development and good fruit yield. But the plants are highly sensitive to water logging and so avoid "Wet Feet" conditions in papaya.

Manures and fertilizers: The papaya crop is heavy feeder so application of FYM @ 10kg, neem cake @1 kg, urea 600g, superphosphate 1.4 kg, and muriate of potash 700g, is required every year to obtain good yield of papaya. The nitrogenous fertilizer should be given in three split doses *i.e.*, at the time of sowing along with phosphatic and potassic fertilizers, vegetative and flowering stages.

Pests and diseases: Damping off, collar rot and stem rot and papaya mosaic are the main diseases which can affect the growth and yield of the fruits. They can be controlled by taking suitable control measures by treatment the seeds with *Trichoderma viride* @ 3-4 g kg^{-1} of seed, Thiram @ 2g kg^{-1} of seed and Rogar 2 ml per litre of water, respectively. Aphids and red spider mites are the main insects which can be controlled by spraying Rogor @ 2 ml per litre of water and Dicofol @ 2ml per litre of water, respectively.

Harvesting and yield: Papaya plants begin bearing fruits after one year of their transplanting. The fruits are harvested when they are mature enough and about to turn from green to yellowish in colour. Although the yield varies from place to place depending upon the soil fertility and various cultural practices followed the average yield of papaya is 350 q ha^{-1}.

Inter-cropping: During pre bearing stages of the crop, short duration vegetables like cabbage, cauliflower, chillies, onion, tomato can be grown as inter-crop which not only keep the weeds free but also generate extra income to the orchardists or farmers. The papaya based agri-horticultural systems are self sustainable where solar energy can be harvested properly, soil resources can be utilized efficiently and cropping intensities can be increased (Saran *et al.*, 2014).

Among the intercrops, turmeric, turnip and suran have been found to be best and most profitable under partial shade of papaya. Economic analysis in terms of benefits: cost ratio is presented in Table 16.1.

Table 16.1: Income from Papaya Based Inter-cropping Systems

Inter-cropping System	*Yield (q/ha)*	*Total Income (Rs./ha)*	*Production Cost (Rs./ha)*	*Net Returns (Rs./ha)*
Papaya sole crop	917.0	2,75,100	1,00,000	1,75,100
Papaya seed production	0.40	16,00,000	1,28,000	14,72,000
Papaya + Turnip +Suran	900+25+195	3,92,500	1,72,000	2,20,000
Papaya + Lentil +Suran	900+8+195	3,83,500	1,70,000	2,13,500
Papaya + Turmeric	900+30	4,01,750	1,86,000	2,15,7500

Source: Saran *et al.*, 2014

References

Bagde, A.B. 2009.Tips for constipation. *Ayurveda*.1(12): 13.

Biswas, P.K. and Kerimenla, S. 2012. Papaya-A potential fruit crop of humid tropics. *Agrobios newsletter*.10(10): 33-34.

Chavan, M.D. 2012.Papaya-a miracle fruit. *Agrobios newsletter*.17(2): 48.

Pathak, R. and Singh, A.K. 2000. Advances in nursery management of banana and papaya. *Intensive Agriculture*. 38(3-4): 18-19.

Prasad, M. 2010. Potassium for you. *Ayurveda*.1(18): 16-17.

Saran, P.L., Chaudhary, R. and Solanki, I.S.2014. Papaya Agro-horti systems. *Agriculture Today*. 17(2): 40-41.

17

Peach (*Prunus persica* L. Batsch)

Common name: Aarru

Family: Rosaceae

Introduction

Peach is a highly popular fruit crop grown in the temperate regions of the world. In India, peach cultivation is mostly confined to the mid hill zone of the Himalayas (Jamwal, 2011) extending from Jammu and Kashmir to Khasi Hills at an altitude of 1500 to 2000 m.

Low chilling cultivars are also grown in sub-mountainous zones of Punjab, Delhi, Haryana, Western Uttar Pradesh and Jammu and Kashmir. The main peach growing areas in Jammu region are Poonch, Doda, Udhampur, Rajouri and Kathua districts. It is also found in Ladakh region of Jammu and Kashmir.

Peaches constitute favourite table fruit because of good nutritive value of the fruit. Processed peaches like canned, dried, frozen, preserved jam; juice and beverages are also very popular products. Peaches like other stone fruits also have medicinal values.

Tea made from the leaves acts as an excellent kidney cleanser. Fruit contains high percentage of water, and is an excellent laxative. It is helpful in eliminating toxins, and usually included in the weight-loss programmes. Cooked and purred peach fruit is extremely helpful in cases of stomach ulcers, bowel inflammations and colitis.

Medicinal Values

Peaches are rich sources of vitamin A (1330-IU) which checks night blindness. Similarly they are rich source of ascorbic acid or vitamin C which checks scruvy and provides resistance to the body. Peaches are high in potassium and low in sodium and are thus, excellent food for people on low salt diet. In addition, they contain Vitamin B and minerals like calcium, phosphorus and iron. Iron is essential for the pregnant women, whereas, calcium and phosphorus are important for the development of bones. The leaves and bark have medicinal value with properties as demulcent, sedative, diuretic and expectorant action. Now-a-days, peach leaves are also recommended in treatment of irritated digestive tract. Fresh peach stimulates digestion, regulates bowel and alkalinizes blood stream. Peach kernel oil is known to stimulate hair growth, and is commonly used as a moisturizer and for massage. Peach seed contain a toxin, hydrogen cyanide which in smaller proportion stimulates respiration and improves digestion but cause respiratory failure and can lead to death if taken in excess.

Botanical Description

The tree is deciduous, medium-sized (upto 3-5 meters). Leaves are linear, oblong and lanceolate, 7 to 16 cm long, pinnately veined. Flowers are pale-pink to purplish and five-petalled, appear in early spring (during March-April). Fruit is a drupe, sweet-tasting, yellow or white flesh, with hairy or smooth skin with a single brown-red seed inside.

Cultural Practices

Climatic requirements: Peach is an important stone fruit grown successfully over a wide range of climate. However, it does well in wet and cool climate, with cold winter and dry summer. Peaches do not require cool nights to develop red skin colour like apple. The best climatic condition for satisfactory tree growth and production of fruit, exist in the mid-hill zone. The limiting factors in peach cultivation are minimum winter temperature, chilling hours, spring frost, hail storm, high humidity and desiccating winds during summer. In sub-tropical areas of the North, the low chilling cultivars require 50-300 hours of chilling during the winter for proper flowering and fruiting (Jamwal, 2011).

Soil requirements: Peach tree thrives best in deep, well drained, light soils (sandy loam and loam). The soil must be fertile and sites previously planted should be avoided.

Varieties: *Elberba, J.H. Hale, Alexander, Babcock, Veteran* are the main peach varieties grown in the hills. On the other hand, *Early Amber, Early Grands, Flordasun, Honey Sweet, Sharbati, Shan-e-Punjab* and *Saharanpur Prabhat*, are the most suitable varieties of peach grown in sub-mountainous regions and plains.

Propagation and rootstock: Rootstock used for peach is wild apricot, plum and peach. Plum stock can be used when peaches are to be planted in slightly heavier soil. Seeds should not be taken from peach cultivars whose fruits mature early in the season, as their germination percentage becomes low. As peach seeds do not germinate readily so they must complete a rest period of about 60-70 days or 3 months prior to germination satisfactorily and produce vigorous seedlings (Singh *et al.*, 2004; Jamwal, 2011). Peach is commonly propagated by T or ring budding done during autumn season (ending October or November). Plants are spaced at a distance of 4-6 m in pits of 1m x 1m x 1m size dug about one month in advance. One year old planting materials are preferred.

Training and pruning: At the time of planting, the trees are headed back at the height of 60-65 cm from the ground level and allow growing only 3-4 branches. In subsequent years all surplus branches that have risen on the trunk are removed. Tips of outsides branches are cut back to allow the scaffold branches to grow in an upward and outward direction.

Manures and fertilizers: Well rotton FYM@ of 40 to 60 kg tree^{-1} should be applied. Fertilizer dose in terms of N, P_2O_5 and K_2O @ of 55-65 kg, 55-65kg and 110-135 ha^{-1} for bearing orchard is recommended.

FYM should be applied during December. All of phosphatic and potassic fertilizers should be applied during December. Half of the nitrogenous fertilizer should be applied before normal bloom while the rest should be applied 3-weeks after fruit set.

Irrigation: Peach trees should be irrigated 2-3 times during summer for getting good yield. In sub-tropical regions more frequent irrigations are required as the weather becomes warmer from 2nd fortnight of March. Irrigation may be applied in such areas at 8-10 days interval in the month of March-May.

Harvesting and yield: The peak harvesting period of peach in sub-tropical plains is between last week of April to mid June, while in hills it varies between June –July.

As all the fruits do not mature simultaneously and 3-4 pickings are required at 4-5 days interval. The yield may vary from 50-125 kg tree^{-1} depending upon the cultivar and orchard management.

References

Jamwal, J. 2011. Peach a highly popular fruit. *The Kashmir Times*. 68(22): 136 (Saturday, January 22)

Singh, R. K., Verma, S. K., Singh, H., Bhardwaj, P. N., Regar, R and Arya, R.R. 2004. Ways to boost the cultivation of stone fruits in Uttranchal. *Indian Farmers Digest*. 37 (5-6): 5-8 and 21.

18

Phalsa
(*Grewia asiatica* DC.)

Common name: Falsa

Family: Tiliaceae

Introduction

Phalsa is a fruit of South Indian origin (Jamwal, 2008; Vyas *et al.*, 2013). But now it is grown in Pakistan, Bangladesh, Nepal, Srilanka, Thailand, Laos, Cambodia and Vietnam. In the dryland areas where either growing of farm crops is not possible or their yields are below the economic threshold due to constraint of water. In such areas growing of drought tolerant fruit trees like Phalsa, ber, custard apple, Jamun etc, is much possible.

In the Indian sub-continent, Phalsa is highly sought in the hot summer months as its fruits can be made into a cooling, refreshing drink 'Phalsa sharbat'. As a matter of fact, it can be considered to be the third favourite fruit after mango and peach due to having medicinal values.

Nutritional Values

Phalsa fruit is a good source of carbohydrates and dietry fibre, calcium, phosphorus and potassium as well as sodium (Vyas *et al.*, 2013). Vitamin A and Vitamin C are also found in good amount. Other vitamins present in the fruit are Vitamin B_1, Vitamin B_2 and Vitamin B_3 (Table 18.1).

Table 18.1: Nutritional Value of Phalsa per 100g Fruit

Nutrient	*g*	*Nutrient*	*g*
Carbohydrates	21.1	Potassium	372.0
Dietary fibre	5.5	Sodium	17.30
Protein	1.6	Vitamin B_1	0.02
Fat	0.1	Vitamin B_2	0.26
Calcium	136.0	Vitamin B_3	0.82
Phosphorus	24.2	Vitamin C	4.4

Source: Vyas *et al.*, 2013

Medicinal Values

Fruit is astringent and stomachic. The unripe fruit relieves inflammation and is used to treat respiratory, heart and blood afflictions as well as in fevers. An infusion of the bark is provided as a demulcent, febrifuge and treatment for diarrhoea. The root bark is used in treating rheumatism. Its leaves are utilized in skin eruptions as they are known to have antibiotic action.

The fruit and the juice have been employed since centuries to treat liver and gall bladder problems as well as to purify blood and to control blood pressure and cholesterol level (Vyas *et al.*, 2013). The fruit is known to assist preventing coughs and colds. For skin problems, one should soak the bark overnight, then pound it and apply the pulp directly on to the affected area. 'Phalsa Sharbat' is good to help sunburn victims and to treat sunstroke.

The juice is found to be digestive and alleviates stomach ache in addition to normalizing heart rate and maintaining blood pressure. It brings on urine incase of stoppage of urine, removes thirst, protects against heat stroke, vomiting, nausea and uneasiness due to heat. The fruits are a minefield of antioxidants and are useful in reducing cancer risk. They also remove headache, pimples and burning sensations in the body, acidity, leucorrhea, anaemia and remove skin dryness.

The leaves relieve all types of inflammations of the skin like cuts, burning boils and eczema. For this purpose, just make a paste of previously soaked leaves in water (soaked a night before). Apply this paste on the affected area. The leaves are antibiotic in action.

Other Uses

The bark is used in the sugar making process to purify the sugarcane juice from which it is prepared. This is because, the bark is mucilaginous. The wood of

Phalsa tree is fine grained, cream coloured, strong and flexible, and such it is used in making archer's bows, spear handles, poles and baskets.

The leaves are used as cattle fodder and the root bark is used to help the people who suffer from rheumatism. The stems of the Phalsa tree and bark can be made into ropes, baskets and are harvested for fuel.

Botanical Description

Phalsa can be grown as small tree or found wild, may grow as a straggly looking bush and attains a maximum height of 4.5m. Bark is rough, branches long and drooping. Leaves alternate, simple, deciduous, broadly cordate to ovate, margins toothed and with pubescence. Inflorescence is axillary cyme with 3-5 flowers inclusters. Fruit is a globose drupe. Fruit is purple to almost black in colour when gets ripened with sweet to sour taste. It is very delicate and perishable.

Cultural Practices

Climatic requirements: It is a fruit of tropical to sub-tropical areas and grows up to an altitude of 900m. It can withstand light frost which can cause only shedding of leaves.

Soil requirements: Phalsa can be grown in any kind of soil *viz.* sandy or clay provided sufficient amount of organic matter is present. However, its best growth can be seen in soils having loam or loamy texture.

Propagation: Seeds of this fruit tree are the usual means of propagation which germinate within 15 days. One year's old seedling after uprooting from seed beds, are transplanted in the well prepared fields during February to March by maintaining 1mx1m distance or 2mx2m with 2500 plants/ha.

It can also be propagated through clumps or cuttings. However, the cuttings make very slow growth. As vast variation exits in nature and single clonal selection can help in establishing desired types.

Pruning: Phalsa fruit tree requires heavy pruning every year as fruits are borne on new growth (in axils of leaves)

Harvesting and yield: Fully ripe fruits are harvested during May or June, and are immediately sent to the market as they are highly perishable. Average yield per plant is 9 to 11 kg in a season (Jamwal, 2014).

References

Jamwal, J. 2008. Fruit cultivation Periurban areas. *The Kashmir Times.* 65 (107): 15 (Thursday, April 17)

Jamwal, J.2014. Useful fruits that is not very common. Personnel communication. Trikuta Nagar, Jammu, Jammu and Kashmir.

Vyas, M., Chauhan, A.K., Nathawat, N.S. and Maheshwari, R. 2013. Phalsa-A fruit of dynamism and exuberance. *Agrobios Newsletter.* 12 (4): 88-89.

19

Pineapple (*Ananas cosmus* L. Merr.)

Common name: Ananas

Family: Bromeliaceae

Introduction

India has a wide variety of climate and soils on which a large variety of horticultural crops such as fruits, vegetables, ornamental and plantation crops, medicinal and aromatic plants are grown. As for instance, farmers of North Bengal adopted large scale cultivation of pineapple in the foot hills of the Himalayas in the districts of Darjeeling, Jalpaiguri and Dinajpur in the early eighty's and now many farmers are growing pine apple on their waste lands. Similarly, considerable area is under cultivation in Nagaland and other North Eastern Hill states (Khate, 2013). It is because pineapple as a fresh fruit has got good demand all over the world. In addition to this, preserved pine apple has excellent market both inside and outside the country. Slices, chunks, juice, jam, jelly are favourite pineapple products all over the world. Pineapple also possesses number of medicinal properties.

Medicinal Values

Pineapple is such a kind of fruit, which in addition to having good taste has the ability to fight number of internal diseases. This fruit is a rich source of many enzymes, antioxidants and Vitamin C. Owing to the presence of Vitamin C, it helps in curing the wounds of liver and stomach as well as swelling. While presence of Vitamin C vigourates the body, it also helps in curing scurvy disease of gums (Anonymous, 2010). Pine apple juice is capable of curing body's internal injuries

and wounds, which has recently been investigated by the scientists in London, United Kingdom (Anonymous, 2010).

Botanical Description

Pineapple is tropical, herbaceous monocots perennial (1.0 to 1.5 meters) plant which has a short, stocky stem with tough, waxy leaves. After flowering, the individual fruits join together to form coalesced multiple berries, known as pineapple. After the first fruit is produced, side shoots (called 'suckers') produced in the leaf axils of the main stem are used for propagation. The stem produces spike-like inflorescence after 12 to 20 months, up to 15 cm. This bears over 100 spirally arranged, trimerous flowers, each subtended by a bract. Flower colour vary from lavender, through light purple to red.

Cultural Practices

Climatic requirements: Cultivation of pineapple is extremely suitable in the areas having relatively high relative humidity and mild temperature.

Soil requirements: Sandy and sandy loam soils with slightly acidic conditions (pH 4.5-6.5) and high content of organic matter is good for pineapple cultivation. Water should not accumulate in pineapple orchards and proper arrangement for drainage must be ensured.

Selection of planting material: Planting material for propagation is obtained from the plant itself, but the material need to be collected from healthy, vigorous, disease free plants which can fruit early.

(a) **Ground suckers:** They develop from the underground portion of the stem and have no fixed time of their emergence. These suckers are, however, vigorous and have characteristic of early fruiting.

(b) **Shoot Suckers:** Develop from axillary buds at the base of leaves and produce early fruiting in plants.

(c) **Slip suckers:** They develop from the fruit stalk and are smaller in size and late in producing fruits.

(d) **Crown suckers:** They are produced on the top of the fruit and are extremely late in producing fruits (2 to 3 years after planting).

Planting method: Plough the land well, collect stubbles and weeds and divide it into plots of convenient size. Depending upon the amount of organic matter present in the soil, add 25 to 50 tonnes of well rotten FYM or compost ha^{-1} and mix it well with the soil.

Planting can be done throughout the year, except rainy season. The best season of planting is after the monsoon. The most appropriate time of planting in North Bengal is mid October to mid December, whereas, in South Bengal it varies between mid September to mid November.

Manures and fertilizers: It is best to apply fertilizers on the basis of soil test values. However, in average situation approximately 25 g N, 15g P and 25 g K may be applied plant^{-1} year^{-1}.

Weeds control: Heavy weed infestation affects the number, size and quality of pineapple fruits. Thus, timely control of weeds is very essential. Hand weeding is generally done in pineapple. Rows of plants are first cleaned by small hand tools. The space between the rows is first cleaned by sickles or spades. Hand weeding should be done 6 to 7 times in a year.

Although application of suitable herbicides may be economical to control weeds yet their use is prohibited due to preference being given to organic food production.

Harvesting and yield: Fruits and suckers of pineapple are available from the second year of planting. In the second year, pineapple yields about 18,000 fruits ha^{-1} and about 10,000 suckers.

References

Anonymous. 2010. Juice of pineapple cures internal wounds and injuries of body-research. *The Hind Samachar* 67 (4): 10

Khate, A. 2013. Horticulture scope of diversification. *Agrobios News letter*.12 (1): 42-43.

20

Plum

(*Prunus domestica* L.)

Common name: Aloo bukhara

Family: Rosaceae

Introduction

Among the stone fruits, plum holds a conspicuous and prominent position due to its wider adaptability. There are many species and varieties of plums which are well adapted to various types of climate (Singh *et al.*, 2004). They are nutritious low calorie food. One medium sized plum possesses 36 calories and provides cellulose, pectin Vitamin C and potassium vis-a-vis anthocyanins. Owing to the presence of the three components in the plums, they are characterised to have medicinal properties.

Medicinal Values

Occurrence of antioxidant pigment *i.e.* anthocynin protects against cancer and heart diseases (Sawhney, 2005). Plums are rich sources of Vitamin A and Vitamin C which check night blindness and scurvy disease, respectively. However, plum may cause allergic reaction in people who are allergic to apricots, almonds, peaches, cherries, aspirin. Plums have amygdalins which break into hydrogen cyanide in stomach and causes cyanide poisoning if taken in excess amount.

The laboratory tests carried out in Texas (USA) revealed that the extracts of plum and peach can help to fight breast cancer cells (Anonymous, 2010). The extracts determined that two specific phenolic acid components; cholorogenic and trichlorogenic were responsible for killing the cancer cells while not affecting the normal cells. These two components are very common in fruits, but the stone fruits such as plums and peaches have especially their high levels (Anonymous, 2010).

Botanical Description

Trees are small to medium in size, with rough bark and spreading branches. Like apple, plum is also slight shade and direct sun tolerant plant and grows in light intensity range of 600-2400m candles (Sharma and Mir, 2001). The chilling requirement of plum is less than apple. Leaves are ovate, with acute or obtuse tips, short petioles and crenulated margins. Flowers are white, smaller, and have longer pedicels, mostly borne in umbel-like clusters of 2-3 individuals on short spurs. Honey bees are the major pollinator. Fruit is a drupe, oval to round in shape.

Cultural Practices

Climatic requirements: Irrespective of varieties of plums, these are well suited for cultivation between 700-2000 m elevations. The chilling requirement of European plum for opening of flower buds is slightly less than apple. The chilling requirement of Japanese plums are even less than those of European. Thus, plums can do well in relatively warmer climatic conditions. Rainfall during the growing season reduces production and causing fruit cracking.

Soil requirements: Plum can be grown with success on a variety of soils but deep, well drained loamy soils with pH of 5.5 – 6.5 are the best for most of the plum varieties. When plums are to be planted on sloppy surface, middle of the ridge would be an ideal for planting.

Propagation and planting: The common root stock used for propagation of plum is 'wild peach', 'apricot behmi' (a natural hybrid of almond). However, large proportion of plums are propagated on peach root stock, which are suitable for light textured, whereas, in heavy soils the apricot root stocks are likely to give good results.

Generally, budding and grafting is used for commercial propagation of plum. The seedlings or rooted cuttings are budded during May-June. Both T-budding and ring budding have given good results.

The grafting is done by tongue method with the advent of spring when the sap flow of the rootstock starts by using dormant scion stick. The cleft grafting can also be practised. After the scion has grown for one year, the plant can be planted in the field at the distance of 4-5 m depending upon the varieties.

Training and pruning: Training in plum is an important operation which provides strong framework for easy management besides entrance of light and proper aeration inside the plant. Plum trees are required to be trained by modified leader system.

Fruit bearing: Plum bears fruit on one year old wood and on spurs on older wood. Only light pruning is given to bearing plum. A certain amount of thinning of branches is essential to admit sunlight.

Manures and fertilizers: Application of 25-30 kg FYM per tree during December-January is recommended. Besides, 50gN, 25gP and 60gK is recommended for every one year age of plant upto 10 years. Nitrogen is applied in the split dose; once before flowering and other half one month later.

Irrigation: Two or three irrigations applied during April to June have been found to be highly useful.

Harvesting and yield: Harvesting of plums is generally done during the months of July to August, When more than 50 per cent of the fruits have ripened. The yield of the fruit varies from variety to variety. However, on an average the yield of the plum varies from 1.5 to 2.0 tonnes ha^{-1}.

References

Anonymous. 2010. Plums can help fight breast cancer. *The Kashmir Times*. 67(154): 12 (4th June, Friday)

Sahwney, D. 2005. Flavours of the month. *Ayurveda*. August Issue, 15-16.

Sharma, J. P and Mir, A. A. 2001. Dynamics of cold Arid Agriculture. Kalyani Publishers, Ludhiana-New Delhi, Noida (UP)-Hyderabad-Chennai Calcutta (Now Kolkata)

Singh, R. K., Verma, S. K., Singh, H., Bhardwaj, P. N., Regar, R and Arya, R.R. 2004. Ways to boost the cultivation of stone fruits in Uttranchal. *Indian Farmers Digest*. 37 (5-6): 5-8 and 21.

21

Pomegranate (*Punica granatum* L.)

Common name: Annar

Family: Punicaceae

Introduction

Pomegranate grows up to 12 m height with deciduous foliage. Pomegranate being a delicious juicy, palatable and favourite table fruit of tropical and sub-tropical climate is commercially grown worldwide. Although it is known as sub-tropical crop yet can be successfully grown between 600 m to 1800 m in mid hill of Uttrakhand (Verma *et al.*, 2005) and Jammu and Kashmir.

A native fruit crop of Iran (Persia) is extensively cultivated in Mediterranean countries. Kabal and Kandhari *anar* has its own identity for its seedless red aril. In India, it is grown from Kanyakumari to Kashmir. But commercially it is grown only in Maharashtra. However, with the introduction of soft seeded varieties, its cultivation is rapidly expanding and plant can withstand different soil and climatic stresses.

The pomegranate fruit is highly nutritious due to the presence of carbohydrates, proteins, fats, minerals and vitamins.

Nutritional Value

The pomegranate fruit is highly nutritious since it contains lot of water (78 per cent), carbohydrates (14 per cent), protein (1.6 per cent), fat (ether extractable 0.1 per cent), mineral content (7 per cent), fibre (5.1 per cent), calcium (0.1 per cent), phosphorus (0.7 per cent), Fe (0.3 per cent), nicotinic acid (0.2 mg 100 g^{-1}), riboflavin (100 mg 100 g^{-1}) and Vitamin C (16 mg 100 g^{-1}). Pomegranate fruit is a rich source of potassium, copper, zinc and manganese also (Suguna, 2012).

Medicinal Values

Pomegranate fruits and even rind have been reported to possess number of medicinal properties. The juice of pomegranate is used for pharmaceutical purposes. It is utilized for treating dyspepsia (indigestion) and is considered to have therapeutic value and help in treating leprosy. Because of the presence of tannin content in bark, leaves, a mature fruit and fruit rind are given as astringent to haul diarrhoea, dysentery and hemorrhages.

A decoction of the flowers is gargled to relieve oral and throat inflammation. Chinese mention pomegranate juice as longevity drug. It contains a powerful antioxidant, falconoid; more effective in fighting heart disease than those known in tomatoes and red wine. Role of pomegranate in counteracting the proliferation of human breast cancer (Bhagat, 2007) has also been proved.

Verma *et al.*, 2005 and Gupta (2011) have also outlined important methods to use pomegranate in curing following ailments/disorders:

1. **Anaemia:** Taking orally the endocarp of *Punica granatum* twice daily for about week cures anaemia.
2. **Cough:** Grind the leaves of *Punica ganatum* and extract the juice and use of 5 ml in each time twice in a day for 4 to 5 days with a little honey.
3. **Dysentery:** Pound the leaves and extract the juice and take 5 ml of the juice twice in a day with a little honey.
4. **Infertility:** Grind roots of *Punica granatum* into paste and intake with a little butter once in a day till cure. Use the paste before sunrise or sunset after 6 hours of the first day of menstruation.
5. **Tuberculosis (TB):** Grind the leaves of *Punica granatum* with the leaves of *Syzygium cumini* and extract the juice, add ginger to it. Take the juice thrice daily till cure.

Apart from the above benefits, it has also proved beneficial to get relief from the redness and heat in the eyes and the fever. For this purpose, its juice is taken which also helps to relieve heat, acidity and uneasiness very affectively (Suguna, 2012). Pomegranate helps in improving the appetite, palpitations and counters the weakness of the intestines. It also helps in increasing haemoglobin level and in weight gain. One can enhance his taste buds by taking 100 ml of pomegranate juice with one spoon of honey and little rock salt. This is very effective in improving appetite. Antioxidants in pomegranate can help to prevent premature aging.

As per recent finding, pomegranate could reduce the risk of breast cancer owing to occurrence of a phytochemical 'Ellagic acid' which prevents cancer cells from developing.

Pomegranate health benefits run deep to bones which reduce the damage to cartilage in case of those suffering from arthritis (Anonymous, 2013). It acts like an oxygen mask *i.e.*, pomegranate juice pumps the level of oxygen in the blood and prevents blood clots by reducing cholesterol due to its antioxidant properties.

Cultural Practices

Climatic requirements: Although pomegranate grows well under tropical to sub-tropical climatic conditions yet it can be successfully grown upto an altitude of 600 m to 1800 m in mid hill of Uttrakhand. It also grows best in semi-arid climate where cool winter and hot summer prevail. In cool climate of Uttrakhand, the tree is deciduous, whereas in sub-tropical/tropical conditions it remains evergreen or partially deciduous (Singh *et al.*, 2005). The tree requires hot and dry weather during fruit development and ripening. It cannot produce sweet fruits unless the temperature is sufficiently high for a long period. Humid climate lowers the quality of fruits and increases incidence of fungal diseases. It can tolerate frost to a considerable extent. As quality of fruit is adversely affected by humid conditions so dry belt of warmer valley and foothills are very suitable for its cultivation.

Soil requirements: Pomegranate can be grown on a wide range of soils. However, it prefers a well drained sandy loam to deep loamy sandy soils. In light soils, the quality and colour development is good but poor in heavy soils. Plants can tolerate sodicity (exchangeable sodium percentage upto 7 per cent) and salinity (upto 9 per cent). Acidic soils require liming for obtaining optimum fruiting.

Varieties: Wild variety of pomegranate is very acidic and as such is of little value except as souring agent (*anardana*). The double flowered pomegranate which do not bear fruits are grown in parks and gardens for their beautiful red flowers. The important varieties of pomegranate are:

1. **Ganesh (GBI):** It is a selection from Alandi released for commercial cultivation in 1936 and renamed as Ganesh in 1970. Its fruits are medium sized with yellow, smooth surface and red tinges. The seeds are soft with pink aril. Ganesh variety has revolutionized the cultivation of pomegranate in Maharashtra.
2. **P26:** It is a seedling selection of Muscut. This variety is superior to Ganesh in yield and fruit rates. Its fruits are large in size, rind yellowish-green with red tinge, fleshy, aril pinkish-white with moderately hard seeds and sweet juice.
3. **Mridula:** A seedling selection from open pollinated progeny raised from F1 progeny of Ganesh Gul-e-Shah Red. Its fruits are medium sized, rind smooth, dark red in colour. It has blood red arils with very soft seeds, juicey and sweet in taste. It is significantly superior cultivar to other commercially grown pomegranate.
4. **Ruby:** It is hybrid from a three-way cross between Ganesh Kabul Yercadu and Gulaba Rose Pink which has soft red arils with good flavour. The plants are dwarf, specific bearer providing uniformly red fruits.
5. **Yercadu:** Its fruits are medium sized with easily peelable rind. The seeds are soft with attractive deep purple aril.
6. **CO1:** It is also high yielding selection. Its fruits are medium sized with attractive rind, soft seeds, high pulp content and sweet taste.
7. **IIHR:** A soft seeded selection with mean fruit weighs of 225 g. It provides fruit with a heavy bearing capacity. It is a selection from open pollinated seedlings.

Propagation: As pomegranate plants occur widely, are usually grown from seeds. But to get true to type cuttings of mature wood with 3-4 nodes are required. Rooting in cutting takes place in 90 days. The pomegranate can also be propagated by air layering or gootee layering done during rainy season and/or during November-December. Application of 1,000 ppm IBA (Indole Butyric Acid) gives profuse rooting.

Planting: Being shallow rooted crop, pomegranate can be planted in 0.6 0.6 0.6 m well manured pits at a distance of 4 4 m. To obtain good field survival, plants require sufficient sunshine. The planting site with south facing slope is selected. Hillocks or open terraces are suitable for its cultivation in hilly areas.

Manures and fertilizers: For getting good sized fruit, regular application of fertilizers is required. Too much nitrogenous fertilization promotes vegetative growth only at the expense of flowering. Application of fertilizers @ 500 g N, 250 g each of P_2O_5 and K_2O per plant (4 to 5 year old bearing plant) per year is optimum. Apart from these fertilizers, application of 20 kg well rotten farm yard manure for optimum economic yield is also recommended. In young fruits, there is fruit cracking. This disorder is due to boron deficiency but in full grown trees, it is due to moisture imbalance or due to irregular irrigation or rains (Verma *et al.*, 2015). To control this, following control measures are required to be followed:

1. Cultivate tolerant pomegranate like 'Jalore Seedless', 'Khogand', 'Badana Bose'.
2. Spray borax @0.1 per cent or calcium hydroxide on leaves and fruits after fruit setting.

Irrigation: Pomegranate responds well to irrigation. Regular watering, especially in basin system will result in smooth flowering and fruiting. Water requirement is high at maturity. During winter season, flowering is less due to short day length.

Pruning: Trees require less pruning. It is done only to give shape to the trees and removal of extra suckers (water sprouts), and dead and diseased twigs.

Pests and diseases: In hilly terrains, moist weather and erratic rainfall are common phenomenon, which leads to incidence of pests and diseases in fruits. The serious pests are pomegranate butterfly and bark eating caterpillar which cause drastic reduction in fruit yield.

Application of systemic insecticides 0.03 per cent after fruit set should be done, 2-3 times depending upon infestation. To check fruit rot application of Dithane M-45 or Capton at an interval of 15 days is optimum. If organic insecticides are available, they should be preferred instead of systemic insecticides.

Harvesting and yield: Normally it takes 4 to 5 years to start fruiting after planting. In hills, the pomegranate trees bear fruits throughout the year but bulk of the fruits are harvested in two periods *viz.*, June-July and September-October. Well coloured fruits should be harvested before they are dead ripe. Fruits withstand transport well due to hard rinds on them. Maximum bearing per tree is 25-40 fruits, weighing 5-6 kg.

Grading, packing and transportation: Fruits after harvesting and removing from the field at proper time should be sent to the market after grading and proper packing. For packing of pomegranate, a wide variety of containers such as wooden boxes, baskets made from bamboos and corrugated fibre board boxes are used to minimize damage likely to be occur during long transportation.

References

Annonymous. 2013. Health benefits of pomegranates. *The State Times*. 19 (264): 5 (October, 4, Friday).

Bhagat, K.C. 2007. Pomegranate: production and protection practices-1. *The Kashmir Times*. 64 (317): 13.

Gupta, R. D. 2011. Pomegranate helps digestion. *Ayurveda*. 1(20): 18.

Suguna, N. 2012. Power of pomegranate. *The Daily Excelsior*. 48 (172): 4 (Sunday Magazine, June 24)

Verma, H.S. Singh, S.K., Bhardwaj, R.K., Regar, P.N. and Arya, R.R. 2005. Pomegranate: a hardy fruit plant. *Indian Farmers Digest*. 38 (1) : 33-35.

Verma, S. K., Jambamma, S and Lavanya, T.N. 2015. Fruit cracking: physiological disorder in fruit crops. *Indian Farmers Digest*. 48(2): 24-25.

22

Walnut
(*Juglans regia* L.)

Common name: Akhrot

Family: Juglandaceae

Introduction

Walnut belongs to the state of Jammu and Kashmir in India and the tree is indigenous to Kashmir (Putto, 2005). The tree is known by the vernacular name as 'Vontdun' (hard walnut). It is because under ordinary circumstances one is unable to break the shell.

According to Kalhan Pandit, fruit culture including walnut in Kashmir valley dates back to the time of King Nara (1000 BC) as has been mentioned in the Rajtarngni. During the period of King Lalitaditya (700 AD) and Raja Harash (1089 AD) this indigenous fruit culture flourished (Motial, 1976). Walter and Lawrence, the Settlement Commissioner of Jammu and Kashmir State (1887-1894) has documented in "The Valley of Kashmir" an enough evidence to establish the heritage status of walnut for the state which now stands emerged a dollar earner not only for Kashmir but for the country on the whole.

Apart from social and religious significance of walnut, consumption of nuts offers lot of medicinal benefits.

Medicinal Values

Despite walnuts containing up to 60 per cent fat, these are mainly composed of polyunsaturated fatty acids (especially linolic and linolenic acids) in a proportion seven times higher than that of saturated fatty acids. Latest research

proves that unsaturated fatty acids exert a reducing effect on the cholesterol production. Hence, the consumption of walnuts does not produce cholesterol but even reduces its production (Singh, 2009). The positive nutritional advantage of walnuts in lowering blood cholesterol, thus, comes from high levels of mono and poly unsaturated fatty acids and tocophenols.

As per recent studies it has been proved that those who consume walnuts on a regular basis manage to add up an extra two and half years to their average life span (Munshi, 2014). Apart from offering benefits of unsaturated fats, the walnuts also supply vitamins, minerals and antioxidants.

Other Uses

Walnuts shells are used as fuel in Kashmir, whereas, in USA these are used as energy sources for generation of electricity.

Social and Religious Significance

In Kashmir, walnut is known as doom, it has been rated as high value nut fruit. Walnut forms an indispensable ingredient of "Heart" Sheoratri (Shivratri) puja in Kashmir, wherein, a gagar (the earthen/metallic pot, symbolizing Devi Parvati) are filled with walnuts and water. The walnut has significance, as its four part kernel is believed to be representing Dharma, Artha, Kama and Moksha. It is believed to be a symbolic of the Vedas-Rigveda, Atharva Veda, Yajurveda and Samaveda.

After three days worship (puja), the walnuts are distributed as *Prasad* on Amavassia among relatives and neighbours by each house hold for eight days till Shakul Paksh Ashtami (teela atham) symbolizing end of winter in Kashmir and start of spring coinciding with sowing a few selected soaked walnut in the backyard to have walnut trees for future generations.

Botanical Description

Walnut is monoceous (staminate and pistillate flowers separately on same plant) fruit crop. For proper pollination overlapping of pollen dehiscence and stigma receptivity is also highly essential. The non-coincidence between the flowering periods of male and female flowers is the major cause of low yield in the walnut.

The walnut seedlings have long juvenile period and come to bearing after 10-12 years. The juvenility can be reduced by planting clonally propagated saplings. Root pruning of walnut trees at a radial distance formed is recommended which affect the quality of nuts as well as kernel.

Cultural Practices

Climatic requirements: Walnut grows in sub-temperate to temperate climates. In Kashmir, it is found almost in every village from an elevation of 1650 to 2250 m (Putto, 2005). Frost or snow during flowering destroys young flowers and new shoots, thereby affecting crop production. The production of a good quality walnut crop is, however, dependent on altitude, temperature fluctuations and humidity/moisture during the main fruit development stage.

Soil requirements: Deep, silty loam or loam, well drained soils, with high organic matter are most suitable for cultivation of walnut. Sandy soil or alkali sub-soil should be discouraged for cultivation of walnut. The accumulation of water is harmful for growth and development of trees and also affects the quality of nuts. Trees grown on shallow soils with scant moisture content remain stunted and are very often affected by sunburn.

Varieties: Recently, two promising varieties *i.e.*, 'Hamdan' and 'Suliaman' have been released by SKUAST-K, Srinagar having a good nut and kernel quality characters with improved yield (Jamwal, 2015).

Propagation: Walnut trees can be raised either from seeds or through budding or grafting methods. The walnut trees are propagated from seed although grafting is not uncommon. Young walnut plants raised from seedlings come to bearing at the age of 10-15 years but optimum production is not expected until 20-25 years age and continue up to 100 years or more. The three varieties of walnut *viz., Kagazi, burzal* and *Wantu* are mostly reproduced from seed.

The seeds are sown in March which fully germinate in May. After two years, the young trees are planted in well prepared fields. It has been calculated that a walnut tree will yield half its bearing capacity in six years from sowing and will be in full bearing in ten years after sowing. Walnut cultivation in India has no serious infestation of insects and diseases (Yousf, 2006).

Harvesting and yield: The nuts are ready for harvesting in August to September when browning of packing tissue and splitting of hull takes place and natural drop of some nuts occur. Walnuts are harvested by striking the branches with a long pole. Piling should be avoided to avoid the mould incidence. Hulling and drying the walnuts should be as soon as possible after harvest. The nut crop is dried in the open air and sun and depending on weather conditions it may take from 5-8 days to dry. Average productivity is 1.4 tonnes ha^{-1}.

References

Jamwal, J. 2015.Status of walnut in Jammu and Kashmir (Personnel communication).

Motial, V.S. 1976. Introduction assessment and performance of apple and pear varieties in Jammu and Kashmir. In: Research report on pome, stone and nut fruits. All India co-ordinated Fruit Improvement Project, Fruit Research Workshop Ranikhet, Uttar Pradesh India, pp.46

Munshi, T.K. 2014. Food that helps you defy age. Personnel communication with Prof. (Dr) R.D. Gupta, 258, C- Sainik Colony Jammu.

Puttoo, B.L. 2005. Walnut: A heritage tree of Jammu and Kashmir. *Asian Agri. History*. 9(1): 59-62.

Singh, C. 2009. 5-Good foods for cholesterol *Ayurveda.* 1 (15): 18.

Yousuf, M.2006. Status of Walnut in Jammu and Kashmir. *The Kashmir Times*. 63 (185): 13.

23

Wild Date Palm (*Phoenix dactylifera* L.)

Common name: Khajur

Family: Arecaceae

Introduction

The desert of India occupies an area of about 27.83 m ha, spreading over four states namely Gujarat, Rajasthan and parts of Haryana and Punjab. In these areas, where irrigation facilities (underground or canal irrigation) are available, the choice of growing date palm may be appropriate. Its cultivation is most suitable in Kacch region of Gujarat and western region of Rajasthan, especially Bikaner, Jodhpur, Jaisalmer and Barmer areas (Srivastava and Misra, 2007).

The date palm fruits have been recognized very popular among the people because they are very delicious and nutritious in nature. Moreover, they are rich source of minerals, vitamins and carbohydrates. Owing to this reason, India is importing several thousand tonnes of soft dates (*Pind khajur*) and dry dates worth millions of rupees every year from Middle East countries. Date palm also possesses medicinal uses.

Medicinal Values

The dried date palm, known as 'Chuhhara', is considered a dry fruit. It is a rich source of Fe, Ca, P and K. It is useful for curing cough and sciatica pain. It quenches thirst of water and relieves in high temperatures. It checks excessive urination, cures dysentery and helps in rising teeth of children. For this purpose, date palm rubs are given to the children for 3-4 times. It also matures gums. It is useful for heart patients. Taking of 3-5 date palm and 10 leaves of neem tree, gives benefit to sugar

patients. To remove the redness of eyes and swelling, the seed of date palm is rubbed and applied.

It is also beneficial to tuberculosis patients by giving its water. Cholera and constipation is also cured by taking its water. Due to presence of tannins, many dates in green stage are astringent in taste. Its juice was the favourite of Mahatma Gandhi being rich in minerals and vitamins.

Other Uses

The leaves of date palm are used for preparation of paper baskets and brooms. Wine is also prepared from dates. Vinegar, liquid sugar, date juice from the date pulp and concentrate protein from dates are manufactured on large scale.

Botanical Description

Leaves are 3m long and leaf crown grows to 10 m wide. Inflorescence grows to 1m, flowers white, unisexual. Fruit is single seeded which turns to purple red on maturity. Data palm is a 10-15m tall tree. It is highly cross pollinated due to its dioecious nature. Two to three males are enough to pollinate 100 female plants. About 2-3 strands of male flowers are inserted between the strands of female flowers.

Cultural Practices

Climatic requirements: The date palm is very exacting in its climatic requirement which according to Arab saying should grow with "its feet in water and its head in the fire of the sky". The successful cultivation of date palm requires a long summer day as well as night with low temperature.

Soil requirements: Soil should be well developed having good water holding capacity and proper drainage. Sandy loam soils are the best suited for date palm cultivation. Date palm tolerates high salinity and sodality.

Varieties: 'Halawy', 'Medjool', 'Zaheedi', 'Shamran', 'Barhee', 'Hayani', 'Khalsa', 'Zagloul', are the main varieties of date palm, which can be grown depending upon the choice of the cultivators and rainfall conditions (Srivastva and Misra, 2007).

Planting: In commercial plantation, the multiplication of date palm is done through ground off shoots which have at least 8-10 kg average weight and well developed root system. These should be planted either in rainy season (August-September) or in February-March.

Planting is done in well prepared and leveled field either at 6 m or 8 m spacing from row to row and plant to plant. Pits of 1 m^3 size should be filled with a mixture of well decomposed compost or farm yard manure. Rainy season (July-September) is the ideal time for planting.

Irrigation: Young date palm trees require frequent irrigation for their better growth with frequency of irrigation from 15-30 days during winter and 7-10 days during summer.

Mulching: Using local weeds such as Bui (*Aerva bersica*), Shania (*Crotonalaria burhia*) and branches of desert shrub Ker (*Capparis decidua*) as organic mulches help in conserving soil moisture during the period of fruit growth and development.

Manures and fertilizers: Well rotten farm yard manure or compost should be applied to the trees in the beginning of winter season. A bearing tree should be applied 40-50 kg of farm yard manure. If there is deficiency of nitrogen, the nitrogenous fertilizer should be given after fruit set *i.e.*, in the month of March-April.

Fruit bearing: After 4-5 years of age, trees start flowering. However, good fruit bearing may start at the age of 10-12 years.

Harvesting and yield: The fruits of date palm are harvested at 'doka' stage particularly of variety 'Medjool'. The dates harvested at doka stage, have 70-80 per cent moisture. They have very poor keeping quality. Therefore, these fruits should be marketed soon or may be cured or may be processed.

The number of bunches per palm varies from 15 to 20 and fruit yield from 50 to 300 kg.

References

Srivastava, D. and Misra, K.K. 2007. Date Palm Management. *Indian Farmers Digest*. 40(1): 29-32.

24

Apricot
(*Prunus armeniaca* L.)

Common name: Khubani, khurmani, chulli, shara, sari

Family: Rosaceae

Introduction

Apricot is one of the important fruit of mid hills and dry temperate regions of the country. It is known as 'chulu', 'sahara' and 'zardalu' in the local dialect found in different regions of Uttrakhand, Himachal Pradesh, Jammu and Kashmir (Anonymous, 2007; Singh *et al.*, 2005). Kargil and Leh districts of Ladakh region in Jammu and Kashmir are rich in diversity of apricot. Although the fruits of wild apricot are unfit for table purpose due to high content of acids and low sugar content, yet their sweet kernels resemble with almond in taste. As such they are extensively used by the locals as replacement of almonds. Not only this, bitter kernels of wild apricot are used for extraction of oil which is edible and medicinal in value (Singh, 2005).

Medicinal Values

The apricot oil consists of unsaturated fatty acids such as oleic (74.3 per cent) and linoleic acid (21.6 per cent) which constitutes up to 90 per cent of total oil yield and has good nutritional and pharmaceutical importance.

Apricots are a good source of carotene (vitamins A), C and E, calcium, iron potassium, phosphorus and carotenoids. The lycopene found in apricot helps to protect eyes and prevent heart disease as well as certain cancers especially skin cancer (Singh, 2013). Presence of fibre in apricots helps to relieve constipation. Seeds of apricot are stomachic, demulcent and laxative.

The seeds of wild and domestic apricot contain a chemical which, when ingested, breaks down into the highly toxic hydrocyanic acid and cause serious illness.

Other Uses

Traditionally pulp of wild apricots is not used for eating. In Leh and Kargil, where wild apricot exist on larger area, after separating the kernels from fruit, pulp is mostly used as animal feed. Now, with the advanced technologies efforts have been made to utilize fruit pulps for manufacturing the following products:

1. **Dried apricots:** Dry apricot is a commercial product in Leh and Kargil districts of Jammu and Kashmir and tribal area of Himachal Pradesh. However, the apricot grown in low and mid-hill regions being low in sugar content cannot be dried directly. But the osmotic dehydration technologies developed by Indian Institute of Integrative Medicine (CSIR), Jammu are not being used to dry the apricot. In this technique, fruits are washed and halved to remove the stones and dipped in hypertonic solution containing "B" sugar syrup for 6-8 hrs. After draining of syrup the fruits are dried in a mechanical dehydrator or in open sun to a constant weight.
2. **Appetizer:** The wild apricots have been successfully utilized for manufacturing of appetizers.
3. **Fruit chutney:** Fruit chutneys are prepared form pulps of wild apricots either singly or in combination with tomato, strawberry and apple.
4. **Fruit powder:** The dried slices of wild and cultivated apricots after removing stones are ground to fine powder, which can be utilized for preparing number of products. Prior to drying, the fruits are required to be dipped in 0.5 per cent solution of boiling sodium hydroxide and halved to remove stones, followed by exposure to sulphur fumes (5g/kg) for 3-4 hours.
5. **Preparation of Jam:** The pulp of wild apricots can be used for preparing jam. The technology for preparation of jam is available and is being used extensively for manufacturing of apricot jam in Leh and Kargil. The sugar can be replaced by non-nutritive sweeteners like saccharine for making product suitable to diabetic patients.

Botanical Description

Botanically apricot is identical to peach. It grows maximum up to height of 10-15 meters but smaller in size if planted in the planned orchards. Bark is reddish brown. Leaves are oval to elliptical to round to sub-cordate, 5 to 7.5 cm long, often pubescent (fuzzy-haired) on the underside around the leaf veins. The flowers are solitary, white to pinkish which appear much before the emergence of leaves. The fruits are hairy, orange or yellowish orange with soft flesh. Fruits had a smooth and hard, flattened stone which contain a kernel inside.

Cultural Practices

Climatic requirements: The apricot growing region is mostly confined to the mid hills, upto an altitude of 1200-3000 m, where winters are cold enough to bring the trees out of rest and to the cold arid regions of Himalayas where the rainfall is low (Singh *et al.*, 2004). The long cool winter (800 chilling hours below 7°C), and frost free and warm spring are favorable for fruiting. Average summer temperature (16.6°C-32.3°C) is suitable for better growth and quality fruit production. Spring frost causes extensive damage to the blossoms, which are killed when temperature falls below 40°C. An annual rainfall of about 100 cm well distributed throughout the season is good for its normal growth and fruiting (Anonymous, 2007). The early blooming makes it more susceptible to frost damage than peach trees, and thus it is generally not planted as far north.

Soil requirements: Apricot can be grown in most of the soils but requires deep, well drained, loam soils for its best growth. However, it can be grown on loamy sand to sandy loam soils also. The tree growing in wet soil can depress physiological activity of leaves and overall growth of the plant. The pH of the soil should be 6.0-6.8.

Varieties: 'Suffaida', 'Charmagh-3', 'Shakarpara', 'Parachinar' and 'Ladakhi' are some of the varieties of apricot. The fruits of these varieties are highly attractive, sweet and delicious.

Propagation and planting: It is propagated through seeds and rarely through vegetative means. Apricot can be propagated well on peach and plum root stock. Its propagation is done in the same manner as that of peach. Both 'T' and ring budding can be done during the months of May and June, and grafting operation is done in February when the plants are in full dormancy. The planting in the field is done 4-5 m apart during early spring, while in dormant stage.

Wild apricot seeds require cold stratification at 4-7° Celsius for a period of 72 hours for better germination. The growth inhibitory activity of the seed-coat disappears with continuous moist cold treatment. The germination can be hastened by removal of shell from kernel after stratification. Seed stratification and gibberellic acid/kinetin treatment also hasten the germination. The seeds are soaked for 24 hours in 500 ppm GA or 5 ppm kinetin before sowing (Anonymous, 2007).

Planting: Wild apricot is planted during the dormant season (ending December to mid-March), but early planting gives better establishment. Pits of 1 m x 1m x 1m size are prepared one month before planting. They are filled with a mixture of soil and 50-60 kg well-decomposed farmyard manure. About one kg SSP and 10 ml Chloropyrophos solution (10 ml/10 litres of water) is also added to each pit. On flat land, a regular layout system such as square or rectangular is followed while on the hill slopes, contour system is generally practised. The plant spacing varies with the soil, climate and vigour of cultivar. The plants are generally planted at a distance of

6m x 6m. High density planting is still to be standardized with proper training and pruning system, and with the use of growth retardants. One year- old, healthy and disease-free plants are planted in the middle of the pit. The soil is pressed gently so that roots are set. Watering should be done immediately to establish close contact between roots and soil. After planting, tree basin is mulched with 10 cm thick hay mulch to conserve soil moisture. In summer, watering should be done as and when required (Anonymous, 2007).

Manures and fertilizers: Application of 25-30 kg FYM per tree during December-January is the recommended dose. Application of N,P and K @, 50, 25 and 60 g, for every year age up to 10 years is also recommended (Singh *et al.*, 2004) through urea, DAP and muriate of potash. Fruit cracking physiological disorder in apricot is sometimes observed due to B-deficiency (Verma *et al.*, 2015). To overcome this problem, use solubor or other boron containing fertilizer.

Harvesting and yield: Fruits are harvested during ending June to August when more than 50 percent fruits get ripened since apricot trees are large and cylindrical so cannot be handpicked easily. The method of harvesting is, shaking of limbs or beating with long sticks. This method, however, causes damage to the tree as all the fruits fall down irrespective of the stage of ripening. Ripe fruits, sometimes, split open and others get bruises. The fruits should be harvested in morning hours and direct exposure of fruits to sun should be avoided during grading and packaging. Department of Horticulture has introduced nets for collecting the fruits while harvesting but the technique is yet to be picked up.

Tree starts bearing at the age of 4-5 years and continues to bear well for 50- 60 years. The full bearing occurs at about 10-15 years when it yields about 85-100 kg fruits per tree. The stone yield varies from 12-17 percent of fruits and the kernel yield ranges 3.14-4.81 kg/tree. The yield of a full-bearing well maintained tree varies from 120-150 kg (Anonymous, 2007). The average yield of apricots has been worked out be 3.8 tons ha^{-1} under Leh and Kargil soil conditions of Ladakh, Jammu and Kashmir.

References

Anonymous, 2007. Wild Apricot. National Oilseeds and Vegetable Oils Development Board (Ministry of Agriculture, GoI) Gurgaon. pp.17.

Singh, C. 2013. Top 10 healthiest fruits. *Ayurveda* 1(22): 16-17.

Singh, N. K., Yadav, V. K and Kumboj, P. 2005. Utilization strategies of wild apricot to improve the socio-economic conditions of hill farmers. *Indian Farmers Digest.* 38 (3): 25-29.

Singh, R. K., Verma, S. K., Singh, H., Bhardwaj, P. N., Regar, R and Arya, R.R. 2004. Ways to boost the cultivation of stone fruits in Uttranchal. *Indian Farmers Digest.* 37 (5-6): 5-8 and 21.

Singh, S.S. 2005. Fate of the Medicinal Plant Resources of Jammu and Kashmir. Directorate of Environment and Remote Sensing Jammu, J&K State.

Verma, S.K., Jambamma, S. And Lavanya, T. N. 2015. Fruit cracking: Physiological disorder in fruit crops. *Indian Farmers Digest.* 48 (2): 22-23.

Shrubs

1

American Aloe (*Agave americana* Linn.)

Common name: Century plant, keora

Family: Asparagaceae

Introduction

Keora plants are native to North America and are adaptable to colder climate in the Pacific North-West and even Canada. In India, keora is an exotic which stands naturalized in tropical and sub-tropical regions of the country. It is cultivated from bulbils in gardens.

Roots, leaves and gum of keora are used in preparing many medicines in Ayurvedic system (Singh, 2005).

Medicinal Values

Roots are diuretic and anti-syphilitic. Sap is laxative, diuretic, emmenagogue and anti-scorbutic. Administration is done in the form of decoction or leaf juice and gum from the roots. Juice of fresh leaves is also anti-syphilitic. Leaf juice is useful for external application to bruises and contusions. Gum exuded from leaves and root is most effective in toothache. Pulp of leaves with sugar is popular home remedy for gonorrhea. Leaves are also used as tonic. Leaf poultice is effective on swellings and acute pains.

Botanical Description

It is a long leaved, succulent plant that naturally forms a rosette and produces a flower spire of attractive cup shaped gloom. The plant is drought tolerant and perennial making it ideal for mature arid gardens. It produces attractive flowers and then the plant dies leaving behind pups or offsets. They have large taproots and do not transplant well. Plant has several surface roots. The plant is 7.5m in length and 2.5 m in breadth. The life cycle of the plant is 8 years.

Cultural Practices

Climatic requirements: Wide variety of climates is capable of supporting agave. It is a warm season crop and requires full sun.

Soil requirements: Agave plant needs gritty soil, but should be well drained. In heavy clay soil, mix sand and organic manure to improve its texture. However, sandy loam, loamy sand and loam textured soils have been found to be best for the growth of agave.

Propagation: Grown up suckers can be planted during rainy season which are usually propagated from bulbils or suckers. In case of bulbils, they are first sown in mother beds at close spacing @ of 5,000 bulbils per bed of 1x20m size. After 6 months, these seedlings are planted in transplant beds of size 20x1m at the rate of 500 plants per bed. In 2nd stage they are kept for 3 months. After 9 months, suckers weighing to ½ kg and 22 to 30 cm height are ready for planting.

Field planting: In main field, they are planted at space of 2x2 m in pits of 30x30x30cm size. Planting is usually carried during rainy season (July to September).

Harvesting and yield: The leaves get ready for harvesting from 3rd year onwards. The older leaves having length not less than a meter are harvested in 3rd year. The life cycle of the plant is upto 8 years. From 3rd year onwards leaf yield of 75-100 tonnes per hectare could be obtained. The leaves may be collected during winter and spring seasons when sufficient leaves are available on the plants.

References

Khajuria, G. L. 2005. Protecting medicinal and aromatic plants. *The Kashmir Times.* 62 (339): 13.

Singh, S.S. 2005. Fate of the Medicinal Plant Resources of Jammu and Kashmir. Directorate of Environment and Remote Sensing, Jammu, J&K State.

2

Black Cardamom (*Amom subulatum* Roxb.)

Common name: Badi Elaichi

Family: Zingiberaceae

Introduction

Large Cardamom or black cardamom is considered to be the queen of spices. Large cardamom is also called Greater Indian cardamom and Thuto Elaichi in Nepali. In Sanskrit it is known as Elagreator. In India, cultivation of large cardamom was originally started in the Eastern Himalayan region, especially in Sikkim and Darjeeling hills as well as in certain parts of Uttrakhand (Pant *et al.*, 2013). Now it is being grown in almost all the evergreen forests of Western Ghats. Besides, large cardamom also grows well in forest areas having cooler climate. Guatemala, India, Srilanka, Tanzania, Vietnam, Laos, Cambodia, New Guinea are the major cardamom growing countries. Now, the demand for large cardamom in the export market has increased like small cardamom (Sandhu and Khera, 2010).

Medicinal Values

Large cardamom is not only used as a spice and in aroma therapy, but is also used as medicine in a number of Ayurvedic preparations. It serves as a diuretic agent as helps the body to release excessive fluid out of the body and assists in treating of kidney stone, lymphatic, swelling gonorrhea, liver disorders and high blood pressure. It is used as a carminative and prevents the formation of gas in the gastro-intestinal tract. As a stomachic, it stimulates gastric digestion and sharpens appetite.

'Curcumin' present in large cardamom acts as an anti-oxidant, which protects the heart by preventing the deposition of cholesterol from narrowing the arteries, called atherosclerosis that is the main cause of heart attack. Large cardamom is used to combat nausea and vomiting and is often useful for migraine.

Many healers, particularly of South Asia have successfully using this herb since centuries as an antidote for snake venom and scorpion. It is also used as remedy for throat and respiratory trouble. For this purpose, the seeds of this herb are used both as preventive vis-à-vis curative measures for throat troubles, congestion of lungs and in the treatment of lungs and pulmonary tuberculosis. Essential oil present in large cardamom is called cincole. Cineole acts as an antiseptic agent *i.e.,* it can kill bad breath bacteria. Cineole is often used aromatherapies as a remedy for fainting.

It is concluded that large cardamom is a versatile spice having a wide range of culinary and therapeutic uses. It is an indispensable part of our daily life. It possesses excellent food value. Analysis of the herb shows that it is low in fat and high in protein content and vitamins A, B and C. The seeds contain 2-3 per cent essential oils which is an essential factor in food items and in the preparation of certain medicines. Cardamom oil is also used in perfumes and as carminative, stomachic, diuretic and cardiac stimulant besides a remedy for throat and respiratory trouble.

Botanical Description

Large cardamom is a perennial herb, with sub-terranean rhizomes with leafy shoots, growing up to 1.8 to 2.5m height, and panicles. Stem is a pseudo stem which is called tiller. Inflorescence is spike with 30 to 40 flowers per spike. Flowers are yellow, bisexual, zygomorphic. Fruit is capsule, achinated, maroon in colour with seeds which are whitish in immature stage and dark brown to black at mature stage. The leafy shoots are mostly grayish to reddish in colour and cylindrical in shape depending upon the variety. Leaves are unicotate with parallel venation.

Cultural Practices

Climatic requirements: The crop grows well under the shade of forest trees at altitudes ranging from 1000-2000 m with annual rainfall of 3000-3500 mm.

Soil requirements: Deep and well drained soils with a loamy texture are best suited for its cultivation. The soils having a pH range from 4.5 to 6.0 are ideal. Even though the crop can be grown in undulating and steep terrains, but moderate slope gives good crop.

Varieties: Some of the popular cultivars under cultivation include 'Ramsey', 'Sawney', 'Golsey', 'Varlangey', 'Seremna' etc.

Land preparation: The land selected for planting is cleared of all under growth, weeds etc. Pits of size 30 x 30 x 30 cm are prepared at a spacing of 1.5 x 1.5 m. Pits are left open for weathering for a fortnight and then filled with topsoil mixed with cow dung compost/FYM @2-3 kg per pit. Pit making and filling operation should be completed in the third week of May before the onset of pre-monsoon showers.

Propagation: Propagation is done through seeds and suckers. Propagation through seeds enables production of large number of seedlings. Plants raised from seeds need not necessarily be high yielders even if they are collected from very productive plants due to cross-pollination. On the other hand, planting through suckers ensures true to type with a high production if they are collected from high yielding, disease free plants.

Planting: Planting is done in June-July when there is enough moisture in the soil. Suckers/seedlings are planted by scooping a little soil from the centre of the pits and planted up to collar zone. Deep planting should be avoided. Staking is needed to avoid lodging from heavy rain and wind and mulching is done at the plant base.

Manures and fertilizers: Well decomposed FYM/compost @ 5kg/plant is applied twice in a year in April-May and August-September.

Irrigation: Plants cannot thrive well under water stress conditions. In the first year of planting watering is required at least at 10 days interval during dry months. Higher yield can be realized in irrigated plantations.

Intercultural operations: Earthing up with a thin layer of fertile soil should be done at the base of clump covering up to the collar region.

Harvesting and curing: The crop is ready for harvest when the seeds of top most capsules turn brown. Mature tillers are cut at a height of 30-40 cm from ground and left for another 10- 15 days for full maturity. The harvested spikes are heaped and capsules separated and dried. The cured capsules are rubbed on wire mesh for cleaning and removal of calyx (tail). Traditionally cardamom is cured in *Bhatti*, by direct heating. Improved curing techniques are presently available. Curing is done till moisture content of the produce is brought down to 12-14 per cent level and gives metallic sound while shuffling. Yield ranges from 200-250 kg ha^{-1}.

Packaging and marketing: The properly dried capsules should be allowed to cool and then packed in polythene lined jute bags. The bags may be stored on wooden platforms to avoid absorption of moisture and avoid fungal growth on the stored produce.

References

Ghatani, . 2011. Large cardamom: The queen of spices. *Agrobios News Letter*. 10(5): 38-39.

Sandhu, S. and Khera, K. S. 2010.Cardamom aromatic herb. *Agrobios Newsletter*. 10(1): 29-30.

Pant, M., Kumar, G., Kumar, A. and Kushwaha, H.S. 2013, High value spices as profitable crops of Uttrakhand. *Indian Farmers Digest*. 46(9): 29-32.

3

Castor Bean Plant (*Ricinus communis* L.)

Common name: Arandi, eranda

Family: Euphorbiaceae

Introduction

Castor is cultivated around the world for its non-edible oil seed. India is the largest producer of castor seeds (62.4 per cent) followed by china (19.2 per cent) and Brazil (12.7 per cent). India is the world's biggest grower country dominating the global trade with a share of more than 80 per cent. Gujarat, Rajasthan and Andhra Pradesh contribute 96 per cent of the total castor seed production in India (Roopa and Bharathi, 2014). Apart from its cultivation, the plant is also found as wild in the fields, gardens, villages, along road sides, near habitations, and wastelands in the hills (Chauhan, 2006).

There is no doubt castor seed is poisonous and two or three seeds when taken have been found to possess fatal effects (Chauhan, 2006) but its oil has a number of medicinal uses. Similarly leaves of the castor have lot of medicinal values.

Medicinal Uses

Decoction of leaves is used as purgative, lactagogue and emmenagogue. Roots are useful in rheumatism, lumbago and sciatica. Seed contains about 30 per cent of a fixed oil which is used as purgative, 10-20 drops. Externally it is rubbed in itch, herpes, eczema and cleaning application for wounds, ulcers, sores. Diluted with bland oil, it is effective in rheumatism. Seeds are also purgative. Leaves applied on breast increases secretion of milk. It is a local remedy for scabies eczema. Mixed with butter milk, it is given in dyspepsia and diarrhoea. Twigs are used as tooth brush (*datoons*). It strengthens gums, cures bleeding and gumboils.

According to Chaudhuri (2006) root, leaf, seed and oil are the major parts for medical use. Castor oil is marketed in all medical stores as well as grocery also. According to age, disease and patient conditions, it is used in a dose of 1 to 3 tea spoonful (5-15m) in constipation, rheumatism etc.

Other Uses

Castor oil is used in making contraceptive jellies. It has also been strongly recommended as an illuminant in lamps. The extract of its leaves has shown its positive results as insecticide.

Botanical Description

Arandi (erandi or eranda) is an evergreen plant which grows commonly on wastelands. It is also cultivated in tropical and sub-tropical areas in hedge and as medicinal plant. A herbaceous, annual, fast growing plant which reaches to a height of 0.9–3.0 m. The long-stalked glossy leaves are alternate, 15 to 45 cm in length with 5-12 deep lobes. The leaf colour varies from dark green with a reddish tinge to dark reddish purple and bronze at maturity. The male flowers are yellowish-green in colour, having prominent creamy stamens, while the female flowers bear prominent red stigmas at the tips of the spikes. The fruit is spiny, greenish capsule containing large, oval, shiny, bean-like highly poisonous seeds. The green capsule dries and splits into three sections, causing seed dispersal.

Cultural Practices

Climatic requirements: Arandi requires tropical to sub-tropical type of climate. However, it is mostly cultivated in areas having tropical and warm temperate climate up to an elevation of 2000m.

Soil requirements: Although Arandi can be grown in light (Sandy, sandy loam and loamy sand) to heavy soils (clayey, clay loam) yet the Kharif crop requires to be protected from excessive moisture during periods of heavy and continuous rains in heavy soils by providing drainage.

Time and method of Sowing: Arandi grows as a main season crop and the time of sowing varies from region to region. Generally June to July and ending September to October are the most probable months for sowing of this plant.

Propagation is mostly resorted by the seeds.

Irrigation Schedule

***Kharif Castor*:** The first irrigation is given as pre or post sowing irrigation to ensure better germination, emergence and stand establishment. The subsequent irrigations are provided with 15-20 days interval depending upon soil and climatic conditions. The most critical period for moisture is the terminal bud initiation to full flowering of primary spike (Roopa and Bharathi, 2014). Furrow irrigation system is ideal for castor sown in ridge and furrow type of seed bed, as this is the simplest system and also very efficient, especially on undulated lands or underdeveloped topography.

***Rabi Castor*:** First irrigation is required to be given about 50 days after sowing or around full flowering at primary spike. Subsequent irrigation should be furnished with an interval of 20 days. Crop of 180 days duration requires around 6 irrigations for obtaining optimum yield. At times of deficient water supplies, first irrigation must be given at primary spike development or secondary spike initiation followed by one or two irrigations during flowering and seed formation stages.

Harvesting and yield: The duration of the crop in the annual type of castor varies from 4 to 9 months, but generally the duration of the crop is about 6 months.

Flowering and fruiting mostly takes place from May/June to August. Flowers are large in terminal branches. Fruits are prickly capsules marked into three parts. Seeds are oblong having crustaceous seed coat. The average yield of the castor is about 20-90q ha^{-1}.

References

Chaudhuri, S.2006. Cheaper c's. *Ayurveda.* April issue, pp.12.

Chauhan, N.S.2006. Medicinal and aromatic plants of Himachal Pradesh, Second Edition. Indus Publishing Company FS-5, Tagore Garden, New Delhi-110027.

Roopa, T.K. and Bharathi, K. 2014. Irrigation scheduling for both *Kharif* and *Rabi* castor (*Ricinus communis* L.). *Agrobios Newsletter*. 12(12): 30.

4

Chaste Tree (*Vitex negundo* L.)

Common name: Bana, nirgundi

Family: Verbenaceae

Introduction

Bana is found throughout the greater part of India. It is a very common shrub of sub-tropical Himalayas growing on the forest fringes and in agricultural lands along the banks of streams, rivers and canals at an altitude of 1500m (Singh, 2005; Chauhan, 2006). All parts of the shrub viz; roots, leaves, bark and flowers are used in medicines.

Medicinal Uses

Leaves of the shrub are used as febrifuge, diuretic, emmenagogue and expectorant. Roots are demulcent, diuretic and expectorant. Flowers are astringent and cardiac tonic.

Bana is, in fact, an alternative, anodyne, antiarthritic, anti-parasitic, appetizer, aromatic, astringent. It has been credited with curative effects in asthma, urinary troubles, tonsillitis, sciatica, rheumatism etc.

Other Uses

Dried leaves are used to protect stored grains and the smoke of the leaves repels mosquitoes (Chauhan, 2006). Twigs are used to keep away evil spirits.

Botanical Description

Bana is a shrub, with thin grey coloured bark. Branchlets are four angled, densely white, tomentose. Leaves are digitately divided, leaflets ovate, lanceolate, whitish beneath. Flowers are blue in colour borne in cymose panicles. Berries black in colour and are equivalent to the size of the pea.

Cultural Practices

Climatic requirements: Plant requires tropical to sub-tropical and sub-temperate climatic conditions. It mainly grows ascending to an altitude of 1500m in the outer Himalayas (Siwalik or Shivalik hills).

Soil requirements: It prefers to grow in light textured soils ranging from sandy to loamy sand and sandy loam. This plant species can be best cultivated in moist soil conditions.

Propagation: The shrub can be propagated through branch cuttings and root suckers.

Flowering and fruiting Time: March to April is the flowering time where fruits set in from June to September.

Yield: Fresh leaves yield about 0.05 per cent essential oil.

References

Chauhan, N.S.2006, Medicinal and Aromatic Plants of Himachal Pradesh. Indus Publishing Company, FS-5, Tagore Garden, New Delhi – 110027.

Singh, S.S. 2005. Fate of the Medicinal Plant Resources of Jammu and Kashmir. Directorate of Environment and Remote Sensing, Jammu, J&K Government.

5

Curry Tree (*Murraya koenigii* L. Sprengel)

Common name: Kari patta

Family: Rutaceae

Introduction

Curry patta is a handsome perennial (sometimes deciduous) ornamental shrub with lush green leaves. The use of curry leaves in Indian cuisine has been referred to in thousand years Kannada Texts (Menon, 2011). It is used as a spice. Interestingly curry patta (*Kadi Patta*) or curry leaves also known as *meethi neem* comes from a plant that grows wild in the Himalayan foot hills and other parts of India. Apart from India, this plant also grows in various parts of North Thailand and Srilanka. (Menon, 2011).

It is remarkable to note that this shrub is not browsed ordinarily by the goats except when there is scarcity of fodder plants/tree leaves or forage crops. Although curry patta grows throughout the country yet in Himalayan region it grows mostly in the Siwaliks up to 1000m altitude (Gupta, 2012). In Siwaliks of Jammu this shrub generally grows in the sub-mountainous tract locally called the *Kandi belt* of Jammu ranging from 300 to 1050 m above mean sea level (Gupta, 2012). Besides it also grows in the *Kandi* belts of Punjab, Haryana, Himachal Pradesh and Uttrakhand.

There is no doubt that curry patta does not play a big role in North Indian cuisine where one is used to fresh coriander in everything. But for south Indian, curry patta is a vital ingredient in recipes for chicken, mutton, fish, chutneys, vegetables etc. The name itself in Tamil is pronounced as "Kariveppulai" (Kari-curry, veppu-neem and lai-leaf) which means "leaf" used to make curry. It is present in almost

all the dishes of Tamil Nadu. The south Indians also make delicious dry chutney with roasted *Karri Patta,* roasted *chana daal* and many vegetables. It is because curry patta has lot of medicinal values which have been described here under vis-à-vis a brief description of the curry patta and its cultural practices.

Medicinal Uses

Its medicinal uses run into a long list although in literature it is mentioned only for few. All parts of the plant are used. The leaves are cooked with many preparations. Chewing of leaves is indicated to give relief in gum troubles. Twigs are used as tooth brush (Datoon). It has many uses in home remedies (Gupta, 2012).

Ayurvedic and Unani Systems of Medicine use curry patta in many ways including herbal tonics. The leaves have the quality of regulating the functions of the stomach. It has some aromatic oils that are antioxidant in nature. Fresh leaves taken with few pepper corns early in the morning are believed to bring down blood sugar level (the west is clamming breakthrough in research regarding properties of curry leaves that help control diabetes while Ayurveda has advocated that for centuries).

An infusion of roasted leaves is used to stop vomiting. Chew a few leaves in butter and add them to butter milk along with cumin and rock salt. This drink is very useful herbal remedy for nausea. Chewing five curry leaves empty stomach in the morning helps to regulate peristaltic movements and gas formation improvement.

Apart from adding distinctive flavour to many dishes, the leaves of the bush are also employed as herbal tonic in Ayurvedic medicines. Use of the leaves is well known to boost the blood circulation. It also has antimicrobial, hepeto-protective, hypo-cholestrolemic qualities and delays premature graying. The leaves of the shrub can too be used as a sedative and hair tonic. Curry leaves are commonly sizzled in hot oil, often with mustard seed and them simmered with fish or vegetables (Natrajan, 2010).

Botanical Description

Plant is an evergreen, gregarious shrub regenerating from seeds. It is an aromatic perennial tree which is sometimes 5 m tall and 15-40 cm in diameter. It was grown in private gardens in India for centuries, but started being commercially produced only recently.

The shrub, in fact, is ornamental with lush green pinnately compound leaves. The leaves are about 15-30 cm long bearing 10-20 leaf lets nearly 3-5 cm long and lanceolate type. This evergreen shrub mostly regenerates from the seeds naturally and makes dense undergrowth in the forests.

Cultural Practices

Climatic requirements: Curry patta is a warm season shrub and it much harmed by frost. It grows well both under tropical and sub-tropical climate conditions. In Siwaliks, it mostly grows in its sub-mountainous areas known as *Kandi* belt.

Soil requirements: It grows practically on all kinds of soils except very light and heavy, and waterlogged ones. Loamy sand and sandy loam soils suit the best for its growth.

Planting time: Although it can be planted both during spring season and rainy season yet its best growth can be seen when curry patta is planted during the rainy season. It is already stated that this shrub regenerates from the seed in the Siwalik Hills. Thus, plant seeds annually producing a bumper crop which germinates readily and densely making a green carpet in the form of seedlings.

Harvesting and yield: The main season of availability of leaves is July-August. The leaf yield after 1st year of planting is 200-400 kg per hectare whereas, total yield upto 20 tonnes per hectare per year can be realised.

References

Gupta, R.D. 2012. Karri Patta in curry, cure. *The Daily Excelsior*. 48(294): 3 [Magazine, Sunday, October, 21]

Menon, P. 2011. The goodness of curry leaves. *The Kashmir Times*, Jammu 68(189): 2 [Sunday Magazine, July 10]

Natrajan, P. 2010. Curry Leaves: Kari Patta. *Tattvalika*. 33(9): 40-41.

6

Indian Barberry (*Berberis aristata* DC.)

Common name: Rasaunt, rosant

Family: Berberidaceae

Introduction

The plant is found throughout the country having altitude in the range of 1600 to 3400 m. In Jammu region of Jammu and Kashmir, it is mostly confined to Kud, Batote, Chenani, Bhadarwah and Kishtwar areas (Singh, 2005). In Himachal Pradesh, it is present in Chamba, Kinnaur, Sirmour, Kulu, Mandi and Shimla. However, due to over exploitation, population in the wild has declined considerably (Chauhan, 2006). All parts of the shrub like wood, root bark and its extract known as "Rasaunt', have been used as medicines since long, which are briefly described hereunder.

Medicinal Uses

The medicinal properties of *Berberis aristata* are just analogous to *Cucurma longa*. The plant and the 'Rasaunt' are utilized as alternative and deobstruent for curing skin diseases, diarrhoea, jaundice and eye infections. A decoction of the root bark is used as a wash for unhealthy ulcers. Rasaunt, mixed with honey is an useful application to aphthous sores (Chauhan, 2006). Berberine (Alkaloid) is used in the form of salts *viz.*, berberine hydrochloride and berberine sulphate, for curing enlargement of the spleen, malarial infections, dyspepsia, diarrhoea, anorexia, vomiting and intestinal catarrh.

Botanical Description

Rasaunt is a thorny deciduous shrub about 2-5 m in height. It is characterized by pale rough, furrowed bark, leaves stalked, obovate or elliptic in the axils of simple or branched spines, flowers yellow in colour in compound racemes with stout pedicels. Racemes are longer than the leaves, fruits 1.8 cm x 0.6 cm ovoid, grey-white when young but become bluish when ripen. Wood and roots are yellow in colour from inside.

Cultural Practices

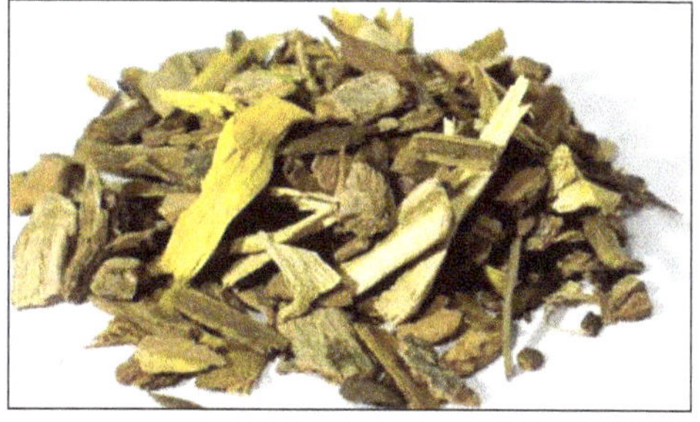

Climatic requirements: The shrub mostly grows from sub-temperate sub-humid climate to high hill temperate wet climatic zones.

Soil requirements: The best growth of the shrub takes place in sandy loam soils having neutral to slightly acidic soil conditions and medium to high categories of available nitrogen; low to medium phosphorus and medium potassium. However, it can also be grown in gravelly sandy loam to fine sandy loam and neutral soil. The status of available nutrients (N,P,K) should be high.

Propagation: Owing to the increased activities in agriculture including horticulture to feed the looming human and livestock population in north-western Himalayan States (Uttrakhand, Himachal Pradesh, and Jammu and Kashmir) like other parts of India, the habitat of Rasaunt has been ruthlessly damaged. The plant has managed to survive in inaccessible slopes, banks of streams and rivers.

Its propagation, however, is mainly through seeds. Propagation through cuttings has also been found suitable. Rainy season *i.e.* June to July or July to August is an appropriate time for its sowing. Generally, no manuring is done. Field after sowing should be kept free of weeds.

Flowering/fruiting, usually occurs during May to July/August to September depending upon the climatic conditions. Average sized plant yields about 650 g fruits in 4 pickings.

References

Chauhan, N.S. 2006. Medicinal and Aromatic plants of Himachal Pradesh, Second Edition. Indus Publishing Company FS-5, Tagore Garden, New Delhi-110027.

Singh, S.S. 2005. Fate of the Medicinal Plant. Resources of Jammu and Kashmir. Directorate of Environment and Remote Sensing, Jammu, Govt. of J&K.

7

Indian Ginseng (*Withania somnifera* L. Dunal)

Common name: Ashwagandha, asgandh, asgand

Family: Solanaceae

Introduction

Ashwagandha is one of the species, which has been prioritised by the National Medical Plants Board for its commercial cultivation. Ashwagandha is used as a substitute for Ginsing. It is known by many other names in Indian languages (Table 7.1). Ashwagandha is also called Rape seed/Winter cherry, Nagori, Asand, Bajiganda (Sharma *et al.*, 2006).

It is an important medicinal plant used both in traditional and modern system of medicines. In Ayurveda, all parts of the plant *i.e.*, roots, leaves, fruits and seeds are used in medicines for curing number of diseases. Presently, its use has been widely promoted by pharmaceutical companies for aphrodisiac, anti-aging and nervine tonic properties. Western herbalists refer Ashwagandha to as Ayurvedic Ginsing because of its reputation for increasing energy, strength and stamina like ginsing or ginseng, a herbaceous perennial plant, native to eastern America (Taj and Kumar, 2013).

Presently, it is commercially cultivated in more than 5,000 ha area in the north-western regions of Madhya Pradesh and the adjoining areas of Rajasthan on low fertility soils. The estimated annual production of ashwagandha roots in India is about 2000 tonnes against annual requirement of about 7,000 tonnes. Ashwagandha roots have tremendous demand in the Indian market as well as international market.

Table 7.1: Vernacular Names of Ashwagandha

Sl.No.	Language	Name
1.	Hindi	*Ashwagandha, Ashwgandha, Asgandh*
2.	Gujarati	*Asan*
3.	Ghoda	*Asor*
4.	Kannada	*Hiremaddinegida*
5.	Malayalam	*Amukkiram*
6.	All other Indian languages	*Ashvagandha*

Source: Sharma *et al.*, 2006

Distribution

This species grows well naturally throughout the drier parts and in the sub-tropical and semi- temperate regions of India upto 1500 meters above mean sea level but also being cultivated in certain regions. It is widely available in Mandsaur and Bastar districts of Madhya Pradesh, all over the foothills of Punjab, Himachal Pradesh, Uttarakhand and Jammu region of Jammu and Kashmir (Pandey and Chilana, 2004).

It is a plant of wastelands, bunds and other neglected sites and grows on soils which are not suitable for other plants. The plant occurs in drier regions naturally and is also cultivated. In India, it is cultivated in warm and arid climates. It also occurs in Middle East countries and parts of Africa. It is point to mention that Ashwagandha does not establish its roots in the hilly and mountainous areas. It requires light and well-drained soils for the formation of thick roots. Thus, it is ideally suited for cultivation in *Kandi* belt of Jammu as well as that of Punjab, Himachal Pradesh, Haryana and Chandigarh. There are 23 species of *Withania* that occur in the dry parts of India, North Africa, Middle East and the Mediterranean region (Das and Palai, 2011).

Chemical Composition

Ashwagandha contains a variety of alkaloids and withaniols (0.13 to 0.68 per cent). Nicotine, somniferine, somniferenine, somine, thiamine, withanamine are important ones. Withaniols and pseudo-withanine are therapeutically very important. The leaves of this plant contains anaferine but are not marketed commercially. The roots contain upto 50 per cent alkaloids, while the roots of wild plants are rich in 3-trophyl trigolate. Withaferin-A (steroidal lactone) have anti-bacterial, anti-tumor anti-arthritic and anti-inflammatory properties. Besides this, a pale yellow crystalline substance having antibiotic properties has also been isolated from the leaves of this plant. It contains a resin, fat and colouring matter. A new alkaloid withasomnine, phenyl-1-5-trimethylene pyrazole has been isolated (Butt, 1999).

Medicinal Values

All parts of plant are used in medicines and the fruit has the property of coagulating milk. However, roots of Ashwagandha are well known Indian drug used in Ayurvedic and Unani systems of medicines in the treatment of chronic joint pain, general debility, mental disorders and gynaecological problems.

(i) It is used as a sedative, a diuretic, a rejuvenating tonic, an anti-inflammatory agent and "adaptogen" (endurance enhancer).

(ii) It is useful in leucoderma, constipation, sexual weakness, insomnia, nervous disorder, and tuberculosis.

(iii) Ashwagandha can be used to cure asthma, senility, cardiac disorders, as well as painful swellings.

(iv) It is being widely used by the pharmaceutical companies for its anti-aging and nervine tonic properties.

(v) It has the ability to relieve from the stress and preveting rheumatism.

(vi) It acts as a narcotic and removes functional obstructions of body.

(vii) The root powder is applied locally on ulcers and inflammations.

(viii) The antibiotic and antibacterial activity of the roots and leaves has been experimentally proved (Jain, 1991).

(x) Ashwagandha provides remedy for dropsy, dyspepsia, hiccup, joint pains, reduces neurosis and inflammations in the lungs.

(xi) World Health Organisation (WHO) has recommended list of 42 medicinal plants in the Modern System of Medicines and Ashwagandha is one of them.

(xii) Some Ayurvedic preparations made from Ashwagandha are Zero Forte (DRDO) which is useful for persons working in high altitude areas such as Leh. Shepherd 10 (DRDO) is useful against frost bite and Stress Cum (Dabur India Limited) is beneficial against stress.

(xiii) Ashwagandha is one of the simple and very effective herbs which can strengthen the brain and leave a soothing effect on the irritated nerves to give a feeling of well being to a depressive patient (Vatsyayan, 2010).

Nowadays, ashwagandha and its extracts are the component of many invigorating herbal teas, powders, tablets and syrups. Traditional ways to use Ashwagandha against various diseases have been shown in Table 7.2.

Side effects: In at least two published clinical trials of *Withania somnifera*, the side effects experienced by *W. somnifera* treated individuals were not significantly different than side effects experienced by placebo treated individuals (Dass and Palai, 2011). However, there has been a report indicating it can stimulate the thyroid and lead to thyrotoxicosis in some people.

Botanical Description

It is a small or medium sized shrub growing upto 1.5 m. The stem and branches are covered with minute star-shaped hairs. Leaves upto 10 cm long, ovate in shape with hair like branches. Flowers are small, pale green in colour and about 1 cm in length. Few flowers are borne together in short axillary clusters. Fruits are 6 mm

Table 7.2: Methods of Using *Ashwagandha* as a Remedial Measure against Diseases/Disorders

Sl.No.	*Name of the Disease or Disorder*	*Method or way to control*
1.	Inflammation associated with rheumatological condition such as arthritis.	About 100-200 mg standardised extracts or 1-2 teaspoons of liquid extract twice a day should be taken.
2.	Stress induced exhaustion.	As stated above *i.e.* 100-200 mg standardised extract or 1-2 teaspoons of liquid twice a day should be taken.
3.	To boost red blood cells and sexual performance *vis-a-vis* rejuvenating effect.	Aged persons (50 to 60) when given 3 g of Ashwagandha powder daily for 3 months, showed significant increase in red blood cells and improved sexual performance also.
4.	To create immunity and relaxation in an individual.	100 to 200 mg standardised extract twice a day provides better immunity to an individual.
5.	As tonic and energetic to invalids.	Powder of Ashwagandha roots, Shatawar roots and Kawanch seeds in equal proportions is given to the weak persons.
6.	Stress Cum capsules	Available in the market in Ayurvedic formulations and taken once or twice a day to relieve stress on brain.

in diameter, red in colour and globose and smooth in shape. These are enclosed in the inflated and membranous calyx.

Cultural Practices

Climatic requirements: Semi-arid type of climate is considered the best for its growth. Excessive humidity and shady conditions slow down its growth. Hot and dry type of climate is most suitable for its cultivation. The semi-arid sub-tropical areas receiving 600-750 mm rainfall are suitable for its cultivation. The crop requires dry season during its growing period, however, 1-2 late winter rains are conducive for the proper development of the plant roots.

Soil requirements: Ashwagandha can be grown on variety of soils ranging from gravelly loamy sand to sandy loam soils. Soil should be well drained with medium fertility and normal soil reaction (pH 6.5 to 7.5). The plant has shown very good growth on sandy, dry loose textured soils (Butt, 1999). It is a kharif crop and grows well in sandy loam oils having pH in the range of 7.5 to 8.0. Water logging is injurious to roots and adversely affects the growth of the crop (Jamwal and Jamwal, 2011).

Varieties: Asgandh 20, Asgandh 134, WSR, Rakshita and Poshita are some of the recommended varieties.

Land preparation: Ashwagandha is usually grown on marginal lands without irrigation facilities. The fields, on which food crops cannot be taken profitably, are used for its cultivation. Land should be well pulverized by ploughing 3-4 times and planked twice to make the tilth fine and weed free. After leveling, it is divided into beds of suitable size.

Method of sowing: Ashwagandha is cultivated through seeds by broadcast method. It can also be raised through transplanting. The seeds should be placed about 1 cm deep in the soil which take about 12-15 days to germinate. Seeds are sown 25 cm apart in rows to facilitate hand weeding.

Seedlings from nursery are transplanted by maintaining 60 cm distance from plant-to-plant and 60 cm from row-to-row. Seedlings can also be planted at 30 cm spacing and ridges are prepared at 60 cm spacing (Jamwal and Jamwal, 2011). The crop can also be raised by transplanting but such crops yield more fibrous roots which are not desirable from commercial point of view.

Seed rate: About 10 to 15 kg seed is required for an hectare of the land by broadcast method of sowing. For nursery sowing, 5 kg seed is sufficient for a hectare. Seed should be treated with thiram or diathane M-45 @3g/kg before sowing to protect the seedlings from seed borne diseases.

Time of sowing: Sowing is usually done during rainy season *i.e.* ending June to July or August. Plant nursery can be raised in mid-July and transplanting should be done during mid of August. However, depending upon rains, the transplanting can be done upto first week of September.

Intercultural operations: The plant does not need much after care except some weeding during initial stages. In direct sown crop, thinning should be done after 25 to 30 days of sowing by maintaining plant population 30-60 plants m^{-2} (about 3.5 to 6 lakh plants/hectare). Closer plant spacing due to competition facilitates more growth of tap root than horizontal spread which is desirable. One weeding cum hoeing is essential to the crop about 2 months after thinning. Over population should be avoided.

Manures and fertilizers: Generally, no chemical fertilizer is applied if sufficient quantity of farm yard manure (10 to 15 tonnes ha^{-1}) has been applied. However, if 35 kg each of N and P_2O_5 ha^{-1} is applied, through urea (44 kg) and DAP (66 kg), the crop gives good yield (Sharma *et al.*, 2006). Application of 15 kg each of nitrogen and phosphorus ha^{-1} has also proved beneficial for getting higher yield (Jamwal and Jamwal, 2011).

Irrigation: Irrigation is not required, if rainfall is regular. However, if irrigation facilities are available, root development and flowering stages are the two critical periods when the crop should be irrigated. During the post monsoon period, life saving irrigation is given as and when required. If crop is grown for leaves, then irrigation is required in summer months.

Pests and diseases : Plants are affected by many diseases such as seedling mortality (damping off), seedling blight, leaf blight, seed rotting and die back. The most serious is the seedling mortality which reduces the plant population drastically

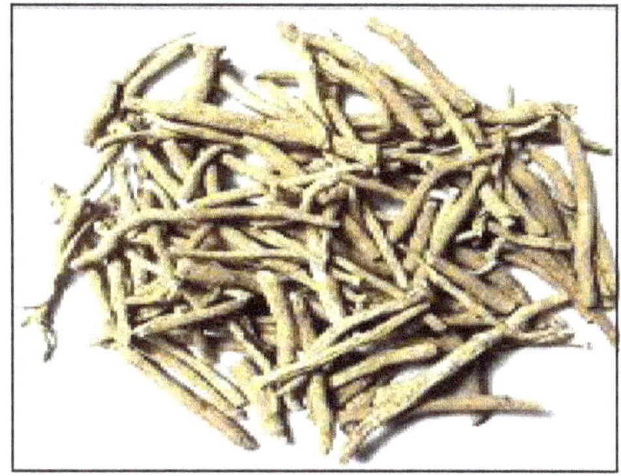

and ultimately the root yield. Seedling mortality becomes serious under high temperature and humid conditions. This disease spread and the damage is increased if heavy rains follow sowing. Seedling blight and leaf blight can be controlled with seed treatment (Thiram or Deltan at 3-4 g/kg seed) coupled with one or two sprays of 0.3 per cent Fytolon, Dithane Z-78 or Dithane M -45.

Major insect affecting crop are aphids. These can be controlled by using *neem* based botanic pesticides. A combination of 0.5 per cent Malathion and 0.1-0.3 per cent Kelthane as foliar spray at 10-15 days interval keeps the infestations under control.

Harvesting/storage: The crop is ready for harvest in January-March after 150 to 170 days of sowing. Roots of the plants should be dug in early February to mid March when the fruits turn reddish and the lower leaves have completely dried. The entire plant is uprooted for roots which are separated from the aerial parts by cutting the stem 1-2 cm above the crown. These are thoroughly cleaned to get rid of soil. The roots should be cut into 7.5 to 10 cm long pieces and then dried in the shade. Berries are hand plucked separately. They are dried and crushed to take out the seeds.

The roots are then graded according to their thickness. The thicker ones are graded better and fetch higher price and different market grades are as under:

Grades	*Characteristics*
Grade A	Root pieces upto 7 cm in length with 1-1.5 cm in diameter, solid, cylindrical and brittle. The external surface is almost smooth and a pure white non-woody xylem inside.
Grade B	Root pieces upto 5 cm in length with a diameter of 1 cm, solid, brittle and white from inside.
Grade C	Solid root pieces upto 3-4 cm in length with 1 cm diameter
Lower grade	Small root pieces, semisolid or hollow, very thin and yellowish inside.

The roots are wrapped in gunny bags and must be stored in well ventilated cool godown for a long period to maintain the quality.

Yield: The duration of the crop is 6-9 months. The usual yield is 300-500 kg roots ha^{-1} and 50-75 kg seeds ha^{-1} (Pandey and Chilana, 2004), whereas, 1000 to 1500 kg yield of dried roots ha^{-1} Sharma *et al.* (2006). Thick roots are found in wild and do not preferred due to high starch content. The cost of production of Ashwagandha being less *i.e.* Rs. 10,000 to 12,000 ha^{-1} and gross income Rs. 39,000, with total net profit of Rs. 25,000 to Rs. 27,000 can be realized even if cultivated on marginal land

(Sharma *et al.*, 2006). The market price of dried roots of ashwagandha varies from Rs.40-50 per kg.

References

Butt, M.S. 1999. Propagation of "*Withania somnifera*" (The Indian ginsing, Ashwagandha) field trial in Kalakote. N.T.F.P. Project during 1998-1999. *Haryali* (Quarterly Newsletter), No. 11-17 (October, 1997-June, 1999), pp 87-92.

Das, G and Palai, S. 2011. *Withavia somnifera*. *Agrobios News Letter*. 10 (6): 66-69

Jain, S.K. 1991. Toning up the system. Nature Care. Indian Express Magazine. April 14, 1991.

Jamwal, J. 2005. Ashwagandha, the Ayurvedic ginseng. The Kashmir Times. 62 (72) : 10.

Jamwal, J and Jamwal, S. 2011. Green gold of dryland farming-Ashwagandha. *Agrobios Newsletter*. 10(5): 30.

Pandey, S.T. and Chilana, K. 2004. Grow Ashwagandha : A stress buster crop. *Indian Farmers Digest*. 37 (12) : 26-27 and 43.

Sharma, K.K., Sambyal, K.S., Sharma, N.K. and Gandotra, S. 2006. Agronomical practices for the cultivation of high demand plant species of medicinal and economic importance. Directorate of Command Area Development, Talab Tillo, Jammu, Jammu and Kashmir State.

Taj, A. and Kumar. S. 2013. Ginseng-amedicinal plant. *Agrobios News Letter*. 11(12): 57.

Vatsyayan, R. 2010. When winter makes you feel depressed, your diet can help. *The Tribune*. 130 (338): 12.

8

Indian Senna (*Cassia angustifolia* Vahl)

Common name: Senna, sanay, swarnpatri

Family: Caesalpmiaeeae

Introduction

Senna was first reported to be introduced in Tirunelveli district in Tamil Nadu during the mid eighteenth century from European Countries. As such, Indian produce as a whole is referred to as Tirunelveli Senna. These days, Senna is grown in number of places in Tamil Nadu including Tirunelveli and Ramanathapuram Districts. Apart from Tamil Nadu, Senna is also being cultivated in Gujarat and Rajasthan. Senna is a wonder plant, the leaves of which are bluish coloured, while the pods are golden yellow in colour.

Senna is cultivated as an annual crop in Tamil Nadu and Rajasthan, usually in marginal soils requiring warm and dry weather throughout the growing period. Three pickings are generally practised. Well dried material maintains a light green to greenish yellow colour. An average dry leaf yield of 6 to 7 q ha^{-1} under rainfed conditions and 15-20 q ha^{-1}under irrigated conditions is harvested.

The most efficient of all vegetable laxatives either as crude or in formulations, senna is widely used for its laxative properties and cultivated as a medicinal plant.

Medicinal Values

Laxative properties of senna make it to formulate into convenient pharmaceutical forms *i.e.* tablets, syrups or additives to be taken with tea or coffee. Additional

advantages of this plant are its cathartic properties and the fact that it does not cause grips to occur. India now ranks as the main producer and exporter of senna leaves and pods.

The laxative nature of the senna is mainly due to the presence of chemicals namely 'Sennosides' A, B, C and D. The demand of senna leaves and pods are very high and increasing due to their use in Ayurvedic preparations, herbal teas as well as bakery products and other home preparation all over the world.

Senna is amongst the top export oriental medicinal plant from India. The 'Sennosides' present in this plant are used as laxatives. The dried leaves, pods and flowers and herbal concentrates are the items being exported. In the European market, it is sold as herbal tea. During 2001-02, about 8,240 tonnes leaves were sold in the International Markets (Jamwal, 2007).

Botanical Description

Indian senna is native to India, Arab and tropical African countries. It is cultivated in India, Pakistan, Egypt and Sudan. It is a small, erect shrub about 2-3 feet high. The stem is smooth and pale green, with long, spreading branches, bearing opposite 4 to 5 pairs of leaves. Leaves are lanceolate or obovate, unequally oblique at the base, pale yellowish-green in color with pointed apex about 1 to 2-inch in length and about 0.2 to 0.3-inch in breadth. The plant bears small and yellow flowers in racemes at the branch tips or leaf axils. The pods are broadly oblong, contain about 5-7 seeds dark brown in colour.

Cultural Practices

Climatic requirements: Senna is sun loving crop which needs bright sunshine for its good growth. Frequent heavy rains, cloudy weather and severe cold are harmful for the successful growth. An average rainfall of 25-30 cm is sufficient to get good yield.

Soil requirements: Senna plants can be grown on a variety of soils and have a great tolerance for salinity. Sandy loams, red loams including coarse gravelly and alluvial soils with loam and clay loam texture are more favourable for this crop. The crop is also grown in poor marginal lands having soil pH range from 7.0 to 8.5 in Tamil Nadu. However, much higher yield can be obtained in well drained and fertile soils (Ramamoorthy *et al.*, 2012).

Sowing time: Generally two sowing times for senna have been recognized. For irrigated crop, February to March and for rainfed crop, July to November is the best period for its sowing. It can be grown as a pure crop as well as mixed crop with gram, ginger, chillies and cotton. Line sown crop at 30 cm x 30 cm and 45 cm x 30 cm gives maximum yield.

Seed rate: The optimum seed for growing senna crop is 27-30 kg ha^{-1} for rainfed conditions and 15 kg ha^{-1} for the irrigated crop (Ramamoorthy, *et al.*, 2012). For quick germination, the seeds should be soaked for 3 hours and mixed with natural manure before sowing. Prior to mixing the seeds with manure, the surface of the seeds should be abraded by rubbing with coarse sand.

Manures and fertilizers: Which ever quantity of organic manure is available with the farmers, must be added before sowing of the crop. The application of 50 to 60 kg of N, 40 kg P_2O_5 and 20 kg of K_2O ha^{-1} is found suitable to produce optimum yield. The fertilizers should be placed 4 to 5 cm below the rows to make it easily available to growing plants.

Intercultural operations: Usually senna requires two weedings- cum- hoeing operations. The first is given at 25-30 days at the time of thinning and second at 50 days after first weeding.

Harvesting and yield: The plants start flowering 60 days after sowing and the first flush is removed to encourage the vegetable growth. First harvesting should be done at 90-100 days after sowing. Fully grown bluish coloured leaves and golden yellow coloured immature pods are stripped manually. The second harvesting is taken up about 30 to 45 days after the first harvest and the third harvesting can be done after 30 to 45 days from the second harvest. At the final harvest the plants are plucked along with their roots.

The 'Sennoside' content of dark green leaves is high. When the buds are yellow and mature and when the leaves are dark green, the 'Sennoside' content is highest (3.33 per cent).

The harvested crop should be spread in a thin layer in an open field to reduce its moisture. Further, drying is done in well-ventilated drying sheds. It takes about 10-12 days to dry the produce completely. The pods are threshed during drying to remove the seeds and is baled under hydraulic pressure and wrapped in gunny bags for export.

On an average, good crop gives 15 q ha^{-1} of dry leaves and 7 q ha^{-1} pods under irrigated conditions. The yield under rainfed conditions is 10 q ha^{-1} of leaves and 4.9 q ha^{-1} of pods.

References

Jamwal, J. 2007. Medicinal/aromatic plats and value addition. *The Kashmir Times* 64(338): 13 [December 6, Thursday].

Ramamoorthy, K. Shingare, P and Kalaichelvi, K. 2012. Senna: A wonderful medicinal plant. *Agrobios News Letter*. 10(11): 42-43.

9

Karonda
(*Carrisa carandus* L.)

Common Name: Karonda, karunda

Family: Apocynaceae

Introduction

Karonda also known as Christ Thorn is an evergreen spiny shrub or small tree. It is widely grown in South Africa, Malaysia and India (Swami and Yumnam, 2010). In India, Karonda is grown on a limited scale in Rajasthan, Gujarat and Uttar Pradesh. In the Siwaliks of Jammu (Jammu and Kashmir), Punjab, Harayana and Himachal Pradesh it grows wild, but now a days, its population has been reduced considerably like girna (*Carrisa spinarum*). Karunda, in fact, is a drought resistant shrub that apart from Siwaliks also grows in Bihar, West Bengal, the Western Ghats, Karnataka and the Nilgiri hills (Kumar, 2014).

Fruits are sub-acidic and sweet when ripe, and rich in vitamins and minerals. Owing to the presence of vitamins and minerals, the fruits of karonda have medicinal values.

Medicinal Values

Karunda is widely used in Ayurvedic System of Medicine and by tribals throughout India (Kumar, 2014). In addition, a few American countries including Kenya and Ethiopia have history of utilizing the fruit for medicinal purposes. All parts of the plant are used in traditional medicines. Tribes in the Western Ghats use the fruit as a blood sugar stabilizer and as a guard against liver damage. Various groups have utilized the fruit as a remedy of anaemia, biliousness, fungal infections, diarrhoea, wounds, skin infections, intestinal worms, fevers, ear infections, intestinal worms, fevers, scabies and also used as anti- scorbutic. Extract of the plant has been found to possess very strong antioxidant activity, anticancer, anti-inflammatory, anti-diabetic, anti-hyperlipidemic (Kumar, 2014).

As the fruit is a rich source of iron and contains a fair amount of Vitamin C, so it is very useful for treating anaemia and has anti scorbutic (counteracting scurvy) properties (Jamwal, 2007). The latex is used for preparation of chewing gum besides, having some medicinal properties.

Other Uses

Fruits can be used in pickling, chutney and for culinary purposes. It also makes excellent jelly. Latex can be used for making rubber.

Botanical Description

Karonda is a genus of about 20-30 species of shrubs or small trees which grow between 3m and 10m tall with spiny branches. The leaves are waxy and oblong, 3-8cm long. Flowers are 1-5cm diameter with five flowers have a fragrance reminiscent of Gardenia. The fruit is a plum shaped berry, red to dark purple black depending upon the species.

Cultural Practices

Climatic requirements: Karonda is very hardy in nature and can be grown in varied climatic conditions. It is a drought tolerant plant.

Soil requirements: It thrives in a wide range of soils ranging from gravelly sandy loam to loamy sand. It can also be grown in silty loam to clay loam soils.

Propagation: It is commercially propagated from seed in the month of August onwards. Karonda can also be propagated by grafting and cuttings though inarching and budding has also been found suitable for vegetative propagation.

Planting is done with the first shower of monsoon at a distance of 1.5m. Plants raised from seeds come into bearing after about two years of planting.

Flowering starts in February to March and fruit ripens from June to August or September in north India.

Harvesting and yield: The fruits become ready for harvest after 100-110 days of fruit set when they develop natural colour with a specific gravity around 1.02. Two to three pickings are required for complete harvest of 3-4 kg fruit per tree as minimum yield. An average maximum yield of Karonda varies from about 10-15 kg fruits per tree per year.

References

Jamwal, J. 2007. Other traditional fruits. *The Kashmir Times*. 64(130): 13 (Friday, May 11)

Kumar, P. 2014. Karunda forgotten fruit. *Ayurveda*.1 (26): 10-11.

Swami, G.S.K. and Yumnam, S.S.2010. Karonda: A useful fruit crop for dryland. *Agrobios, Newsletter*. 9 (6): 27.

10

Malabar Nut
(*Adhatoda vasica* Nees.)

Common name: Adusa, basuti, brainkar

Family: Acanthaceae

Introduction

Vasaka is a plant of wastelands, heavily eroded and degraded areas. It makes extensive landscapes in which most of the species remain absent. Vasaka is traditional plant of Ayurveda and stands listed in the Indian Pharmacy. Due to its diversity of application as medicines, it is matchless (Singh, 2005).

Medicinal Uses

Medicinally vasaka has innumerable uses. It is a traditional medicinal plant of Ayurveda and drug '*Vasaka*' is prescribed in bronchitis, asthma, colds and coughs.
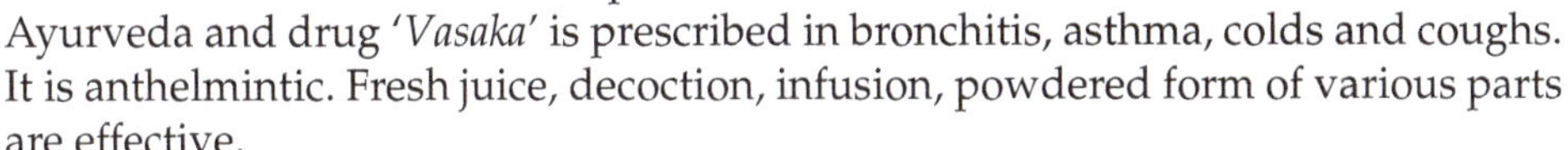
It is anthelmintic. Fresh juice, decoction, infusion, powdered form of various parts are effective.

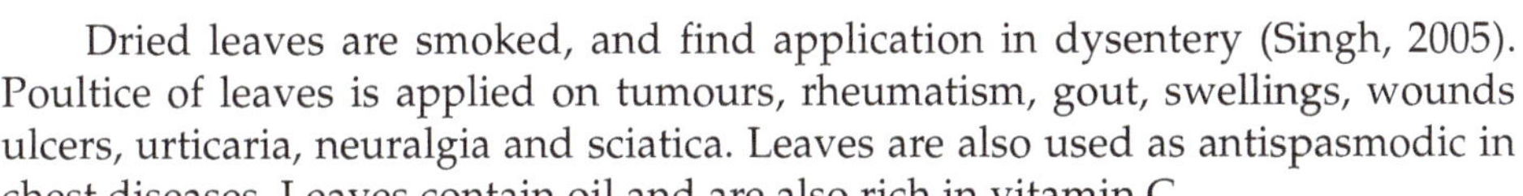
Dried leaves are smoked, and find application in dysentery (Singh, 2005). Poultice of leaves is applied on tumours, rheumatism, gout, swellings, wounds ulcers, urticaria, neuralgia and sciatica. Leaves are also used as antispasmodic in chest diseases. Leaves contain oil and are also rich in vitamin C.

Flowers and fruits are bitter in taste but leaves are mostly used. Flowers are mostly eaten to improve blood circulation. Root is expectorant, antiseptic and antispasmodic. Recent research has shown that extract of vasaka has anti-ulcer activity against those induced by the consumption of ethanol, pylorus ligation and aspirin.

Other Uses

Vasaka is a multipurpose plant for economic and environmental purpose. It covers the bare slopes and protects them against soil erosion. It is a fodder plant for horses and in ripe conditions its leaves are eatan by goats and sheep and also by cattle in fodder scarcity. It is eaten as vegetable by some of the people of Uttar Pradesh. Its flowers render excellent honey forage.

Botanical Description

Vasaka is an evergreen, gregarious stiff, perennial shrub about 2-3 m tall. It is an unarmed shrub with broad, ovate opposite leaves and bi-tipped pink spotted white flowers. Fruit is a dehiscent capsule.

Cultural Practices

Climatic requirements: Vasaka plant responds well to tropical hot climate to sub-tropical. Under sub-tropical climatic conditions, it mainly grows on the slopes of Siwalik hills upto an altitude of 700 to 1300 m.

Soil requirements: It prefers to grow on light textured soils viz; sandy, loamy sand and sandy loam. However, it also grows well in heavy soils-clay loam and clay. Soil should be well drained and fertile for good cultivation of vasaka.

Propagation: Like *Vitex negundo*, vasaka shrub can be propagated through cuttings and root suckers. This species has wide adaptability and can be cultivated best in the moist conditions, open waste lands and near the deciduous forests.

References

Singh, S.S. 2005. Fate of the Medicinal Plant Resources of Jammu and Kashmir. Directorate of Environment and Remote Sensing, Jammu, J&K Government.

11

Marijuana (*Cannabis sativa* Linn.)

Common name: Bhang, charas

Family: Cannabaceae

Introduction

Bhang is widely distributed in the temperate areas of the world and also cultivated in the tropical regions. In India, Bhang is cultivated for narcotic drugs. Apart from its use as narcotic drugs, cardage fibres and seeds are also used for culinary purpose. Various uses of Bhang (dried leaves and flowering tops), charas (resinous aerial exudations collected mainly from female plants) and ganja (dried resinous flowering tops of the female plant free from the leaves) are mentioned below.

Medicinal Uses

Bhang is least narcotic, whereas 'chars' is the main narcotic drug. In action, the drug is anodyne, antispasmodic, anaesthetic, aphrodisiac, appetizer, astringent, digestive, exciting, sleep inducer, sedative, toxic and narcotic (Chauhan, 2006). Seeds are fed to the birds and milch animals as a lactogogue. The semi-drying oil is used as liniment in pains. Paste of fresh leaves is used to resolve tumors. Leaf powder is a useful dressing for wounds. Seeds are roasted and eaten as culinary by the hill people.

Botanical Description

Bhang is an annual tall herb having height upto 5m under cultivable or arable conditions. Stem is slender, grooved finally tomentose. Leaves are alternate, palmately

divided, 5-20 cm long, toothed, rough above pale downy beneath, leaflets 7-11. Flowers dioecious *i.e.,* borne on different plants, are yellowish green. Male flowers are clustered in short, drooping panicles.

Cultural Practices

Climatic requirements: Although Bhang can be planted in diverse type of climate, ranging from tropical and sub-tropical to sub-temperate and temperate climate with mild humidity, yet it can be grown in sub-mountainous areas of Siwaliks.

Soil requirements: The plant is grown on rich loamy soils in well prepared fields.

Sowing time and seed rate: The sowing time varies depending upon the type of climate. In sub-tropical and tropical area, it is sown during *Kharif* (June-July), whereas, in temperate to sub-temperate zones it is sown during rabi season (October to November). The seed rate varies from 50 to 75 kg ha^{-1}.

Thinning: For seed purpose, the plants are thinned out after germination.

Harvesting and yield: The crop is harvested as soon as lower leaves are shed, and the tops of stalks and flowers become yellow. On an average, the yield of the crop is about 5.5 to 7.5 tonnes ha^{-1} based on dry stems. Bhang is the chief source of narcotic drugs which, if abused can play havoc with the human lives (Singh, 2005).

References

Chauhan, N.S. 2006. Medicinal and Aromatic plants of Himachal Pradesh. Indus Publishing Company FS-5, Tagore Garden, New Delhi-110027.

Singh, S.S. 2005. Fate of the Medicinal Plant Resources of Jammu and Kashmir. Directorate of Environment and Remote Sensing, Jammu, Govt. of J&K.

12

Milkweed (*Calotropis gigantea* L.)

Common name: Ak, madar, arka

Family: Apocynaceae

Introduction

Ak is a very common shrub which is found throughout India in association with *C. procera*. Both these shrubs are generally confined to the stream beds and wastelands and also in the fallow agricultural lands upto 700m elevation (Singh, 2005). All parts of *C. gigantea* Linn viz; roots, leaves, latex and powder of the flowers have medicinal properties which are as follows:

Medicinal Values

Root bark is used as dysentery cure. It is utilized as a substitute for Ipecac and serves as diphoretic and emetic. Tincture of the leaves is used in intermittent fevers. The latex of the shrub is of great medicinal importance. It is used against skin diseases and is also prescribed in rheumatism, spleen complaints, indigestion, epilepsy and toothache.

Powdered flowers are useful in colds, cough, asthma and indigestion. Recent reports indicate that extract of roots and leaves is anti-cancerous and used in lupus, tuberculosis, scorpion sting and night blindness besides leprosy and rheumatism. Plant is also useful in smallpox, leg and chest pain, dropsy, tetanus, scabies, ringworms, cholera, and pneumonia.

Leaves and stalks possess calotropin and calotropagenin. Latex contains uscharin (0.45 per cent), calotoxin (0.15 per cent) and calactin (0.15 per cent). From the stem bark, root bark, flowers and latex, β-sterol, taraxasterol, α and β amyrins, triterpenoids have been isolated.

Botanical Description

It is a small to large shrub, bark ash coloured, wrinkled and its leaves are sessile and sub-sessile, thick, ovate-oblong, acute covered with white tomentum. *C. gigantea* and *C. procera* are the two most common species in the genus. The flowers are about 1.5 to 2 inch (3.8 to 5.1 cm) in size, with umbellate lateral cymes. Flowers are white without fragrance in *C. gigantea* and pink fragrant with purple spot in *C. procera*. The root bark occurs in short, corky, deeply vertically, cleft pieces (Chauhan, 2006). *C. gigantea* grows to a height of 8 to 10 ft (2.4 to 3.0 m) while *C. procera* grows to about 3 to 6 ft (0.91 to 1.83 m). The seeds are compressed, broadly ovoid, with a tufted micropylar coma of long silky hair.

Cultural Practices

Climatic requirements: It is found throughout India in warm and dry climatic conditions.

Soil requirements: Like Datura, it can be grown in coarse textured soils ranging from sandy ravine to gravelly loamy sand/sandy loam. Its frequent growth can be seen in the waste lands, stony lands near the banks of the streams, rivers and road sides.

Propagation: Propagation of the shrub can be done by the cuttings which are raised in soil/sand culture.

Seeds after making free from silky white floss can be raised in nursery during warm weather (April-May) conditions.

Flowering and fruiting occurs throughout the year, but chiefly it takes place in the cold season *i.e.*, during the months of November, December and January.

The leaves, flowers and roots are harvested through out the year. On an average, the yield of cultivated crop in Thailand is 1.1 tonnes of dry leaves and 56 kg latex per hectare.

References

Singh, S.S. 2005. Fate of the Medicinal Plant Resources of Jammu and Kashmir. Directorate of Environment and Remote Sensing, Jammu, Govt. of J&K.

Chauhan, N.S. 2006. Medicinal and Aromatic Plants of Himachal Pradesh, Second edition. Indus Publishing Company FS-5, Tagore Garden, New Delhi-110027.

13

Mugwort
(*Artemesia vulgaris* L.)

Common name: Common wormwood, chitra, jhau

Family: Asteraceae

Introduction

Chitra is native to temperate Europe, Northern Africa and Alaska and has naturalized in North America. In India, this species along with others, is found in Western Himalayas, Assam, Western and Southern India. The Western Himalayas covers the states of Uttrakhand, Himachal Pradesh and Jammu and Kashmir. In Jammu and Kashmir, the species along with *Artemisia annua* are mostly found in cold arid zone of Ladakh. It is remarkable that before the second world war species like Kuth (*Saussurea lappa*), the *Artemisia vulgaris* was used to be harvested 100 tonnes a year from the forests of Kashmir including Ladakh (Annonymous, 2003). Chitra besides having medicinal uses also acts as an active weed, so one must be aware that it may be a noxious weed and invade the uncultivated areas.

Medicinal Uses

Researchers suggest that applying a lotion containing mugwort, a methnol

directly to the skin relieves itching in number of burn victims, stomach problems, diarrhoea, cramps, slow digestion, vomiting, epilepsy, low energy and anxiety. The anti-malarial compound 'artemisinin' has potential activity against premaquine resistant strain of *Plasmodium*. It kills parasitic worms internally. It is not recommended for use during pregnancy.

Botanical Description

It is cold loving plant and survives in freezing temperature. The plant is found on fallow grounds and near roadsides and on field bunds. Chitra is a tall-growing, fragrant, herbaceous perennial plant with angular and purplish hue stems, growing 1-2m tall with woody root. Leaves are simple, alternate, smooth dark green on the upper surface but white on the under-surface. Size is 5-20cm long, pinnately lobed with dense white tomentose hairs on the underside. Erect stem often has a red purplish ting. Flowers are small (5mm) arranged in long, terminal panicles. They are radially symmetrical with yellow or dark red petals. It flowers during July-September.

Cultural Practices

Climatic requirements: It is a very hardy plant and survives well even in freezing temperature.

Soil requirements: Although this plant generally grows on nitrogen rich soils, yet it can occur in weedy and un-cultivated areas and waste lands. It can tolerate both slightly acidic and alkaline soil conditions, the ideal pH of the soil in the range of 6.6 to 8.5. It can be easily grown on a well-drained, loamy soil, preferring a sunny locations and a moist soil.

Cultivation Practices

i) **Growing from seeds:** Broadcast seeds on surface of soil in well prepared fields @ 120 kg ha^{-1}. This should be done at least about 56 days before the occurrence of last frost. It takes about 21 days to germinate the seeds. Established plant is drought tolerant and is an aggressive and invasive plant as it spreads by underground rhizomes and by re-seeding. Seeds germinate best with a cold treatment.

ii) **Transplanting seedlings:** Seed are sown in the surface soil during late fall to early summer. Prick out the seedlings into permanent beds in summer. About 36 days old seedlings are transplanted. Ensure that temperatures are mild and all chance of frost have passed before planting out as it is a hardy plant.

Harvesting and yield: Harvest dried seed heads during late spring. Gently rub between hands to remove seeds, carefully winnow away chaff from seeds. Each year the crop is harvested during early August. Harvest the stems with leaves, tie with string, hang upside down in a cool room out of direct light when dry, and pull off the leaves.

On an average, accumulated the dry matter yield varies from 2.76 to 5.74 tonnes ha^{-1}.

References

Annonymous. 2003. Conservation and sustainable utilization of Medicinal plants, Jammu and Kashmir State. State Level Planning Committee, Jammu and Kashmir Forest Department.

14

Rose
(*Rosa centifolia* Linn.)

Common name: Gulab

Family: Rosaceae

Introduction

Rose is considered to be one of the nature's beautiful creations, which is universally acclaimed as the 'queen of flowers'. No other flower is a better symbol of love, adoration and other virtues than rose. In fact, rose is the best and the most popular of all garden flowers throughout the world and has been growing on this planet earth since long. In India, the rose was referred to as 'Turunipushpa' in Sanskrit literature, and did adore not only the royal palaces but also in the 'ashramas' of saints (Kumar and Singh, 2009).

Cultivation of rose is a highly remunerative enterprise, particularly the domestic cut flower production market. Not only this, it has become the part and parcel of life, being connected with all phases of life right from birth to death. In India, rose cultivation picked up with the distillation of roses like rose water popularly known as *gulab da arak, pankhuri,* 'gulrogan' etc. as mentioned in Ayurveda by Charka about 100 AD. All the species of rose are medicinal in nature.

Medicinal Values

Jangli gulab (*Rose moschata*) is used for washing of sore eyes with its decoction (Singh, 2005). Rose extracts and rose oil are well known for their emollient and hydrating properties, helping skin stay soft and supple. The water content contains a bit of tannin, which aids in skin toning (Anonymous, 2011). While rose oils add radiance to complexion, rose extracts are rich in Vitamins A, C, D and E and have flavonoids that act as anti-oxidants for the skin. The stimulating action in rose petals helps to fight early aging, enabling a soft and youthful complexion.

The perfumes of various species of rose find a vast application in other medicines and soft drinks. When rose water applied, ice cold to skin after sun exposure, it helps to sooth the burning sensation. Preservative of rose known as 'gulkand' is used in improving digestion.

Other Uses

The different ways in which rose can be utilized for large and small scale plantings consist of as bush, climber, hedge, edge and pot plant, in hanging basket etc. Rose sustains small perfumery units on cottage scale. It can form a big economic resource. As one farmer in Punjab has earned between Rs. 1,25,000 to Rs 2,00,000 from each ha of rose cultivation (Sharma, 2010) with one litre of rose oil fetching Rs 3.75 lac.

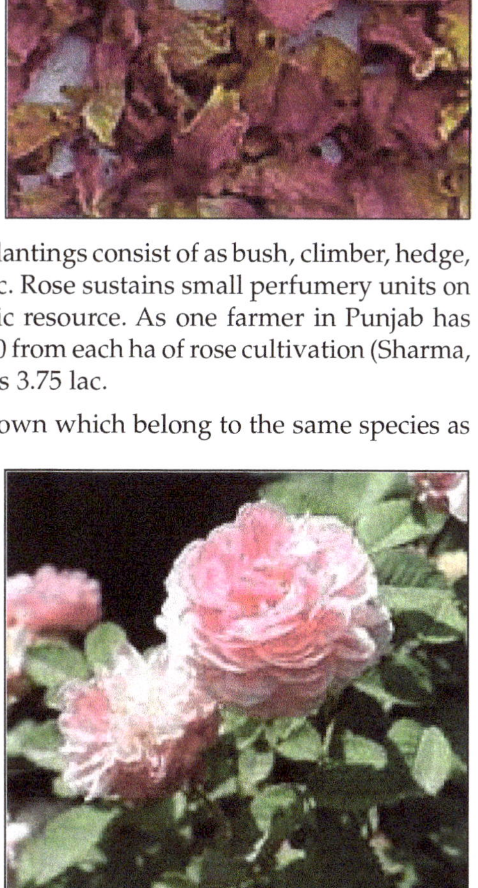

In Europe, edible roses are being grown which belong to the same species as non-edible ones but have to be fertilized and sprayed organically (Anonymous, 2010). Mostly, their petals are used for preparing novelty dishes to attract customers from New York to Barcelona.

Botanical Description

Rose can be cultivated throughout India. Rose is a small, erect shrub in appearance growing upto 1.5–2.0 m. Plant is considerably branched, with hooked prickles and bristles, compound, alternate, grayish green leaves with five leaflets. The flowers are round and globular, usually pink or red and fragrant but vary in hues, form and size.

Cultural Practices

Climatic requirements: The roses require sunshine and free ventilation (Kumar and Singh, 2009). The ideal temperature for growing of rose lies between 15 to 25° Celsius. Low temperature produces flower with short neck (part of the stem between the bud and the upper most leaf), while higher temperature induces long neck. Growth is slow at day length below 12 hours, while increased light intensity increases the quality and production.

Soil requirements: The soil must be loose and structurally stable down to a depth of around 50 cm so that the optimum feeding of air and water is guaranteed with long standing cultivation. Loamy soils are the best for its cultivation. Roses flourish in soils which contain a generous amount of organic matter with good drainage conditions. A slightly acidic soil is preferred by roses.

Preparation of beds and planting distance: The width of the bed varies according to the width of the green house. Normally beds are prepared 1.0 to 1.6m wide and between 30 to 40 m long. The plants are planted in two rows with spacing of 15-18 cm apart. The planting density ends up at 60,000 to 70,000 plants ha^{-1}. The land should be ploughed deep 2-3 times and brought to a fine tilth. Raised beds of 30 cm height should be prepared leaving 30 cm between two beds. Well decomposed farm yard manure or compost should be added to the order of about 50 tonnes ha^{-1}.

Planting time: The best planting time varies from September to October in the northern plains and most of the other parts of the country. However, planting of rose flower can also be done during February-March.

Method of planting: Six to eight months plants are bud grafted on plants having a long root system. It is essential to dig the pits sufficiently deep so that entire roots are not bent upwards. While refilling the pits make sure that there is good contact between the roots and the soil, but never compact the soil around the plants too firmly.

After planting, the soil around the plants must be kept warm and humid for a period but not too wet. Frequent overhead spray with water will assist early establishment.

Harvesting and yield: Harvesting varies with varieties, stage of harvest, local market or distance of export market. Roses for local markets are harvested when outer petals start curling outwards, while for distant markets they are harvested at tight bud stage, when the bud show full colour but the petals have not yet started unfolding.

Flowers should be harvested during early morning or late in the evening to avoid damage of buds due to high temperature. After harvesting, flowers need to be placed in buckets containing clean water. The time gap between harvesting and placing in water buckets should be as less as possible and must not be more than 15 minutes.

The yield of cut flower ranges between 150-200 per m^{-2} or more.

References

Anonymous. 2010. Edible roses. *The Tribune*. 130(268): 3 (In Spectrum, Sunday, 28th September).

Anonymous.2011. Get wow skin with scented flowers. The State Times. 16(64): 5 (Sunday, 6 March, 11).

Kumar, S. and Singh, D. 2009. Production technology and marketing of rose in India. *Indian Farmers Digest*. 42(11): 12-16.

Sharma, S.P. 2010. This farmer makes roses bloom in a desert. *The Tribune*. 130(93): 6 (Monday, 5 April).

Singh, S.S 2005. Fate of the Medicinal Plant Resources of Jammu and Kashmir Directorate of Environment and Remote Sensing, Jammu, J&K Government.

15

Seabuckthorn
(*Hippophae rhamnoides* L.)

Common name: Chhermang, charma

Family: Elaegnaceae

Introduction

Seabuckthorn is a thorny temperate deciduous shrub distributed throughout the dry temperate regions. The genus *Hippophae* has been derived from the Greek word. Addition of Seabuckthorn's leaves/twigs in fodder resulted in weight gain and shiny coat for horse. Thus, this gave the name to the shrub 'Hippophae' as in Latin "Hippo" means horse and "Phaes" means shine.

The bush has been grown and used in Tibetan, Chinese and Mangolian systems of medicines for more than 1,300 years. However, in Himachal Pradesh it is locally known as "Tsermang" (Tserbang) and known to occur in Lahaul Spiti much earlier than China, where people have made its use as a fuel, fodder, fruit, oil and medicine. Seabuckthorn, also called "Wonder plant" and "Ladakh gold" has numerous medicinal and nutritional properties besides helping in soil conservation and nitrogen fixation.

There are 6 species and 12 sub-species native over a wide area of Europe and Asia and it grows naturally in many temperate parts of China, Mongolia, Russia, Britain, France, Finland, Germany, Romania, Nepal, Pakistan and India. In India, Seabuckthorn is mostly grown in cold arid zone of Ladakh (Jammu and Kashmir), Lahaul and Spiti, Kinnaur and Pangi areas of Chamba (Himachal Pradesh), parts of

Kumaon and Garhwal mountains (Uttrakhand) and certain areas of North-Eastern region(Sanjay *et al.*, 2006). Although Seabuckthorn is distributed over a large area naturally in Ladakh and Lahaul-Spiti, yet potential of the shrub especially for its medicinal uses has not been fully exploited so far (Gupta, 2008). It is known by different trade names like 'Leh Berry', 'Ladakh Berry', 'Power Berry' or 'Sindhu Berry' (Dwivedi *et al.*, 2006).

The Ministry of Environment and Forests and the Defense Research and Development Organization (DRDO) have launched an initiative to increase area under its cultivation in cold arid ecosystems in the N-W Himalayas. This also superimposes the objectives of Green India Mission- a part of the National Action Plan on Climate Change.

More than 90 percent or about 1.5 million hectares of the world's Seabuckthorn resources can be found in China where the plant is exploited for soil and water conservation purposes.

Nutritional Value

China has already recognized Seabuckthorn as food stuff besides medicinal plant without any toxicity and side effects. The berries are rich source of nutrients and biologically active components. The berries contain essential amino acids, fatty acids and bioactive substances. Due to presence of more quantity of Vitamin C, it is called 'King of Vitamin C' (Wali, 2006).

The oil has been approved as organic food in China. The Vitamin C, Vitamin E and Vitamin K contents range from 300-1600, 162-255 and 100-200 mg 100^{-1}g of the berries, respectively. Vitamin A, Vitamin B_1 and B_2 are also present though in small amount (Gupta, 2008).

The fruit/berries contain more than 100 natural nutrients, 18 amino acids and 8 vitamins, 24 minerals (Fe, Cu, Mn, Ca, Mg and others), oils, lipids, carbohydrates and high quantity of antioxidants. Field Research Laboratory has registered a patent for processing seabuckthorn juice in Ladakh.

Berry pulp consists of nearly 20-25 per cent of oil, whereas, mature seeds contain 8-20 per cent oil. The oil has medicinal and nutritive properties and is suitable for use in medicine and cosmetics. The tender and healthy leaves are a rich source of dietary fibre, and can be incorporated along with the fruit pulp into biscuits. They can also be used for making herbal tea and as an ingredient of cake for animal feeds.

Medicinal Values

Every part of Seabuckthorn has some pharmacological effect. The people of China were probably the first to use seabuckthorn as drug. Medicinal uses include therapeutic, anti-arsenic, anti-aging anti-bacterial, anti-inflammatory etc. Recent studies have shown that various products have definite effects on blood pressure, respiratory infection, heart problems, gastric ulcers, cancers, burns, blood deficiency and some brain disorders. Its products include foodstuffs, beverages, medicines, cosmetics, healthcare products, chemicals, industrial material and so on. Fed with it, a sheep or goat produces more mutton and cashmere (Ali and Kaul, 2011).

Its berries have some pharmacological effect on the lungs, stomach, spleen and blood circulation. Berries have also many good products for children, lactating and expecting mothers, senior citizens, sportsmen, and people living and working at higher altitudes. It is because berry is a natural energizer and enhances body endurance against scurvy and other diseases as well as vagaries of high altitude climate. Fruits can be used to make pickle, jams, lotions, fruit wines and liquors. It provides a nutritious beverage rich in Vitamin C and carotenes. Berry oil either taken orally or topically is believed to be a skin softener. Berry juice is a common drink rich in Vitamins C and E, carotenes and fatty acids.

Barks and leaves are used for treating diarrhoea, and dermatological disorders. It has been proved to possess good anti-inflammatory, anti-senility and anti- radiation properties because the bark of the stem and fruit certain 5-hydroxy tryptamine, a rare occurrence in the plant kingdom. This compound acts as a neurotransmitter and controls human emotions, harmone levels, neuro-endocrine disorders, age related problems, loss of memory and inhibits cancers cell growth. The flavonoids present in berries have antioxidant properties. Local practioners called 'Amchis', use it in Tibetan System of Medicine.

The oil was regularly used by the former USSR cosmonauts as vitamin supplement during 1960-1970's to protect against radiations in space travel. It has also been used in the beauty cream due to its anti-wrinkle and anti-skin sclerosis properties. Seabuckthorn oil improves the micro-circulation of capillary vessels, nourishes skin and hair, cures scabies and dry skin.

Other Uses

Seabuckthorn plays a significant role in the conservation of wildlife by providing habitat for a number of wildlife species. Being symbiotic nitrogen fixer, it enriches the soil fertility and can be used as fuel and fodder.

Use in Soil Conservation and Landslides

The seabuckthorn because of its nodulated, fibrous, long and well developed root system of *Frankia* is an excellent plant species for waste land reclamation. *Frankia* being unicellular but having very thin and slender type of mycelia in their cells are having soil binding properties and the species has proved a boon to control soil erosion in cold arid regions. Seabuckthorn is not only helpful in binding loose soil with its extensive root system, controlling the soil erosion, but also retains surface runoff and checks landslides (Rongsen, 1992; Dhyani *et al.*, 2007).

Botanical Description

Seabuckthorn is a deciduous shrub that typically varies 0.5- 6.0 m in height and equivalent spread, but may reach up to 18.0 m in central Asia. It is very hard that rapidly develops extensive roots capable of fixing atmospheric nitrogen. Branches are dense and stiff, and very thorny. The leaves are pale silvery-green, lanceolate, 3-8 cm long and < 7 mm wide. It is dioecious, with separate male and female plants. The male plant bears brownish flowers which are pollinated by wind. The female plant bears orange cloured berries 6-9 mm in diameter, soft, juicy and rich in oils. The berries are an important winter food resource for some birds also. Flowers generally appear during late April to mid of May *i.e.*, about one year after planting.

Cultural Practices

Climatic requirements: Seabuckthorn requires cold temperate climate, hardy, drought-resistant and tolerant to extreme temperature of – 43° C to + 40° C. In India, it grows naturally in cold arid regions comprising Ladakh province of Jammu and Kashmir, Lahaul Spiti, Kinnaur and Pangi areas of Himachal Pradesh. The plant has an extensive root system, requires full sunlight for good growth and cannot tolerate shady conditions.

Soil requirements: The best growth occurs on deep, well drained, sandy loam soils. Heavy soils *i.e.*, clayey or semi-clayey are not found suitable for its growth. It grows on various types of soils *i.e.*, Chernozems, brown soils, grey forest, peaty loam and peaty swampy soils.

Sowing time: Seabuckthorn is mostly grown during spring season *i.e.*, during the months of April and May.

Harvesting and yield: Harvesting is difficult due to dense thorn arrangement on the branch. A common harvesting technique is to remove an entire branch, though this is destructive to the shrub and reduces future harvests. A branch removed in this way is then frozen allowing the berries to be easily shaken off.

However, the ideal method of harvesting is to shake the branches vigorously between 5-8 am and pick-up the fallen berries. The falling of berries has to do with the temperature making it more difficult as the day progresses. The fruit is too juicy to be plucked as a result of which people cut the branches and then beat them against the floor. This results in loss of yield also. Early morning harvesting gives nearly 70 per cent of productivity but the other method reduces the yield to as low as 30-40 per cent. FRL is also engaged in developing a mechanized harvester that uses vibration technique.

Harvesting period lies between late June and mid of July. Yield of the seabuckthorn per plant has been worked out be 9 kg on an average.

References

Ali, A. and V. Kaul. 2011. Seabuckthorn: A Valuable Resource of The Cold Desert (Ladakh) ENVIS Bulletin: *Himalayan Ecology* 19.

Dhyani, D., Maikhuri, R.K., Rao, K.S., Kumar, L., Purohit, V.K., Sundriyal, M. and Saxena, K.G. 2007. Basic nutritional attributes of *Hippophae rhamnoides* (Seabuckthorn) populations from Uttarakhand Himalaya, India. *Current Science* 92(8): 1148-1152.

Dwivedi, S.K., Singh, R. and Ahmad, Z. 2006. The Seabuckthorn, FRL, Leh Ladakh.

Gupta, R.D. 2008. Grow Seabuckthorn (A wonder Medicinal Plant) for greening cold Desert Himalayas. *Wasteland News*. 23(4): 40-42.

Rongsen, Lu. 1992. Seabuckthorn: A multipurpose plant species for fragile mountains. ICIMOD Occasional Paper No. 20, Kathmandu, Nepal. pp. 1-50.

Sanjay, D., Singh, R. and Ahmed, Z. 2006. *The Seabuckthorn.* Leh: FRL(DRDO) Bulletin.

Wali, P. 2006. Drink Vitamin C. *The Daily Excelsior*. 42 (356): 3 (*Sunday Magazine*, December, 24).

16

Thorn Apple (*Datura stramonium* L.)

Common name: Datura

Family: Solanaceae

Introduction

Native of Tropical America, datura is naturalized in the Himalayas upto 2700m height. It is found in cultivated as well as wastelands. Datura is one of the medicinal plant which acts as the most important source of the life saving drugs for the majority of the world's population despite having a long history of use for causing delirious states and death (Chauhan, 2006).

Medicinal Values

Datura helps in relieving the spasm of the bronchitis in chronic asthma. It is used in treatment of parkinsonism and haemorrhoids. It has proved useful in dysmenorrhoea, neuralgia and sciatica and is given with benefit in cases of nymphomania and in purpuramania with a tendency to suicide. It is beneficial in boils, breast inflammation caused by excessive formation of milk, salivation, travel sickness (sea or air), diarrhoea, enlarged testicles, insanity and itch.

Datura is also used as a useful remedy for heart disorders (Verma, 2013). It relieves cardiac pains, distress, palpitation and aortic disorders. The fruit of datura is a specific remedy for phlegmatic and bilibous type of malarial fever. Datura leaves are useful in the treatment of earache. A preparation made from datura seeds with other ingredients is beneficial in patchy baldness. It is also useful in impotency.

The active ingredients present in the datura are tropane alkaloids which include 'Atropine', 'Hoscyamine', 'Hyoscine', and 'Scopolamine' classified as deliriants or anti-cholinergic. The group of tropane alkaloids comprises more than 200 compounds characterized by a two ringed structure with pyrrolidine and a piperidine ring sharing a single nitrogen atom and two carbon atoms (Verma, 2013).

Botanical Description

Datura is an annual, erect, nearly glabrous, tall robust herb with broad, ovate, coarsely and irregularly lobed and toothed leaves. Flowers are white in colour, single on short, usually axillary stalks, 7.5 to 15cm long and tubular. Capsules are ovoid 3.5x2.5 cm covered with rigid sharp prickles, surrounded at the base by enlarged and reflexed lower part of the calyx. Seeds are black in colour but wrinkled.

Cultural Practices

Climatic requirements: A woody perennial herb grows in climate varying from sub-tropical, sub-temperate to temperate in elevation from 300m to 2500m or upto 2700m.

Soil requirements: Grows on variable types of soils ranging from gravelly sandy loam to loam or loamy sand to gravelly loam.

Propagation: Datura is generally propagated by seeds. Vegetative propagation by stem cuttings has also been found suitable for its cultivation. Seeds of the herb are gathered from the mother plant while the fruits turn yellow. After washing, the seeds are dried in the shade and sown during March-April in well prepared fields. Frequent watering and shade is required for proper germination. The seedlings get ready within 3 months time which can be transferred directly to the fields during monsoon period.

For planting, suitable field having proper drainage should be selected. Pits having dimensions 30cmx30cmx30cm should be dug at least 1 month prior to planting. It should be filled with a mixture of top soil, sand and farmyard manure in the ratio of 1:2:1. The plants begin to yield after about six months.

Naturally regenerating plants are stronger, much branched and bear bold capsules and yield good quality seeds.

References

Chauhan, N.S.2006. Medicinal and Aromatic plants of Himachal Pradesh. Indus Publishing Company FS- per cent, Tagore Garden, New Delhi-110027.

Verma, K.C.2013. *Datura stramonium*-potent plant for Pharmaceutical Industry. *Indian Farmers Digest*. 46(05): 23-25.

Climbers

1

Bitter Gourd (*Momordica charanita* L.)

Common name: Karela

Family: Cucurbitaceae

Introduction

The vegetable, Kerala (bitter gourd), brings about many frowns to people's faces owing to its bitter taste. It grows on a vine and probably is one of the most bitters among all of the vegetables (Singh, 2005). Karela is grown throughout the country up to 1500 m altitude as vegetable crop. It is cultivated during warm season and is so well known to common man and to cultivate. Bitter gourd is a valuable herb gifted to us by mother Earth. It is one of the healthiest vegetable known to mankind (Kumar, 2011).

Its names in Tamil, Telgu, Kannada and Malayalam are: Paavarkai; Kakrkaya, Hagalakai and Kaipakka, respectively.

Nutritional Composition

Bitter gourds are very low in calories but rich in precious nutrients. It is an excellent source of Vitamins B_1 B_2, B_3 and E, magnesium, zinc, phosphorus, manganese, iron, calcium and folic acid. Bitter gourd contains a unique phyto-constituent called 'charantin' which has been confirmed to have hypoglycemic effect (Kumar, 2011). It has insulin like compound known as polypeptide P. Bitter gourd contains plenty of beta carotene.

Medicinal Values

Karela is consumed for health benefits as it has excellent medicinal values. It is an antidote for poison and anti-pyretic *i.e.*, to control fever. It works as an appetizer, and is good for stomach aches and bile juice. Karela serves as laxative also (Natarajan, 2010). Bitter gourd is specifically used in native medicine for diabetes, piles, constipation, blood and respiratory disorders. It is too used to check alcoholism and cholera. It has antipyretic and antibilious property.

Tender shoots and leaves are eaten raw as salad or cooked as vegetables in Java and the Philippines. Leaves are rich in calcium and, thus, make up the deficiency of this nutrient in the body. Seeds yield a clear reddish brown oil used in many household remedies such as rubbing on affected skin. Seed cake is highly rich in protein and is eaten.

Fruits are tonic, stomachic, carminative cooling and are effective in gout, rheumatism, and arthritis and in diseases of liver, spleen. Wild variety of the plant is more effective medicinally and its fruit is a febrifuge.

Seeds of karela are anthelmantic. Fruits and tender leaves are more effective than other parts. Fresh juice of the leaves is also taken to lower down the levels of blood sugar. For all health conscious and health minded people Karela is a panacea. Karela is known to reduce glucose level by improving peripheral glucose uptake (Gupta and Mahajan, 2009). It also generates the activity of pancreas. This vegetable therefore, should be included in the diet of diabetic patients.

Usage of Kerala in South and North India

It is a popular food in south India and can be made as a stir-fry, in gravy with *dhals* and tamarind. This is called *pitlai* and has been adopted from the Maratha rulers who ruled Tamil Nadu centuries ago. In Andhra Pradesh it is cooked into a dish called *pachi pulusu*, a gravy dish with fried onions and other spices.

In the North India, Karela is cooked with onions, tomatoes, red chili, turmeric, coriander and spice powders, salt, and seasoned cumin seed. The whole of the unpeeled bitter gourd is boiled and them stuffed with spices, potatoes and other vegetables in many parts of India and deep fried.

Botanical Description

As already stated, karela is a common climber. Generally two varieties of karela are available *i.e.*, dark green and pale green (Natarjan, 2010). Karela can he long like cucumber or tiny ones little larger than peas. The knobbly exterior should be dark green in colour and must be pliable when twisted. If there are signs of yellowing or orange streaks on the karela, avoid buying, for the vegetable as it has started to ripen.

Cultural Practices

Climatic requirements: It is a warm season crop but can tolerate moderate cold better than other cucurbits. However, it is susceptible to extreme cold and frost conditions. Karela thrives best in warm and humid region.

Soil requirements: Although karela can be cultivated on all types of soils yet it thrives best in sandy loam and silty loam soils. Soil must be rich in plant nutrients and well drained.

Sowing time: The crop is sown in the months of February-March in sub-tropical zone, whereas, in intermediate zone, it is sown in the months of April and May. This crop is also sown in June and July.

Seed rate and method of sowing: Seeds are directly sown in the field, after soaking in luke warm water for 24 hrs and then kept in moist cloth till the seed coat ruptures, 2-3 seeds are planted per hill. Seed rate is 5-6 kg ha^{-1}. Seeds are sown by maintaining row to row distance at 1.5 to 2.0 m and plant to plant 2.0 to 2.5 m.

Manures and fertilizers: Well decomposed organic manure *i.e.,* farm yard manure (FYM) or compost/vermicompost @ 25-30 tonnes or even up to 50 tonnes ha^{-1} is recommended depending upon its availability. Farm yard manure or compost should be applied at the time of final ploughing during land preparation. The manure is required to be well mixed with the soil.

Well decomposed organic manure here refers to having C:N ratio of 10:1 and it should be either brown/dark brown or black in colour and amorphous in nature. Although in addition to farm yard manure or compost, application of chemical fertilizer to the extent of 100 kg N, 50 kg P_2O_5 and 50 kg K_2O is also recommended but the use of chemical fertilizers is not preferred.

Training, inter-culture and irrigation: Plants are trained on supports when they start running, shallow hoeing and regular weeding is done. During summer, crop should be irrigated at 3-4 days interval but in rainy season, one irrigation per week depending up on the availability of the rain, is sufficient.

Harvesting and yield: The vines start fruiting 60 to 75 days after seeding. The marketable fruits are immature with the bright green colour. This colour, however, fades away to a slight whitish yellow or whitish green (Annonymous, 1992).

References

Annonymous. 1992. Directorate of Extension Education (DOEE), S.K. University of Agricultural Sciences and Technology Regional Horticulture Research Station, Udheyawala, Jammu.

Gupta, R.D. and Mahajan, N. 2009. 10 Herbs for diabetes. *Ayurveda.* 1 (13): 9.

Kumar, P. 2011. Better guard bitter gourd. *Ayurveda.* 1(20): 16-17.

Natarajan, P. 2010. Karela (Bitter gourd). *Tattvaloka.* 33(8): 40-41.

Singh, S.S. 2005. Fate of the Medicinal Plant Resources of Jammu and Kashmir. Directorate of Environment and Remote Sensing, Jammu, Govt. of J&K.

2

Bottle Gourd
(*Lagenaria siceraria* L.)

Common name: Ghia, kaddu

Family: Cucurbitaceae

Introduction

Bottle gourd is an important cucurbitaceous vegetable grown both in summer and rainy season. Bottle gourd is also called while flower gourd.

Besides using bottle gourd in number of recipes like *Raita, kofta, halwa,* sweets etc; it has very important role from health point of view. It is because bottle gourd has a lot of medicinal properties with regard to human health.

Medicinal Values

Bottle gourd is easily digestible and mainly prescribed to weak and diabetic patients. It is also proved beneficial for jaundice. Seed oil has cooling effect while applying to head during summer.

Botanical Description

Bottle gourd is believed to be indigenous to Tropical Africa and India on the basis of variability in seeds and fruits. It is monoecious, annual, vine with soft pubescence. Leaves are generally cordate-ovate to reniform-ovate with entire margin but one variant PBOG54, has been found with segmented leaf characteristic and can be used as marker gene for hybrid seed production.

The flowers are white and open in the evening. The staminate flowers are born on long pedicles with three stamens, whereas, pistillate flowers are single with short peduncle and hairy ovary. Since flowers open in evening selfing/crossing must be done after 5.00 pm to late night or early morning.

Cultural Practices

Climatic requirements: Bottle gourd is mostly cultivated in warm and humid climate. Although it can withstand cold conditions but cannot tolerate frost. It thrives well at temperature range of 25-30 ° Celsius. Heavy rainfall might attract insect pests and diseases to the crop.

Soil requirements: Bottle gourd can be grown on all kinds of soils. However, loam and silty loam soils are the best suited for its production provided they are well drained and rich in organic matter. The pH of these soils should be in the range of 6.5 to 7.5.

Land preparation: Field should be thoroughly prepared and ploughed 2-3 times. Soil is, thereafter, well leveled and ridged to sow the seed.

Improved varieties: Number of improved varieties have been developed. These are *Pusa Summer Prolific Long, Pusa Summer Prolific Round, Pusa Naveen, Pusa Meghdoot, Pusa Manjori, Punjab Komal, Arka Bahar, Kalayanpur long Green, Pant Sarkar, Lauki-1* etc.

Seed rate and sowing time: In plains, bottle gourd can be grown twice in a year *i.e.* during summer, sown in February-March and during rainy season, sown in June-July. Seeds with 80 percent germination capacity at the rate of 4-5kg ha^{-1} are good for sowing. The spacing varies with soil fertility and variety. Ridges are prepared at a distance of 3m and plant to plant distance is maintained at 1m.

In Kashmir valley, the seeds are sown in the field with 2-3 seeds per hill with seed rate ranging from 6-8 kg ha^{-1}. In sub-tropical and sub-temperate or intermediate climatic zones of Jammu region, the sowing time is February and June, July and March and April-May respectively. Row to row spacing is 2.25m and plant to plant is 75 to 90 cm.

Manures and fertilizers: To obtain good yield of bottle gourd, it is essential to apply adequate quantity of manures and fertilizers. About 20-25 tonnes of well decomposed FYM or compost should be mixed in the soil at the time of ploughing. Besides this, 60-100 kg N, 50-80 kg P_2O_5 and 30-60 kg K_2O ha^{-1} are required for overall growth depending upon the soil fertility. It would be much better if the fertilizers are applied on the basis of the soil test. One third nitrogen and full dose of phosphate and potash should be added before sowing. Rest of the nitrogen is divided into two equal parts and applied at 4-5 leaf stage and flowering time through top dressing.

Intercultural Operations

Irrigation: During rainy season, crop does not require irrigation, but if there is long spell without rains then irrigation should be provided. During summer, continuous and frequent irrigation is required in 5-7 days interval.

It is always advantageous to keep the field free from weeds, especially in the early stages. Frequent hoeing and inter-culturing of the young plants promote healthy growth and heavy fruiting.

Harvesting and yield: Generally, bottle gourd matures faster, immediately after fruit set. Fruit reach edible maturity stage within a week or slightly later. They should be harvested when are immature. They lose fresh appearance soon if not kept in a cool place. Because of highly perishable nature they cannot be transported to distant markets under ordinary packing.

The yield of the crop varies from variety to variety. The average yield in plains of various Indian states is 250-350q ha^{-1} (Tiwari, 2010). In various regions of Jammu and Kashmir, bottle gourd yields about 150 q ha^{-1} in case of *Pusa Summer Prolific Long* variety whereas, *Pusa Summer Prolific Round* yields 125-150 q ha^{-1}. In Kashmir Valley, the yield of *Shalimar Improved Variety* is 150-200 ha^{-1}.

References

Tiwari, A. 2010. Production technology and improvement of bottle gourd. *Indian Farmers Digest*. 43(4): 29-31.

3

Coral Bead Vine (*Abrus precatorius* L.)

Common name: Rati

Family: Fabaceae

Introduction

It is found in South China, the Pacific Island, Tropical Africa, the West Indies and almost all tropical countries including India in moist areas. It is a very common climber in kandi belt of Jammu, Punjab, Haryana, Himachal Pradesh and Uttrakhand and the Siwalik hills growing in forest and open places and in hedges. All parts of the plant are medicinal in nature, which are described in detail as below:

Medicinal Values

Root of the climber is often used as a substitute of liquorice, known as Indian liquorice. It is found in pieces of varying lengths and is very effective in coughs and catarrhal infections. Roots are diuretic, tonic and emetic and are used in preparations to control disorders like gonorrhea, jaundice and haemogoburic, etc. leaves have a sweetish taste and are eaten raw, even pan and cooked as vegetables. Leaf decoction is given in colic, coughs and common colds. Leaf juice is applied to sores, boils, and swellings. It is also useful in leucoderma, itching and skin diseases (Singh, 2005).

Seeds are purgative, emetic, tonic, aphrodisiac, and are used in nervous disorder and cattle poisoning. They are also used externally in ulcers, skin diseases, sciatica, stiffness of shoulders, joints and paralysis.

Seeds possess major poisonous constituents; abrin, a toxalbumin similar to ricin from castor seeds, a glucoside; abrussic acid, alkaloid; abrine and abarnin. Roots and leaves have precol, abrol and two alkaloids; abrasine and precasine.

Botanical Description

Rati is a much branched climber with a woody stem. Leave are peripinnate 5-8cm long, leaflets opposite and thin, flowers reddish or white in colour, born in racemes. Fruits 3 to 7 cm long, seeds small, are a measure of weighing gold and precious stone. Rati is the smallest unit of weight in the Indian system. Pods oblong, 3 to 5 seeded, seeds ovoid or sub-globse, generally two third scarlet are white and the rest jet black giving appearance of a black eye.

Cultural Practices

Climatic requirements: Rati mostly grows in areas having tropical kind of climate. However, it can also ben grown in sub mountainous areas of Siwaliks locally known as *Kandi*. In Jammu *Kandi,* Rati grows in forest and hedges, whereas, in Himachal Pradesh it grows in Kangra, Una, Mandi and Poonta Sahib range and Nalagarh-Kunihar division below 1200m altitude (Chauhan, 2006).

Soil requirements: Rati requires sandy loam to loam soil for its good growth having soil pH range from 6.5 to 7.5. In *Kandi* belt it can also be grown in gravelly loam to sandy loam soil with good amount of organic carbon.

Cultivation: Rati can be sown during April-May through seeds. Prior to sowing seeds should be soaked in water for about 24 hours to hasten the germination. These seeds are then sown directly in well spaced pits.

As it is a delicate climber, so it requires support for its better performance. Support should be either temporary or permanent. As permanent support the plants of Rati should be grown along fences. It can also be propagated by cuttings of firm shoots (Chauhan, 2006).

References

Chauhan, N.S.2006.Medicinal and Aromatic Plants of Himachal Pradesh, Second Edition, Indus Publishing Company, FS-5, Tagore Garden, New Delhi-110027.

Singh, S.S. 2005. Fate of the Medicinal Plant Resources of Jammu and Kashmir. Directorate of Environment and Remote Sensing, Jammu, Govt. of J&K.

4

Cucumber (*Cucumis sativus* L.)

Common name: Kheera

Family: Cucurbitaceae

Introduction

Cucumber is one of the most ancient vegetable, most popular and familiar to common man, belonging to cucurbitaceous family. It contains most of the vitamins and minerals which one needs every day. Due to the presence of a number of vitamins and minerals; cucumber possesses lot of medicinal values.

Medicinal Values

Cucumber contains enough sugar, B Vitamins and electrolytes to replenish essential nutrients the human body lost, keeping everything in equilibrium (Mohindra, 2010). Among the B Vitamins group, Vitamin B_1, Vitamin B_2, Vitamin B_3, Vitamin B_5 and Vitamin B_6 are the major ones. Vitamins C and E, and folic acid are also present. Iron, magnesium, phosphorus, potassium and zinc are the main minerals, which are present in the cucumber.

Cucumbers have been used for centuries and often used by European trappers, traders and explore for quick meals to thwart off starvation.

Cultural Practices

Climatic requirements: Cucumber is a warm season crop and requires about 50-60 days from seeding to maturity. It grows best in areas with temperature range between 18° and 25°C.

Soil requirements: Cucumber thrives well in wide variety of soils from sandy to clayey provided; these are well drained and rich in organic matter content. However, sandy loam soil is good for early crop, whereas loam, silty loam or clay loam is preferred where heavy yield of cucumber is desirable.

Sowing time and seed rate: February-March is the suitable time for growing of cucumber in sub-tropical areas. In the sub-temperate and temperate agro-climatic zones, March and April-May are the best suitable times for sowing of cucumber. 2.5 to 3.0 kg seed is sufficient for sowing one ha area. Row to row and plant to plant spacing should be maintained at 1.5 to 2m and 0.6 to 0.7m, respectively.

Manures and fertilizers: Application of 25 tonnes of well rotten organic manure (FYM, compost or vermicompost) as basal is beneficial. This should be supplemented with 40 kg N, 25 kg each of P_2O_5 and K_2O ha^{-1}. Add full dose of P_2O_5 and K_2O and ½ N at the time of sowing and the remaining ½ N after 30-35 days of sowing.

Inter-culture and irrigation: Frequent hoeing and inter-culture of young plants promote healthy growth and heavy fruiting. During summer, crop should be irrigated once in a week, depending upon the availability of rains.

Harvesting and yield: Proper stage of fruit maturity should be judged by size and not by the age. The immature edible fruits should be harvested after 40-45 days. Fruit yield is 8-10 tonnes per hectare in 8-10 picklings.

References

Mohindra, V. 2010. 10 home uses of cucumber. *Ayurveda.* 1 (18): 18-19.

5

Indian Tinospora (*Tinospora cordifolia* Thunb. Miers)

Common name: Giloe or guduchi.

Family: Merispermaceae

Introduction

Giloe is a vigorous climber. Its Engish name is *Gulancha Tinospara*. There are about 40 species of giloe found throughout the world, comprising parts of Africa, Southern Eastern Asia, Australia. Out of 40 species, only 4 species are found to occur in India namely: (i) *Tinospora cardifolia* (ii) *T. sinensis* (iii) *T. malabarica* and (iv) *T. tomentosa*. *T. cardifolia* is, however, widely found in the tropical and sub-tropical regions of India ascending an altitude of 300 m to 1200 m above mean sea level. Contrary to this, *T. malabarica* grows upto 1300 m above mean sea level. Both these species are found in association with each other in the warmer and valley areas of Uttarakhand (Gautam and Maheshwari 2007). In fact, *T. cordifolia* is indigenous belonging to tropical areas of India, Myanmar and Sri Lanka. It is an important herb used in traditional Ayurvedic medicines and known by many common names (Table 5.1). In sub-mountain area of Jammu region of J&K especially *Kandi* belt of Jammu, giloe grows upto 700 m above mean sea level.

Table 5.1 : Vernacular Names of Giloe

Language	*Names*
Sanskrit	Guduchii
Hindi	Amrita
Bengali	Gulancha
Gujarati	Galo
Tamil	Shindikodi
Telugu	Tippatigo
Dogri	*Giloe*
English	Tinospora

Source: Gupta and Mishra (2008).

Medicinal Values

It is one of the most useful ayurvedic preparation which acts as tonic and aphrodiasac. It is also anthelmintic, anti-arthritis, anti-periodic, anti-pyretic, blood purifier, cardiac, carminative, digestive, diuretic, expectorant, stomachic, rejuvenating, appetising and anti-inflammatory. Starch from the roots of giloe as well as from stems can be used against chronic diarrhoea and dysentery. The juice of the fresh plant is diuretic and used in gonorrhoea. It is also useful for curing chronic fevers, gouts, vomiting, cardiac debility, skin diseases, leprosy, anaemia, cough, asthma, jaundice, seminal weakness, uropathy and splenopathy. In fact, giloe is famous for strengthening the immune system and helping the functioning of various body organs to remain in a balanced state (Gupta, 2013). As a bitter tonic, giloe has also proved to be an immune modulator, anti-oxidant, rejuvenator and restorative agent (Vatsyayan, 2011). It has been described one among, foremost herbs called "medhya rasayan" meaning boon for the brain and the nervous system.

The recommended dose of whole plant as powder, decoction and extract is as under:

1. Powder – 1-2 g upto maximum 5 g
2. Decoction – 50-100 ml
3. Extract – 1-2 ml

The powder is taken empty stomach with half cup of water in the morning.

The decoction or extract is taken after diluting with half a glass of water early in the morning.

Ayurvedic Formulations

Giloe plant is used for the preparation of various types of ayurvedic products like *Yograja guggulu, Kaishore guggulu, Sanjivani-vati, Amritarista, Sudarshan churana.*

Chemical Composition

The important chemical constituents of the plant are tinosporin, perberillin, palmarin, berberine, tinosporon, hepta consol tinosporic acid adntinosporol. The fresh stem bark yield giloin, giloinin and gilosterol. Hypoglycaemia agent and phenolic lignin have also been isolated from this plant. There is no doubt that giloe is bitter in taste but is packed with alkaloids and glycosides which reduce inflammation and enables the body to fight infections (Gupta, 2013).

Traditional Use of Giloe in Medicines

1. **Skin diseases:** Stem juice taken with neem, haldi and amla is very effective.
2. **Piles:** Juice with butter milk is useful.
3. **Breast milk:** Decoction of the stem is given to improve the quality of breast milk.
4. **Toxins:** It is considered as best herb for clearing micro-circulatory system. Its juice is very effective in removing both exogenous and endogenous toxins. It clears out the brain toxin that inhibits mental function.
5. **Asthma:** The root and bark with whey is used in the treatment of respiratory troubles particularly asthma.
6. **Diabetes:** Juice is taken in high quantities.
7. **Excessive bleeding during menstruation, bleeding after abortion or delivery:** Stem, leaves and roots can be used. About 5 g each of leaves and roots are crushed together to extract the juice. Consumption of two cups for few days after diluting it with water (2 to 3 ml in half a cup of water) till the condition improves.
8. **Malaria and other fever:** Decoction of the stem with *pipli* (*Piper longum*) and honey is taken.
9. **Indigestion:** The juice with honey or the paste of leaves can be taken with butter milk.
10. **Conjunctives and cataract :** In some parts of India the juice is applied inside the eyes.

Other Uses

Every part of this graceful twining, climbing plant has lot of value. Local practioners have used it for centuries as a longevity tonic (Anonymous, 2009).

It is used in cancer prevention, cancer treatment support, high cholesterol and liver protection. It is used as strong anti-aging agent. Many natives use the fruits of *Tinospora* in face care. It has been used to treat convalescence from severe illness, arthritis, food allergies and anaemia. *Tinospora* has adaptogenic effects, a term that indicates it helps the body to adopt to stress. In children it is used in general debility, digestive disturbance, loss of appetite and fever. Different plant parts can be used for curing various diseases (Table 5.2).

Table 5.2 : Use of Giloe Plant for Curing Diseases

Part Use	*Disease that can be Cured*
Stem	Fever, Acidity, Blood purification and Eye disease
Fruit	Rheumatism, Jaundice
Root	Leprosy
Leaves	Rheumatism

Source: Gautam and Maheshwari (2007).

Botanical Description

Giloe is a large extensively spreading, glabrous, perennial, deciduous climber having succulent stems and corky dotted bark with grey-brown or creamy-white colour. The leaves are simple, membranous, alternate broadly ovate, deeply cordate and shortly acuminate. Its flowers are small, yellow or greenish-yellow in appearance when the plant is leafless. They are usually solitary in the female and clustered in the male plant. The fruits are generally drupes and red in colour when they ripe. The drupes are ovoid, glossy, succulent, red and seeds are curved. Usually, fruits bear only one seed.

The flowering in giloe usually starts in the month of March-June and fruit set in the month of July and matures during winter months. The most important part of the plant used for therapeutic purpose is fresh stem, though it is commercially available in the market in the dried form.

Cultural Practices

Climatic requirements: As explained earlier giloe is grown in tropical to sub-tropical climate with an elevation ranging from 300 to 1300 m above mean sea level. In fact, the giloe grows throughout the country in deciduous and dry forests (Verma and Baigh, 2012)

Soil requirements: Giloe can be successfully grown in large variety of soils, ranging from loamy sand/sandy loam or loam to clay loam. However, the soil should be well drained with sufficient moisture.

Propagation

Planting time: Planting is usually done during rainy season (July to August). As it is climber so it requires the support of fast growing species like Jatropha (*Jatropha curcas*) and Moringa (*Moringa oleifera*) for its good growth. However, it can be grown during spring season also.

Propagation through seeds and cuttings: Giloe is propagated both through seeds and cuttings yet propagation through cutting is cheaper and easy as compared to seeds. Cuttings do not require any chemical treatments to enhance the rooting. The cuttings should have atleast two internodes. Propagation through seeds requires some chemical treatments before sowing in the nursery bed.

Irrigation: Initially at the time of planting, watering is required but during winter it can be done once in a week depending upon climatic conditions.

Weeding and hoeing: Regular weeding and soil working is required to obtain good yield.

Manures and fertilizers: If 10 quintal of FYM is added per hectare, there is no need to add chemical fertilizers.

Harvesting and yield: Harvesting is done during November, the stem and leaves are harvested, dried in clear weather conditions. The yield is around 10-12 q ha^{-1} on dry weight basis. The present market rate of whole plant is Rs. 50-60 kg^{-1}.

References

Anonymous 2009. Ayurveda, reinterpreted. *The State Times*. 14 (219): 5 (In Infotainment, Saturday, August 8).

Bisht, P.S., Kumar, R. and Uniyal, C. 2005. Giloe : An important medicinal plant of Uttaranchal. *Indian Farmer's Digest*. **38** (4) : 36-37.

Gautam, S.K. and Maheshwari, G. 2007. Giloe – A valuable medicinal plant: Cultivation and use. *Agrobios Newsletter*. **6** (2) : 42.

Gupta, R. D. 2008. Giloe is Amrita. *Ayurveda*. 1 (10) 20-21.

Gupta, R. D. 2013. Giloe as medicine. *The Daily Excelsior*. 49 (221) : 4 (*Daily excelsior Magazine*, August 11, Sunday).

Gupta, S.K. and Mishra, P. 2008. *Tinospora cordifolia* : An immunization herb. *Agrobios Newsletter*. **7** (1): 21-22.

Vatsyayan, R. 2011. The goodness of giloy. The Tribune.131 (254): 16 (Wednesday, September 14).

Verma, K.C., and Baigh, M.A.2012. *Tinospora cordifolia* (wild): Plant with enormous pharmacological applications. *The Indian Farmers Digest*. 45(1): 8-21.

Herbs

1

Ajowan Caraway (*Trachyspermum ammi* Sprague) Syn. *Ammi copticum* L.

Common name: Ajwain (in Hindi); Yamini (in Sanskrit)

Family: Apiaceae

Introduction

A herb compliant, yields large quantity of seeds. The herb is mostly cultivated in East India, Central India, especially in Indore and its adjoining parts and in Andhra Pradesh-South India. Home remedies made out of ajwain are well known. It is due to the presence of 'thymol' in its seeds which help against number of diseases of the digestive tract and to treat fevers. Use of ajwain in treating various diseases/disorders includes indigestion, flatulence, diarrhoea, cholera, dyspepsia, hysteria and spasmodic affection of bowels. It is a good expectorant in bronchitis. Ajwain is not only used as a spice to add taste to the culinary but also as a major ingredient for different kinds of medicines. Our "Grand Ma" has always used ajwain effectively in managing problems such as vomiting, stomach ailments etc.

Medicinal Values

Oil of ajwain is rubbed over the affected parts to relieve rheumatic and neuralgic pains. Ark of ajwain seeds is given in early stages of cholera, oil and flowers combined in diarrhoea, dyspepsia and flatulence. It is strongly anthelmintic

in the shape of oil known as thymol for treatment of stomach-ache, colic, cough, catarrh. Apart from preventing indigestion problems including acidity, bronchitis and asthma, it is also used for getting relief from tooth ache, backache, joint pains and sore throat as well as to get relief from burning sensation in feet. Crushed seed or leaves is also made into a poultice and applied for colic.

Seeds are generally chewed to correct foul breaths. Roasted seeds are useful in dyspepsia. The paste made from seeds is used to relieve pain in cephalagia. Coriander along with ajwain water is useful in indigestion and other bowel complaints. It is beneficial as carminative and antibilious tonic, refrigerant and aphrodisiac.

Botanical Description

Ajwain is thought to have originated in the Asia Minor and spread to the Indian sub-continent. It is widely grown in arid and semi-arid regions with high concentration of salts. The plant is an annual, erect, small, hairless, branched, cool season herb, which grows up to 2-3 feet (0.6 to 0.9 m) in height. Flowers are tiny, white-petalled, borne terminally in umbels. Fruits are small, aromatic, greyish brown, with oval-shaped seeds one seeded that are ready for harvest by the end of winter or early spring. Seeds are olive green to brown in color, have similar appearance to cumin or caraway seeds with vertical stripes on their outer surface. However, they can be easily distinguished by their elliptical shape and tiny size.

Cultural Practices

Climatic requirements: It loves to grow in colder areas mostly during rabi season. High humidity at the time of maturity is detrimental. Crop is relatively tolerant to dry conditions and grows under wide climatic conditions.

Soil requirements: Ajwain can be grown on all kinds of cultivable soils and even in black cotton soils but the soils having sandy loam or loamy texture are considered the best for its growth. The soil should be rich in organic matter and preferably well drained (Malhotra and Vashishtha, 2008).

Nursery and field preparation: The soils should be thoroughly ploughed with 2-3 deep ploughings followed by planking to conserve soil moisture.

Sowing time and seed rate: Early crop under rainfed conditions is sown in August-September whereas; normal sowing is done in September-October. Seed rate- 2.5 kg for irrigated and 4 kg ha^{-1} for rainfed conditions is recommended. Seed sowing is done in drills 30-45 cm apart depending on conditions. Close spacing is recommended for rainfed crop and deep sowing should be avoided.

Varieties: Mostly the varieties released by National Research Centre on Seed Spices are under cultivation namely; 'Gujrat Ajwain', 'Lam Selection 1 and 2', 'RA 1-80' and 'RA 19-80' with an yield potential of 8-10 q ha^{-1}.

Manures and fertilizers: Apply 10 tonnes of well rotten FYM and plough it into the soil. Apart from this, N, P and K@ 40, 20 and 20 kg ha^{-1} are recommended through chemical fertilizers.

Hoeing and irrigation: Shallow hoeing and regular weedings are required to obtain better yield. Although the crop can grow without irrigation but irrigating crop at 15-25 days intervals produces optimum yield.

Harvesting and yield: The crop matures in about 130-180 days and is ready for harvest in Feb-March. Harvesting is done manually and is stacked for drying and threshed by beating with sticks. Yield of the crop varies from variety to variety and ranges between 4-6 q ha^{-1} to 12-15 q ha^{-1} in case of rainfed or irrigated crop respectively.

References

Malhotra, S.K. and Vashishtha, B.B. 2008. Package and practices for production of seed spices. National Research Centre on Seed Spices (ICAR), Ajmer, Rajasthan 63-70 pp.

2

Black Caraway (*Bunium persicum* Boiss. Fedtr.)

Common name: Kala zeera

Family: Apiaceae

Kala zeera, an aromatic and medicinal perennial herb belongs to genus *Bunium*. The plant is a native of Europe and Western Asia, but is now widely distributed in dry temperate regions of both hemispheres. and species is widely cultivated in many parts of the temperate zones - in Holland, the USSR, Hungary, Poland, Denmark, Germany, Norway and lately also in United States. It has been found to grow at an altitude between 2000-3000 m above mean sea level in the Western Himalayan region, extending up to Baluchistan and Afghanistan (Panwar and Chauhan 1990) and worldwide known by different vernacular names (Table 2.1). It generally grows in forests, grassy slopes and pastures of Himachal Pradesh, Jammu and Kashmir and Uttrakhand. High altitude regions of Kinnaur, Lahaul Spiti, Pangi and Bharmour of Chamba area in Himachal Pradesh and Padder valley and Gurez area of Jammu and Kashmir are the potential areas for its production in India and its cultivation is quite remunerative as one hectare of the crop yields net return of Rs.13, 680-1, 05,277 between 4th -10th year of

its establishment (Panwar *et al.*1993). It is one of the important commercial crops in the backward areas of Gurez, Machail, Tangdar, Kargil, Paddar, Pulwama and some parts of Budgam and Srinagar areas in Jammu and Kashmir. However, in Himachal Pradesh, efforts to domesticate it have been successful and have become a promising cash crop in cold arid zone (Panwar, 1992). The seeds are harvested only during June-July from the wild population from the forests or grazing lands by the migratory people of these areas.

Table 2.1: Vernacular Names of Kala Zeera (*Bunium persicum*)

English	Black cummin
Dutch	Zwarte komijn
Danish	Sort Kommen
French	Cumin noir
German	Schwarzer Kreuzkümmel
Farsi	Zireh e irani, Zireh kuhi
Hindi	Himali jira, Kala jeera, Kala jira,Kashmiri jira, Shahi jira
Italian	Cumino nero
Japanese	Burakku-kumin
Bengali	Kasmiri jira, Sajira
Gujarati	Kal jiru
Sanskrit	Krishna jiira
Urdu	Kalazira, Shah zira
Kashmiri	Kashmir zireh, Koshur zur
Maithili	Sahajir
Nepali	Kalo Jira, Himali jira
Oriya	Kalajira
Punjabi	Kala jira
Kannada	Shahajirige
Spanish	Comino negro
Dogri	Kala jeera, Kala jira

Medicinal Uses

It is an economically important medicinal spice, condiment as well as an aromatic plant. It is a low volume, high value, non-perishable commodity. Its fruit commonly called seed is an expensive spice used for culinary purpose and for flavouring food and beverages. The plant has been used in the Indian System of Medicine since time immemorial. The seeds are regarded as stimulant, carminative and useful in curing of diarrhoea, dyspepsia, fever, flatulence, stomachic, haemorrhoids, amnesia, boils, obstinate hiccups and hoarseness of voice. Seeds, a rich source of thymol are used against hookworm infection and also as antiseptic in many proprietary preparations. The seeds mixed with lime juice are administered in cases of billious nausea in pregnant women. The seeds taken by women shortly after child birth are reported to help secretion of milk. The powder of seed mixed with honey, salt and clarified butter is applied to scorpion bites. The straw left after harvesting of seed is also a rich source of volatile oil (1.25 per cent) and oleo resin. Besides, these it is also considered in dishes, bakery and culinary items.

Besides this, it is used for flavouring alcoholic beverages like different types of sprits and liquors and as one of the important condiment commonly used to add fragrance and taste to non-alcoholic beverages like frozen dairy desserts, candy, baked goods and also in meat products. Its tubers are also eaten as vegetable in J&K state.

Botanical Description

Kala zeera, belonging to family Apiaceae is a perennial, glabrous, branched herb. Roots are tuberous, leaves 2-3 pinnate, segments filiform, lower leaves petiolate, upper sessile. Fruits oblong, 3-4 mm long, dark brown, ridges prominent stylopodium flattened, styles reflexed with inner seed face concave (Raina, 1997). Crop is naturally cross-pollinated. On maturity, it yields 5-10 g/plant mature fruits during June-July and average seed yield of 5-6 quintal per hectare in a well established crop can be realized. The plant is initially propagated through seed and subsequent regeneration is possible through tubers. The above ground biomass of the plant can be pulled up easily on maturity, leaving the tuber underground which remains dormant till March-April. Tuber possesses a productive span of at least 10-12 years.

A long chilling period is essential for the germination of seed and for getting a good crop. The plant thrives well under optimum temperature of 75-80° F and sufficient moisture in the soil. If planted with seed, crop yields after three years provided it is properly nourished.

Cultural Practices

Climatic requirements: The crop is very climate specific and can only be grown in dry temperate regions receiving adequate snow during winter months. Light to moderate showers in the growing season *i.e.* (April- May), bright sunshine, a severe winter with snow up to 1-5 meters and heavy dew is a blessings for the crop. Snowfall after sprouting and even frost is injurious to the crop. But seed and tubers require long chilling periods for better crop. However, crop ready for harvest if marred by a cloudy weather or rains, affects yield, flavour and seed quality.

Soil requirements: It prefers well drained, light textured (sandy to loamy), rich in organic matter.

Preparations of land: The field should be thoroughly ploughed 3-4 times before sowing and planked. Small beds of convenient size are made to facilitate sowing.

Propagation: *Direct seed sowing-* Kala zeera is a seed propagated plant which is sown in Oct.- Nov. and takes 4 years to produce seed. Each seed produces one tuber and tuber does not multiply vegetatively. Seed is sown at shallow depth in rows 30 cm apart and covered with soil.

*Planting of tubers-*Tubers of size 2-3 cm diameter too can be used for raising the crop and give crop in the same year. Same tubers continue to give crop for 8-9 years.

Seed rate: Seeding@1-1.5 kg/ha at first sowing and re-seeding @200g/ha seed in subsequent years, help in maintenance of optimum plant population (1.5 - 2.0 lac/ha). About 1.67 lac tubers are required to plant one hectare area at spacing of 30x20 cm.

Method of sowing: The crop can also be grown as an inter-crop with rajmash, beans, peas, minor millets or saffron in the fields or on slopy areas as understorey crop in the orchards of apple, apricot and almond. Before sowing, seed should be treated with fungicide Bavistin or Dithane M-45/Z-78 @2 g/kg seed.

Manures and fertilizers: The crop is raised traditionally without the application of organic manures, chemical fertilizers and adoption of best agro-techniques. The following doses of manures and fertilizers should be used to get optimum yield:

1. FYM–10-15t/ha
2. Nitrogen–60 kg/ha
3. Phosphorus–30 kg/ha
4. Potash–30 kg/ha

Full doses of FYM, phosphorus, potash and 1/4th dose of nitrogen should be applied before the final ploughing. The balance dose of nitrogen should be splitted into 3 equal doses and applied as top dressing during intercultural and weeding operations at sprouting, bolting and flowering respectively. Since the crop is perennial, therefore, in subsequent years only one light ploughing during Oct.- Nov. with the same recommended dose would be sufficient to harvest good crop.

Intercultural operation: Kala Zeera crop must remain weed free. 3-4 weedings at an interval of 20-25 days are sufficient. Use of weedicide like Stomp@ 5.0 litre or Lasso @ 3.0 litre per hectare can be effective to control weeds. Similarly 2-3 irrigations are sufficient to meet the water requirement of the crop.

Pests and diseases: Following control measures are recommended against insect pest and diseases which cause severe damage to the crop.

Diseases: Diseases like blight and tuber rot can be controlled by treatment of tubers with spraying the crop at a fortnightly Carbendazim @ 0.3 per cent interval with 0.2 per cent Dithane M-45 or Blitox 50 WP from first appearance of the disease. For controlling tuber rot, soil drenching with Brassicol and Bavistin is recommended.

Insect-Pest: The tubers are attacked by white grubs and above ground foliage by hairy caterpillars, army worms and semi loppers. White grubs can be controlled by mixing BHC/HCH 10 per cent Dust or Aldrin 5 per cent dust @ 25 kg/ha in soil during last ploughing. For others spray the crop with 0.5 methyl parathion (Metacid-50 EC) @ 1ml in 1litre of water or 0.05 per cent Endosulfan @ 1.6ml/litre.

Harvesting and yield: The crop raised through seed takes about 4 years for the production of seed. Normally crop matures in July. The crop should be harvested when the seeds turn brown. Over mature crop is prone to shattering and harvesting is done daily in the morning time and stacked in a dry store. The total harvesting period lasts for 10-12 days. The stacked plants should be dried in sun for 2-3 days and seeds are separated by beating with sticks at a protected place. Seeds are separated after winnowing and stored in well dried air tight containers.

Crop yields vary from year to year and on an average from a well established crop about 1.5-2.0 q/ha of kala zeera seed can be produced. The yield levels can still increase provided adequate plant population is maintained. As far as economics is concerned a farmer growing kala zeera under scientific management can earn 30-40 thousands rupees annually per hectare. Market price is variable and ranges between Rs.1500-2000/kg.

References

Gupta, R.K. 2007. Black Cumin status and cultivation in NW Himalayas: An overview. In: *Production, Development, Quality and Export of Seed Spices* (Eds. S.K. Malhotra and B.B. Vashishtha). NRCSS, Ajmer, India. Pp.190-194.

Gupta, S.K. 2013. Status and improvement of Black caraway (*Bunium persicum* Boiss. Fedtsch) in dry temperate region of Jammu and Kashmir. In: *Forest Biodiversity: Conservation and Management* (Eds. Arvind Bijalwan and C.P.Kala). Pointer Publishers, 807, Vyas Building, Jaipur, India. Pp.102-112.

Gupta, S.K. and Ateet Pal Singh 2010. Cultivation of black caraway/kalazeera (*Bunium persicum*) in dry temperate regions of Jammu division. *Green Farming* 1(1): 106-107.

Panwar, K.S. 1992. Kalazira – A low volume high price crop for dry temperate hills. *Indian Farmers Digest,* 25(11): 21-23.

Panwar, K.S. and Chauhan, V. 1990 Agronomic traits differentiation of *Bunium persicum* (Boiss) Fedtsch. under domestication. *Indian Forester* 116(4): 334-335.

Panwar, K.S., Sagwal, J.C., Sharma, S.K. and Kapil Saroch. 1993. Economic viability of kalazeera (*Bunium persicum* (Boiss) Fedtsch.) cultivation in high altitude dry temperate regions of Himachal Pradesh. *Agriculture Situation in India,* 48 (6): 151-154.

3

Black Nightshade (*Solnum nigram* L.)

Common name: Makoy, makoi, gurkkamai, kakamaei, kakmachi

Family: Solanaceae

Introduction

It is a small shrubby plant and grows throughout the dry parts of India up to 800 m altitude. However, this plant grows commonly in all places of Jammu and Kashmir up to 200 m to 800 m elevation. However, makoy is not the part of forest vegetation. Makoy is one of the medicinal plants suitable for cultivation on a commercial scale, especially in Bihar, Delhi, Uttar Pradesh, Madhya Pradesh and Rajasthan. Medicinally makoy has wider applications.

Medicinal Values

Fruits of the makoy are very effective in stomach problems, fevers and eye problems. Leaves of the shrub make a good poultice on swellings and joint pains. Paste of the leaves is used on piles to get relief (Singh, 2005).

Botanical Description

Makoy is an erect herb, 30-60 cm high with white small flowers and green berries (fruits) which turn red or black on ripening. Seeds are discoid, smooth, yellowish.

Flowers and fruits come in the months of August to October. It grows as a weed throughout dry parts of India.

Cultural Practices

Climatic requirements: Usually Makoy is cultivated in tropical and sub-tropical agro-climatic conditions.

Soil requirements: The plant grows in different kinds of soils including dry, stony, shallow or deep soils.

Nursery raising and planting: The seeds are sown during April-May in well manured nursery beds. It takes about 15-30 days to grow seedlings. The seedlings develop in 31-35 days after sowing which are transplanted in rows and 60 x 60 cm apart in the well prepared fields.

Manures and fertilizers: The medicinal crop species have to be grown without chemical fertilizers and pesticides. Organic manures like FYM, vermicompost, green manure etc. may be used as per requirement of the species.

Similarly, to prevent diseases and attack of insect pests, bio-pesticides can be prepared, (either single or mixture) from neem (kernel, seeds and leaves), cow urine etc. are used.

Irrigation: The nursery beds and plantation should be irrigated periodically as and when required; weekly or fortnightly. The plants are irrigated till they begin to produce flowers (Anonymous, 2003).

Harvesting and yield: The crop usually gets ready for harvesting after 1-6 months depending up on the climate and soil. The plants are collected and dried. The dry herbage yield varies from 2-4 tonnes/ha.

References

Anonymous. 2003. Makoy. In: Indian Society of Agribusiness Professionals 1 (w): 3.

Singh, S. S. 2005. Fate of the Medicinal Plant Resources of Jammu and Kashmir. Directorate of Environment and Remote Sensing, Jammu.

4

Black Pepper (*Piper nigrum* Linn.)

Common name: Kali Mirch

Family: Piperaceae

Introduction

Black pepper, often referred to as the king of spice or Black gold is the most important spice in the world. The western world knows black pepper merely as a condiment, whereas, in Asia, especially India, it is also one of the foremost indigenous medicine besides serving as a condiment.

Black pepper is believed to be a native of Western Ghats. In the medieval era, its fame as a spice attracted the traders from all parts of the world. Now a days, black pepper is cultivated in number of tropical countries including India. India is a major producer, consumer and exporter of black pepper in the world with cultivation in an area of 0.211 million ha. It is also cultivated in Vietnam, Indonesia, Brazil, China, Malaysia, Sri Lanka etc. In India, Kerala, Karnataka and Tamil Nadu are the major black pepper producing states, out of which almost 90 per cent is contributed by Kerala. Right from the writings of Charaka and Sushruta to the work of other Acharyas, black pepper has been amply described in almost every Ayurvedic text as it possesses number of medicinal values.

Medicinal Values

Black pepper is used worldwide in cuisine, preservatives, perfumery and in medicines. It is used at all stages of the cooking process and as table condiments. Being a native of Western Ghats, black pepper forms an important gradient of several indigenous medicines of India as well as African and Chinese systems. In Ayurvedic System of Medicine, black pepper is known in different names such as Maricam (killer of poison), Krishna (corrosive), Ooshana (giving burning sensation and vellayam (antihelminthic)) as reported by Alex and Soni (2010).

The whole of the black pepper plant is being used as medicine in various Ayurvedic preparations. It can be used as a stimulant, carminative, digestive, stomachic, nervine, resolvent deobstruent, diuretic and antiperiodic. Black pepper is much employed as an aromatic stimulant against cholera, as an antiperiodic for malarial fever and alternative to arthritis disease. The health benefits of black pepper include relief from respiratory disorders (cough, common cold, anaemia, impotency and heart diseases (Anand and Awasthi, 2013).

Vallue Added Products

The modern analysis of black pepper shows that it contains chiefly resin (chavicine), volatile oil and a few alkaloids. Among various alkaloids present in the black pepper piperine is the most abundant. The presence of resin makes black pepper a stimulant, while volatile oil imparts odour and aromatic taste and the alkaloid piperine gives the febrifuge property. Usually the oleo-resin, piperine, essential oil show many fold increase, about one to two months prior to completing the maturity of the berries.

Apart from the presence of volatile oil, piperine, oleoresin, the black pepper also possesses moisture, protein, carbohydrates, fibre and minerals. It is rich in vitamin-B complex. Black pepper too contains traces of calcium, iron and phosphorus. Black pepper acts as carminative, stimulant, aromatic, digestive, diuretic, tonic and anti-coagulating agent. Thus the curative properties of black pepper have been described in Ayurveda. It excites salivary and sweat glands besides killing intestinal worms and propelling downward movement of abdominal air.

Black pepper has been used for various health problems. Starting from common cough and cold, sinusitis and bronchitis to indigestion, distention of the abdomen, colic and conditions involving sluggishness of liver, black pepper is used singularly and in Indian home is present as an important culinary item. Some simple tips to gain from its medicinal benefits are as follows:

- ☆ To promote appetite and allay distention, a quarter teaspoonful of powders of both black pepper and white cumin (safe zeera) should be taken with butter milk an hour before lunch. As already stated black pepper is beneficial in the treatment of cold and fever. In the case of acute running nose accompanied by headache and bodyache, taking for 2 to 3 days warm milk boiled with pinch of powders of pepper and turmeric is tried as home remedy. Black pepper powder and common salt are an excellent dentifrice which prevent dental caries, foul breath and painful gums.

- After mixing black pepper, dry ginger and *Piper longum* (magh) in equal parts a distinctive combination is achieved which is known as *trikatu*. Having multiple uses like triphala, trikatu is a drug of choice for diseases like sinusitis, bronchitis, indigestion, urticaria, obesity and many other kapha and vata disorders. To be avoided in case of acidity, it can be taken after mixing in honey in dose of 1-2 g twice a day (Patwari, 2005).

Botanical Description

Black pepper, commonly known as Kali mirch is a flowering vine, and is usually dried and used as spice after seasoning. The fruit, known as peppercorn when dried is approximately 5mm (0.5cm) in diameter, dark red when fully mature, and, like all drupes, contains a single seed.

The black pepper plant is a stout, glabrous, climbing shrub with small cordate leaves when young gradually getting larger, latter sending out flowering branches with large leaves and fruits. Leaves are simple, alternate, cordate, varying in breadth, broadly ovate and 5-9 nerved (Alex and Soni, 2010). This is a dioecious/ monoecious plant with minute flowers in spike varying in length. Fruit is globose, one seeded drupe, bright red when ripe. It is pungent and slightly musty in odour. Seed is globose.

Cultural Practices

Climatic requirements: Black pepper can be grown in areas having tropical climate. However, some promising new lines of black pepper have been released by Indian Institute Spice Research, Calicut (Sasikumar and Sarma, 2002).These black pepper lines include 2 hybrids suitable to higher elevations above 900m.

Soil requirements: Black pepper can be grown in soils, which are neither too dry and nor susceptible to flooding. They should be well drained and rich in organic matter. In dry soils, the young plants require watering every alternate day during the season at least for the first three years.

Planting: Black pepper is propagated through rooted cuttings. It develops different kinds of aerial shoots like primary stem or climbing stem, runner shoots which originate from the base of the vines, fruit bearing lateral branches and hanging shoots. Different planting materials will produce different kinds of plants.

In India, generally cuttings from the runner shoots are used for raising rooted planting material. Runner shoots of high yielding and healthy vines in the garden are selected and cuttings with 2-3 nodes are planted either in nursery beds or in bags filled with fertile soil. This traditional method, however, does not provide enough plant materials to meet the requirement.

So, alternative methods of rapid multiplication like multiplication of rooted pepper cuttings using pit method *i.e.* single node from runner shoots are used for rooting. Propagation of rooted lateral or fruiting branches is done as bush pepper.

Micro-propagation of black pepper from nodal explants using tissue culture method is also being carried out (Alex and Soni, 2010).

Irrigation: After planting, the plants require watering in the initial years especially in the soils which are dry. Hence, depending upon the situation, the plants are irrigated twice or once in week.

Weeding and hoeing: The weeds or the competing plants are required to be cleared away. For this purpose, manual weedings and hoeing should be followed.

Mulching: The roots of the black pepper are covered with leaf mulch and manure and the shoots. This will help in saving the irrigation as, the soil water will be stored in the rhizosphere of the roots.

Harvesting and yield: The yield of the plant varies from variety to variety. However, generally the yield varies from 3.5 to 4.6 kg/vine (fresh) in case of HP-105, HP-813 varieties.

In order to prevent spike shedding in advance stages of fruit development, hybrid variety, 'Panniyur -1' should be planted as its tendency of shedding spikes, is comparatively low. Balanced use of N,P,K fertilizers is required to control this disorder (Hnamte and Lungmuana, 2012).

References

Alex, S and Soni, K.S. 2010. Black pepper. The King of spices. *Agrobios Newsletter.* 9(7): 15-16.

Anand, S. and Awasthi, P. 2013. Black pepper- A natural way to stay healthy- *Indian Farmers Digest.* 46(5): 33-34.

Hnamte, V and Lungmuana. 2012. Problems in pollination, fruit set and various disorders influencing the productivity of major spices. *Agrobios Newsletter.* 10(9): 29.

Patwari, H. N. 2005. Nature cure-Black pepper. *The Daily Excelsior*, 41 (231): 3 (August 21, Sunday Magazine).

Sasikumar, B. and Sarma, Y. R. 2002. New black pepper varieties from Indian Institute of Spices Research, Calicut. ICAR News. 8(1): 16.

5

Broccoli (*Brassica oleracea* L.)

Common name: Broccoli

Family: Brassicaceae

Introduction

Vegetables and other plants have been used medicinally since antiquity. Among the various vegetables having medicinal values, green leafy ones like spinach, fenugreek etc; vis-à-vis tomato, potato and broccoli, are worth mentioning. Although broccoli is not very common in our country yet it must be popularized because of its value as good vegetable. As a matter of fact, broccoli is a wonderful nutritious vegetable very popular in foreign countries. It is because, of lot of medicinal values. It quite resembles with the cauliflower except that it has small sized heads with enlarged green flower buds or refered to the flowering top of a cabbage derived from the Italian word, plural of brocoloc (Neethu and Tripathi, 2013). The peculiarity of this vegetable is that when the central head is removed new side shoots emerge which also form small heads and become available for cooking as dish. Broccoli has a wide potential of nutritional and medicinal benefits (Madhu *et al.*, 2015).

Medicinal Values

Broccoli is an excellent source of vitamin C, fibre, iron and calcium. It has been proved that it provides protection against cancer. It purifies blood and increases the immune system of the body. 'Sulphoraphane', a compound found in broccoli is a great solution to boost liver and skin cells which help in detoxification of body

but at the same time this compound can repair sun damaged skin. Broccoli has been singled out as one among the vegetables and fruits that contributes to the significant reduction in heart diseases. Phytonutrient antioxidants found in broccoli also prevent cataract. To obtain the maximum benefits of the aforesaid medicinal values, it is very important to buy right kind of broccoli. Hence, always select firm stems which will have heads dark green purple in colour and buds should always be closed with no signs of yellow flowers (Pandey and Tripathi, 2012).

Broccoli is the superhero of the vegetable kingdom with its rich vitamins and occupies the special status of being one of the most nutritious vegetables around. Nutrient content of broccoli is shown in Table 5.1. The data show that broccoli has much vitamin C content. Broccoli is a big source of vitamin A. Its leaves are completely edible (Neethu and Tripathi, 2013) and contain generous amount of vitamin A. Broccoli has a unique combination of vitamin A and vitamin K. Ample amount of vitamins A and K can keep vitamin D metabolism in balance. Broccoli is a useful natural source of selenium.

Table 5.1: Nutritional Value of Raw Broccoli per 100 g

Nutrient	*Amount*	*Nutrient*	*Amount*
Carbohydrates	6.64 g	Calcium	47 mg
Fat	0.37 g	Magnesium	21 mg
Protein	2.82 g	Potassium	316 mg
Thiamin	0.07 mg	Phosphorus	66 mg
Riboflavin	0.12 mg	Iron	0.73 mg
Niacin	0.64 mg	Zinc	0.41 mg
Vitamin C	89.20 mg	–	
Vitamin E	0.78 mg	–	

Source: Neethu and Tripathi, 2013.

Broccoli is blessed with vitamin B complexes like vitamin B_1 (Thiamin) Vitamin B_2 (Riboflavin), Vitamin B_3 (Niacin), vitamin B_9 (folate, folacin and folic acid). Two carotenoids found in broccoli are lutein and zeaxanthin play an important role in the health of eye. Broccoli is also an excellent source of potassium, magnesium, phosphorous, zinc, molybdenum and selenium which help in maintenance of blood pressure.

Cultural Practices

Climatic requirements: Like cauliflower, broccoli requires cool, moist climate. Dry weather accompanied with low humidity is not suitable for its production.

High temperature generates poor quality curds. The best temperature for its growth is around 20° Celsius.

Soil requirements: Although it can be grown on wide variety of soils yet sandy loam soils are the best suited for its production.

Sowing and transplanting: In nursery, it is sown in the third week of September and transplanting is done towards the end of October. Heads of broccoli become available after 85-95 days of transplanting.

Manures and fertilizers: About 20-25 tonnes of well decomposed organic manure *i.e.*, FYM, compost or vermicompost is the minimum dose for the production of normal heads of the broccoli in the small scale.

Irrigation: Irrigation is only given if there is long gap in the rains.

Harvesting and yield: The plants are cut well below the head so that the stump thus left protects the head from damage during transportation to the market. The harvesting should be done either during morning or evening so that the produce may look fresh. Yield of the broccoli varies from 150-200 q ha^{-1} under normal situations.

References

Madhu, Kockhhar, A., Singh, S and Vishawkarma, S. R. 2015. B for broccoli: A blissful vegetable in human diet. *Indian Farmers Digest*. 48(2) : 11-12.

Neethu, T.M and Tripathi, S. 2013. Broccoli: A healthy snack of necessary nutrients without unwanted kilojoules. *Agrobios News Letter*. 12(1): 37-39.

Pandey, H and Awasthi, P. 2012. Winter vegetables: Treasures of nutrients and medicinal properties. *Indian Farmers Digest*. 45(2): 15-16.

6

Buckwheat (*Fagopyrum esculentum* Moench.)

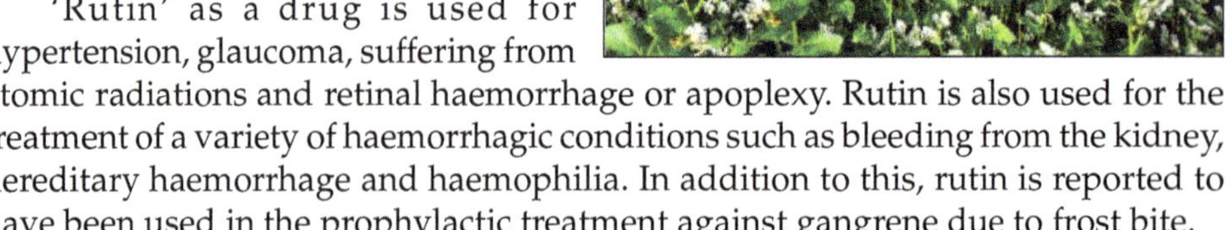

Common name: Kutu

Family: Polygonaceae

Introduction

Buckwheat is an important herb which is commercially grown for rutin. Rutin is mainly found in flowers and leaves which is developed as a drug. The drug can be used for curing a number of diseases.

Medicinal Values

'Rutin' as a drug is used for hypertension, glaucoma, suffering from atomic radiations and retinal haemorrhage or apoplexy. Rutin is also used for the treatment of a variety of haemorrhagic conditions such as bleeding from the kidney, hereditary haemorrhage and haemophilia. In addition to this, rutin is reported to have been used in the prophylactic treatment against gangrene due to frost bite.

Botanical Description

It is an annual herb growing up to 80-160 cm in height. The leaves are alternate, acute and 2.5-7.5cm long, flowers produced in terminal cymes and are pink whitish in colour, fragrant, diamorphic and highly sterile.

Cultural Practices

Climatic requirements: It requires moist cool climate for its optimum growth. A monthly average temperature of $18-30^{\circ}$ Celsius appears to be the most satisfactory. It is a photosensitive crop and hence the period of its cultivation must be carefully chosen for optimum production.

Soil requirements: Buck wheat can be grown on a variety of soils, but it thrives best on well drained sandy or loamy sand soils (Satish and Rajangam, 2012). It does not grow well on ill drained, heavy and acidic soils.

Time of sowing: September to early December is considered the best period for obtaining its good growth and yield.

Land preparation: The land is ploughed 3-4 times and brought to a fine tilth. The entire quantity of FYM is applied at the time of field preparation.

Sowing: Ridges and furrows are opened at 20cm apart and seeds are sown in furrows at 10 cm apart.

Irrigation: If there is insufficient moisture in the soil, the field is irrigated immediately after sowing. Later, the crop should be irrigated at weekly interval during the dry period.

Intercultural operations: Weeds are constantly removed by mechanical means to get good yield of the crop.

Plant protection measures: No major pest and diseases are seen on the crop, hence no plant protection measures are required.

Harvesting and yield: The crop gets ready for harvest after 40 to 50 days of sowing. The crop should be harvested above the ground level when the plants begin to bloom and seed setting to get good rutin production. Delayed harvesting causes sharp decline in rutin content from 3.8 to 6.0 per cent. The stems and leaves contain 88 -90 per cent moisture.

An average fresh herb yield of 50-60 q which in turn may yield up to 5-6 q dry yield ha^{-1}, may be obtained.

References

Satish, G and Rajangam, J. 2012. A rutin yielding buckwheat. *Agrobios News Letter* 11(2): 50.

7

Carrot (*Daucus carota* sub sp. *sativus* Hoffm. Schubl. and G. Matens)

Common name: Gajjar

Family: Umbelliferae

Introduction

Radish, turnip and carrot, are used in India in abundance, but not many people are quite aware that their green parts (leaves) are rich sources of minerals and vitamins. The portion of green leaves discarded as waste material is in fact most nutritious-the darker the leaves, the greater are its nutrients content.

Carrot being a winter vegetable not only adds minerals and vitamins to our diet but also acts as a store house of other nutrients which are considered good for health. Apart from this, carrots are found to contain carotenoids which serve as an antioxidants and help in combating a number of diseases. Recently, it has been found that carotenes prevent the appearance of certain cancers. Here the list of medicinal values of the vegetable carrot is made available. Its cultural practices are also detailed.

Medicinal Values

Wild carrot is more strongly aromatic than the cultivated one. It finds number of uses in house remedies. Infusion of root is diuretic, stimulant and highly effective in urinary problems.

Cultivated carrot finds hundreds of medicinal values. Infusion of roots is effective in thread worm infection. It increases quantity of urine, helps in elimination of urea and uric acid from blood. In diet it maintains nitrogen balance and reduces blood sugar and blood pressure.

Roots are made into poultice in ulcers and swellings. Rich in Vitamins A, C and D, and is a tonic for eyes. Vitamin A is found in three forms such as vitamin A_1, vitamin A_2 and vitamin A_3, (Gupta and Mahajan, 2012). It is rich source of calcium, phosphorus, iron as well as carbohydrates, proteins and mineral matter (Table 7.1). In addition to this, it is good source of thiamine, vitamin, riboflavin and vitamin D. Carrot seed is good for kidney troubles. Leaves and other wastes make excellent cattle feed.

Table 7.1: Ingredients in Carrot

Sl.No.	*Ingredient*	*Per cent*
1.	Carbohydrates	10.70
2.	Protein	0.19
3.	Mineral matter	1.10
4.	Phosphorus	0.50
5.	Calcium	0.08
6.	Iron	1.50

Source: Mittal, 2005.

The scientists have now found out that eating carrots are rendered more attractive (Anonymous, 2012). Their research shows that men and women whose skin has a yellow glow are thought to be particularly attractive and healthy. Healthy and yellow pigments found in the carrot play a key role in giving skin that hue. The study further indicated that just two months of increased consumption could produce visible results.

Carrot has not only proved beneficial to health but its regular consumption also reduces three times heart attack risks (Anonymous, 2010). It has recently been investigated that the presence of β-carotene helps in controlling the cholesterol. Presence of fibre in the carrot saves the human beings from cancer as has been discovered by the Britian scientists, especially lung or mouth. The carrots are loaded with eye, skin and hair enriching nutrients (Mohindra 2014).

Carotenoids found in carrot are antioxidants which help soak up damaging compounds produced by stresses and strains of everyday living life, especially when the body is combating disease. Carotenoids are also important for our immune and reproductive system. Seeds of carrot are aromatic, stimulant, diuretic, carminative and are used in dropsy and kidney diseases.

Other Uses

Carrot is characterized by soothing healing nourishing properties and easily, digestible sugar. Purifies blood, helps semen production and make it thick. Tends to reduce the hazards of infection in child birth, eliminates uric acid from blood, possesses antiseptic properties and for curing chronic diarrhea and colitis. It is also a remedy for gallstone troubles, cleans liver, increases vigour and vitality. Juice, a natural solvent cures ulcers and constipation, jaundice and destroys worms.

Botanical Description

Carrot is grown all over the world and is liked by all. It is generally of yellow colour (Mittal, 2005). However, red, orange and black varieties are also common and all are used for table purpose. Black carrot contains lot of iron and is mainly used for preparing Kanji (an Indian type of decoction) which is used for digestive purposes. Carrot is a root crop, which exists both as wild carrot and cultivated carrot. Wild carrot is more strongly aromatic than cultivated one. In Sanskrit, it is known as *Garijara* in Sanskrit. Cultivated carrot is descendent of a wild ancestor.

Botanically carrot is edible, biennial herb, mainly cultivated for erect taproots growing up to 5-30cm and branched. Leaves are alternate, finely divided, lowest ones larger than upper and in rosette of 8-12 above ground surface. Leaves embrace the stem with a sheathing base. Flowers (white, red or purple) are small (2mm), dense, terminal umbel and appear in May-June. Fruit is flattened.

Cultural Practices

Climatic requirements: The climatic requirement of carrot is quite akin to radish *i.e.,* cool season crop. Although carrot can tolerate high temperature of 18.3 to 23.9 ° Celsius yet its roots develop a good colour under the temperature at 15 ° Celsius. Seedlings are very sensitive to temperature extremes, higher temperature may kill the plant whereas, that below 10 C (50 F) tend to cause the development of longer, paler roots. The plants also require adequate moisture and organic matter.

Soil requirements: Deep, friable and well drained loamy to sandy loam soils are the best for the growth of carrot. These should not become hard after rains or irrigation. The soil must be rich in organic matter and pH in the range of 5.5 – 7.0.

Sowing time: In tropical and sub-tropical areas August to October is the suitable time for sowing of Pusa Kesar variety whereas, for Nantes it varies from October to November.

In sub-temperate zone, the sowing time ranges from August to September for Pusa Kesar variety and September to October for Nantes. Contrary to this, the time of sowing in temperate zone is April to August for these varieties.

Carrot is direct seeded. Prior to sowing, the seeds should be gently rubbed with hand and then sown about 1.5 cm deep on 45 cm apart ridges. The seed rate is 10-12 kg ha^{-1}.

Land preparation: As seed of carrot is directly sown in the field so land must be prepared thoroughly to obtain good tilth after one deep ploughing followed by 2-3 shallow ploughings. Clods are broken and field levelled by planking.

Manures and fertilizers: The carrot crop is a heavy feeder. Apply 30 tons of well decomposed farm yard manure or compost before sowing *i.e.*, while preparing the land. Farm yard manure is supplemented with 60 kg N, 30 kg P and 50 kg of K ha^{-1}. The whole quantity of phosphatic and potassic fertilizers and one half of nitrogenous fertilizer should be applied along with farm yard manure or compost. Rest of the nitrogen should be added in two splits, one after the first hoeing and the second after 30 days of germination.

Intercultural operations: Little hoeing is needed after plants become 8-10cm high. Weeding and hoeing should not be done during dry weather, because the roots are exposed to the dry winds causing loss of moisture and, thereby, arresting growth and damage to the roots.

Thinning: When the plants are well established, thin them to 3-4cm apart to ensure proper development. Thinning should be done in moist soil.

Irrigation: The sowing should be done in optimum moisture conditions. The first irrigation should be given 5 days after sowing and the second 3-4 days, thereafter subsequent irrigations are provided before any wilting appears.

Harvesting and yield: Start digging when the roots are nearly 2.5-3.5 cm in diameter at the upper end. Before harvesting, light irrigation is given. Tops are removed and roots are dug with a spade.

Generally Asiatic type (Pusa Kesar) gives 200-300 q ha^{-1} and European ones (Nantes and Golden Heart) yield 100-150 q ha^{-1}.

References

Anonymous. 2010. Eating carrot reduces 3 – times heart attack. *The Hind Samachar*, Jammu. 4(70): 10 (Friday, December 31).

Anonymous. 2012. Carrots make people more attractive. *The Kashmir Times*. 69 (126): 13 (Tuesday, May 8).

Gupta, R.D. and Mahajan, N. 2012. 14 Dietary Tips to prevent cancer. *Ayurveda*. (21): 10-11.

Mittal, S.R. 2005. Carrot is useful in diseases. *Ayurveda*. December Issue, pp 3-4.

Mohindra, V. 2014.7 vegetables you should eat to boost your disease fighting potential. *Ayurveda*.1 (26): 18.

8

Clove (*Syzygium aromaticum* L. Merrill and Perry)

Common name: Laung

Family: Myrtaceae

Introduction

Cloves are native to the Maluku or Mollucca Islands in Indonesia and used as a spice in cuisines all over the world. They are mostly found in Indonesia, India, Madagascar, Zanzibar, Pakistan and Srilanka.

Cloves have been historically used in Indian cuisine (both North India and South India). In North Indian cuisine, cloves are used in almost all families, rich and poor especially in spicy dishes as an ingredient of a mix named as a "Garam Masala" along with other spices *viz.*, cumin, black pepper, coriander etc. Although it is neither an everyday ingredient for home cuisine and nor it is used in summer very often yet it has a number of medicinal values in Ayurvedic System of Medicine Chinese Medicine and western Herbalism and Dentistry.

The clove is an indigenous plant species of the former Spice Islands, the Mollucca Islands which are now in Indonesia as well as the southern parts of the Philippines Islands. Owing to their commercial value and high sale potential, large plantations of cloves exist in the tropical countries like Tanzania and Madagascar, and to some extent in Brazil and the West Indies.

Medicinal Values

In dentistry, oil of cloves is used as an anodyne (pain killer) for dental emergencies. Cloves are used as a carminative, to increase hydrochloric acid in the stomach and to improve peristalsis. Cloves are also said to be a natural athelmintic. The essential oil is used in aromatherapy when stimulation and warming are needed, especially for digestive problems. Topical application over the stomach or abdomen are said to warm the digestive tract. It also helps to decrease infection in the teeth due to its antiseptic properties.

Cloves can be used internally as a tea and topically as oil for hypotonic muscles, including for multiple sclerosis. This is also found in Tibetan medicine. Some recommended, avoiding more than occasional use of cloves internally in the presence of 'pitta' inflammation such as, found in actule flares of autoimmune diseases. In West Africa the Yoruba's use cloves infused in water as a treatment for stomach upsets, vomiting and diarrhoea. The infusion is called 'Ogun Jedi-Jedi'. Use of clove may reduce blood sugar levels. The buds have anti-herpes virus properties. Western studies have supported the use of cloves and clove oil for dental pain. However, studies to determine its effectiveness for fever reduction, as a mosquito repellent and to prevent premature ejaculation have remained inconclusive.

Cloves have been reported of killing harmful bacteria present in the pancreas and intestines (Mohindra, 2010). It increases the white blood cells and, thereby, the resistance power of the body. It also stops bad breath and works as pain killer in different parts of the body.

According to Chinese medicine, cloves have hot and moving properties that relieve arthritis pain caused by cold and stagnation (Sood, 2009). Cloves contain a phyto-chemical that interrupts the pathways of protein complex in the body that has been linked to inflammatory diseases such as arthritis.

Non-Culinary Uses

The spice is used in a type of cigarette called Kretek in Indonesia. Kreteks have been smoked throughout Europe, Asia and the United States. During 2009, clove cigarettes (as well as fruit and candy flavoured cigarettes) were out lawed in the United States. However, they are still sold in similar form, relabeled as, filtered clove cigars. Cloves are also an important incense material in Chinese and Japanese culture. And clove essence is commonly used in the production of many perfumes. During Christmas, it is a tradition in some European countries to make pomanders from cloves and oranges to hang around the houses. This spreads a nice scent throughout the house and serves as holiday decorations. Cloves are often used as incense in the Jewish practice called Havdala(Kulkarni, 2012).

Other Uses

Dried cloves constitute a key ingredient in Indian masala, spiced tea, a special variation of tea popular in some regions, mostly in Gujarat. In United States, it is often sold under the name of Chai or Chaitea, as a way differentiating it from other types of teas sold in the United States. In Mexican cuisine, cloves are the best known as clavos de olor and often used together with cumin and cinnamon. In American cooking, it is often used in sweet breads such as pumpkin or zucchini bread along with other sweet spices like nutmeg and cinnamon. Due to the Indonesian influence, the use of cloves is wide spread in the Netherlands. Cloves are used in cheeses, often in combination with cumin. Cloves are an essential ingredient for making Dutch speculaas. Furthermore, cloves are used in traditional Dutch stews like hachee (Kulkarni, 2012).

Botanical Description

Cloves are the aromatic dried flower buds of a tree. In Sanskrit the cloves are known as Lavangaha. The clove is an evergreen tree that grows to a height ranging from 8-12m, having a large straight trunk, leaves and sanguine flowers in numerous groups of terminal clusters. The flower buds are of a pale colour in the beginning but gradually become green, after which they turn into a bright red, when they are ready for collection.

Freshly collected cloves are pink in colour, on drying; they turn into rust brown colour. They are 12-16mm in length and resemble with small nails, with a tapered stem. The large end of the clove is the four pointed flower bud. Being extremely hard, it is difficult to grind cloves with a mortar and pestle so an electric grinder is required for this purpose (Srivastava *et al.*, 2012).

Cultural Practices

Climatic requirements: The clove plant is cultivated at different places having a tropical climate around the world. All the regions surrounding the Indian Ocean are good areas for growing clove as they enjoy the optimal tropical soil conditions.

Soil requirements: The best growth of clove has been found in the river bed soils known as alluvial soils. They vary in texture ranging from sandy loam to loam and silty loam to clay loam. The soils must be well drained. As clove does not grow well in water logged soils, so clove in such soils should be avoided.

Propagation: Semi-ripe cuttings of the plant are used in the summer to propagate the plant, while seeds are used during spring to propagate the tree. The young flower buds are plucked and harvested for sun drying and processing two times every year in most of the plantations.

Harvesting and yield: Cloves are harvested when 1.5-2.0cm long consisting of a long calyx, terminating in four spreading sepals and four unopened petals, which form a small ball in the centre (Kulkarni, 2012). The time for the harvest of cloves occurs when the lower parts of the clove flower turn from green to purple. At this ripened stage, all the flower buds are collected and they are then sun dried into the familiar deep reddish brown spice.

Under normal conditions, the yield of the clove has been recorded to the extent of 400 kg ha^{-1}. Normally a full grown tree (> 7 years old) yields 2 kg cloves.

References

Kulkarni, B. 2012. Advantages of cloves. *Agrobios News Letter* 11(1): 48-49.

Mohindra, K. 2010. Floral remedies. *Ayurveda.* 1(18): 14-15.

Sood, S. 2009. 10 Spices that heal cancer, diabetes and more. *Ayurveda.* 1(15): 22-23.

Srivastava, N., Mehrotra, A. and Singh, N. 2012. Clove and clove oil (Eugenol)-Medicinal values. *Indian farmer's Digest.*45 (1): 22-25.

9

Coriander
(*Coriandrum sativum* L.)

Common Name: Dhaniya

Family: Apiaceae

Introduction

Coriander is a strong scented annual herb valued for it its tender green leaves and seeds. Its origin is believed to be Eastern Europe and Asia. Coriander is cultivated not only in India but is grown in Russia, Balkan Slates, Burma, Pakistan and North Africa. In India, coriander is cultivated in all states. However, Punjab, Maharashtra, Uttar Pradesh, Madhya Pradesh, Karnataka are leading states, where it is being grown on a large commercial scale. The cultivation of coriander is so easy that it can be grown even in the kitchen garden for fulfilling the requirement of its fresh leaves and seeds. The oil from coriander seeds is used in various kinds of perfumes and cosmetic products. Above all, coriander is used in traditional medicine for digestive, antispasmodic properties. In fact, coriander is an important medicinal herb. As a matter of fact, the use of coriander as medicinal herb has been reported since 1500 BC. Some of the most important medicinal values of this herb/plant are as follows. The nutritional value of coriander is also given vis-à-vis cultural practices for its cultivation.

Nutitional Value

As far as nutrition value of coriander is concerned, 100g of fresh leaves of this herb provides moisture to the extent of 87.9 percent, protein 3.5 per cent, fat 0.6 per

cent, carbohydrates 6.5 per cent, mineral matter 1.7 per cent, Ca-0.14 per cent, P-0.06 per cent, Fe-10mg. The herb is an excellent source of vitamin A-10,000 to 12,000 IU and vitamin C-150-250 mg. Fruit yields a volatile oil to the order of 1 per cent, fixed oil 1.3 per cent, fatty matter 1.3 per cent, mucilage tannin, malic acid and ash 5 per cent. In addition to this, coriander oil contains coriandrol-an alcohol, 2-d pinene, 1-pinene, geraniol and baborneal.

Medicinal Values

The fruit paste of coriander is used for skin and mouth ulcers. Prior to inventing of tooth paste, coriander seeds used to be chewed as breath sweetener. In medicine, it is used as a carminative, antispasmodic, refrigerant, diuretic and aphrodisiac. It is too used against seasonal fever, stomach disorder and nausea as household medicines.

For relief from the rheumatism pain, pound the seeds and mix with hot water to make a paste and then apply to the affected area. As a gargle, they are useful in thrush and as a poultice; with barley meal added to it they are applied to chronic ulcer and carbuncles. Fruit decoction may be used as eye wash for preservation of the eye sight in small pox and also in chronic conjunctivitis. Coriander is used traditionally for digestive and gastric complaints as well as for cough, chest pain, bladder complaints, leprosy rash, fever and dysentery, externally for headache and oral disorder.

Tea or decoction prepared from seeds of coriander is used for curing stomach upset and relieves flatulence. Chinese herbalists use coriander seeds to treat indigestion and stomach aches. They also suggest that coriander herb can be used to treat influenza in which there is no sweating. Coriander seed oil is an aromatic stimulant and a digestion stimulating agent. It is generally beneficial to the nervous system. It is used in loss of appetite and eliminating bad mouth odour. It helps

in digestion and relief from burning sensation of all over the body. Decoction of coriander with dry ginger in equal amount corrects indigestion, malabsorption etc. (Chaudhuri, 2006).

Coriander is primarily used for curing severe *Streptococcal* skin infections, which caused many deaths before the advent of antibiotic. In modern medicines, coriander is used as a flavouring agent in them and a stomach soothing addition to more irritating compounds. Its main use is in masking foul medicines, especially purgatives, as it has semi gripping qualities.

Coriander has long history as a traditional medicine with cholesterol lowering and blood sugar lowering effects. Several studies have demonstrated reduction of total cholesterol, bad cholesterol and increase in good cholesterol (Khosla, 2013).

Domestic Usage

The whole young plant is used in preparing chutney. Its leaves are used for flavouring curries, sauces and soups. The dry seeds/fruits of the coriander are used in preparing curry powder, pickles, spices and seasonings.

The stem, leaves and fruits have a pleasant aroma. This is the reason why coriander leaves are used in many popular dishes to enhance their flavour and taste. These leaves also have antiseptic and carminative properties. Apart from this, the coriander leaves contain a number of phyto-chemical compounds like quercetin and essential oils.

Botanical Description

The common name of coriander in Hindi is *Dhaniya* or *Dhania*. The coriander plant is an annual herb, with erect stems with many branches, grows up to the height of 30 cm or even more up to 90 cm. The plant is bright green, shining and glabrous. The lowest leaves are stalked and pinnate, the leaflets roundish or oval, slightly lobed. It bears small, white or pinkish purple flowers borne on compound terminal umbels. Fruit is oval shaped to globular, 3 to 4 mm in diameter, when pressed break into two locules each having one seed. Fruit is aromatic, with pale white to light brown seed. The seed clusters are symmetrical and seeds fall as soon as ripe.

Cultural Practices

Climatic requirements: Coriander is mainly cultivated in tropical to sub-tropical regions but even grows in sub-temperate climate conditions. It favours frost free climate, especially during flowering and seed formation. Moderate cool and dry weather at the time of seed formation increases the quality and quantity of the crop.

Soil requirements: Coriander can be grown on different kinds of soils but it thrives best on loamy and sandy loam soils. The soils must be rich in organic matter and free from water logged conditions. It can also be grown in black cotton and laterite soils as well as grey brown podzolic and brown hill podzolic soils.

Sowing time and preparation of field: Preparation of field is done by 2 to 3 ploughings followed by 1 to 2 harrowing. Although coriander can be grown from August-September in *kharif* for leaves, yet the best time for its sowing is October. However, the crop sown in August-September requires great attention against monsoon disorders. Seed rate for cultivation of coriander varies from 15-25 kg ha^{-1} depending upon climate conditions of the area.

Varieties: Commonly grown varieties in India are: S-81, NP-92, Pusa Selection-260, Cm-490, CS-193, 533 etc. T5365 is a high yielding and wilt resistant variety.

NRCSS-Acr-1, Gujarat Corinader-1 and 2, CO-1, Co-2, CO-3, CS-287, Rajendra Swathi, Rajendra Sonla, Sadhva (Cs-4), Swathi (CS-6), Sindhu (CS-2), RCr-41, RCr-20, Rcr-435, RCr-436, RCr-684, Rcr-446 etc. are the other varieties of coriander which are grown on other parts of the country (Malhotra and Vashishtha, 2008).

Method of sowing: Seeds are generally sown by broad casting or in rows, maintaining row to row distance of 30cm and plant to plant of 10cm. Before sowing the seeds are crushed into two parts or halves by rubbing and soaked in water for few hours for getting good germination.

Manures and fertilizers: The requirement of manures and fertilizers varies according to chemical composition of the soil. On an average, it is profitable to apply FYM@10-25 tonnes ha^{-1}. It should be applied at the time of sowing. Farm yard manure must be well decomposed *i.e.* brown or black brown in colour and amorphous in nature. Use of chemical fertilizers should be discouraged in the light of the increased demand of organically grown food by the consumers.

Irrigation: Irrigation depends upon the condition of the soil *i.e.,* moisture holding or retaining capacity of the soil. However, field should be irrigated at an interval of 15-20 days.

Weeding and hoeing: Weeding and hoeing should be done at regular interval to make field weed free.

Harvesting and yield: For garnishing purpose, the leaves and stems are nipped carefully after they attain 8-10cm height. The crop is ready for harvesting a month after the flowering. The seeds should be harvested when 50 per cent of the seed turn yellow. After drying, the seeds are threshed/separated by light beating with stick and winnowing.

On an average seed yield of coriander crop is 12-18 q ha^{-1} or even more up to 15-22 q ha^{-1}. The yield of the leaves varies from 12-15 q ha^{-1}.

References

Chaudhuri, S. 2006. Cheaper 'C' s. *Ayurveda*, April, 2006.

Khosla, I. 2013. Spices of life. *The Tribune*. 133 (276): 3 [The Sunday Tribume Spectrum, 6 October].

Malhotra, S.K. and Vashishtha, B.B. 2008. Package of practices for Productivity of Seed Spices. National Research Center on Seed Spices (ICAR), Tabji, Ajmer, Rajasthan, India.

10

Cumin
(*Cuminum cyminum* L.)

Common name: Safed jira, zeera

Family: Umbellifereae

Introduction

Cumin has been grown and used as a spice since ancient times. Originally cultivated in Iran and Mediterranean region, cumin is mentioned in the Bible in both the old Testament Isaiath 28:27 and in the New Testament Matthew 23:23. It was also known in ancient Greece and Rome. The Greeks kept cumin at the dining table in a container just as pepper is frequently kept today and this practice continues in Morocco. It is now mostly grown in Iran, Morocco and Egypt, India, Syria, North America and Chile.

In India, cumin is culivated in 1,63,688 ha with production of about 6,500 tonnes. Its sizeable quantity is exported to USA, Singapore, Japan, Saudi Arbia and UK for earning foreign exchange. It is mainly cultivated in the states of Gujarat and Rajasthan. In Rajasthan, cumin is mostly grown in the districts of Tonk, Ajmer, Nagour, Pali and Barmer. In Jammu and Kashmir, black coloured cumin (kala Zeera) botanically called *Bunium persicum*, a high value but low volume crop is collected from forest areas in the hilly areas of Paddar, Machil-Kishtwar district and Doda district in Jammu Region and Srinagar, Budgam, Anantnag Pulwama districts in Kashmir Region and Kargil in Ladakh Region (Singh, 2008).

The aromatic flavour in the cumin seed is due to the presence of a volatile oil, the percentage of which varies with the varieties, geographical types and climate. The volatile oil content in Indian varieties varies from 2.52 to 3.0 per cent (Choudhary *et al.*, 2000). The chief ingredient of volatile oil is cumin aldehyde. Besides, curcumin is also present in cumin seed. Both these constituents present in cumin bless this spice with medicinal values.

Medicinal Values

It is no surprise to many spice researchers that cancer rates are lower in India, where cumin is a staple diet. Studies show that cucurmin in this spice inhibits the enzyme that help cancer cell invade healthy tissue and also keeps tumors from developing the new blood vessels that assist them to grow (Sood, 2009).

The fruit contains 2.5 to 4 per cent essential oil, which is composed of cumin aldehyde, perilla aldehyde, cumin alcohol, dipentene and p-cymene. In toasted cumin fruits, a large number of pyrazines has been identified as flavor compounds. Various alkyl derivatives like dimethyl parazine and a sulphur compound was found (Johari, 2009).

Traditional Uses

Cumin seeds are used as a spice in cooking. They are medicinally used in hoarseness of voice, dyspepsia and diarrhoea. Seed powder reduces itching and pain in skin disorders. Clinically it has been tried against jaundice and also used as galactologogue. In debility due to fever, it improves appetite and energy (Johari, 2009). Cumin is used as warming oil and help in relieving muscular pains and also arthritis. In the digestive system, cumin is a stimulant and helps in colic, flatulence and indigestion. For the nervous system, it is a tonic and has a beneficial effect on headaches, migraine and nervous exhaustion. Cumin oil shows in varying degree of anti-microbial and anti-fungal activity at different concentrations and also in combinations with other essential oils. The constituents of essential oil are known to possess carminative properties (Johari, 2009).

Botanical Description

Plant is a small flowering annual usually 30-50 cm in height. Its stem is slender and branched, rarely exceeding 1 foot (30 cm) in height and somewhat angular. Leaves are dark green in colour alternate, compound, divided into long, narrow segments. The flowers are small, pink or white, in stalked umbels and bloom in June and July. Fruits are cremocarp often separated into mericarps brown with light coloured ridges, elongated about 4-6 mm long, 2 mm wide, tapering at ends and slightly compressed laterally. Taste is richly spicy. Seeds are light in colour and with bristles instead of being smooth.

Cultural Practices

Climatic requirements: Dry and moderately cool climate is the best suited for cumin cultivation. Areas having high humidity and frequent rains particularly after flowering stage are not suitable for cumin crop as it increases the incidence of diseases like blight and powder mildew.

Soil requirements: Well drained, loamy soil is the best for cumin cultivation. However, it can also be grown in medium heavy i.e clay loam and moderately saline soils.

Field preparation: The soil should be brought to fine tilth by ploughing the field twice or thrice followed by planking. If soil moisture level is low, a pre -sowing irrigation should be given.

Varieties: 'RS-1' and 'RZ-19' are the main improved varieties. The seeds of the former are bold and hairy. The crop matures in about 110-120 days. This variety being susceptible to wilt must be sown in those areas where the incidence of wilt is less. 'RZ-19' variety is suitable for all areas of Rajasthan. This variety matures in about 125 days. It is a less susceptible to diseases (wilt, powdery mildew and blight).

Sowing time and method of sowing: Crop can be sown from mid November to first week of December. However, to obtain the best results cumin should be grown around 15th of November. Cumin is generally sown by broadcast method. The field after preparation is divided into beds of convenient size. The seeds are then broadcasted and lightly covered by soil by running the iron rakes in the beds. For convenience of intercultural operations the line sowing is better. For line sowing shallow furrows are drawn in the beds at a distance of 22.5 to 25.0 cm with the help of iron or wooden hooks. After putting the seeds in the furrows they are covered by running the rakes lightly in the beds. An optmum plant population can be achieved using seeds @ of 12-15 kg ha^{-1}.

Manures and fertilizers: Cumin crop requires no additional manuring if 10-15 tonnes of FYM ha^{-1} has been applied to preceding kharif crop otherwise 10 to 15 tones FYM or compost may be incorporated into the soil at the time of field preparation. In addition to this, application of 20 kg phosphate ha^{-1} at the time of sowing and 30 kg N ha^{-1} through top dressing in the standing crop either all in single dose at 30 days after sowing (DAS) or in two equal splits-one at 30 DAS and the remaining at 60 DAS is recommended to achieve best results.

Weed control: Because of slow growth and short stature, the cumin crop faces severe weed competition at all stages of growth. At least two weedings and hoeings are essential to check the weeds and for proper aeration of the soil. First weeding should be done at about 30 DAS and second at 50 to 60 DAS. At the time of first weeding, thinning should also be done to remove excess plants.

Irrigation: First light irrigation is given immediately after sowing to wet the surface soil. Second irrigation should be given 8 to 10 days after first irrigation. Subsequent irrigations are given at an interval of 15 to 25 days depending upon the weather conditions and soil type. At the time of grain formation, give the last heavy irrigation.

Pests and diseases: Always use healthy and disease free seeds obtained from wilt free crop. A three crop rotation *i.e.,* cluster bean-cumin, cluster bean-wheat, cluster bean-mustard was found to be a better crop sequence to reduce the wilt disease in cumin. Use of biopesticide *i.e.,* trichoderma @ 4.0 g kg^{-1} during summer also reduced the disease incidence. Similarly certified seed, free from blight and powdery mildew should be used to lessen the attack of these diseases. Moreover, if necessity arises, use of bio-pesticides be made to control the aforesaid diseases.

Harvesting and yield: Cumin crop matures in about 130 to 150 days. Harvesting can be done by uprooting the plants but cutting by sickles is recommended to prevent the contamination of seed with sand or dung which may spoil the quality of the produce. Harvested plants are stacked on the clean threshing floor or tarpolin for drying in the sun. After drying, the grains are separated by light beating with sticks followed by winnowing. The clean seeds are then filled in gunny bags and stored in damp free well aerated store houses.

The average yield of cumin varies from 5 to 8q ha^{-1}. In Rajasthan the yield of cumin is recorded to be about 5q ha^{-1} whereas, in Gujarat it ranged between 7-8 q ha^{-1}.

References

Choudhary, G.R, Sharma, R.K and Jain, M.P. 2000. Production technology of cumin. *Intensive Agriculture.*37(11-12): 19-22.

Johari, R. 2009. Cumin-A spice from Bible. *Ayurveda.* 1(12): 37.

Singh, H. 2008. Medicinal plants of western Himalayas as prevalent in Jammu and Kashmir State and their economic exploitation. (Personnel Communication), Talab Tilo, Jammu.

Sood, S. 2009. 10 spices that heal cancer, diabetes and more. *Ayurveda.* 1(15): 22-23.

11

Deadly Night Shade (*Atropa belladonna* L)

Common name: Belladona

Family: Solanaceae

Introduction

Belladona is a sub-shrub, which occurs mainly in Russia, Italy, Romania, Pakistan and India. It is also naturalized in parts of North-America, where it is often found in shady moist locations with limestone rich soils. In India, it is found in the forests of Western Himalayas, extending from Shimla to Kashmnir. In Kashmir, it is mainly confined to the Lolab Valley forests. However, due to over exploitation, the naturally occurring belladona herb has become a rare species (Khajuria, 2005). Like other temperate herbs, it has lot of medicinal uses.

Medicinal Values

All plant parts, root, leaves and top is used officially. The products from the leaves of the plant are well known to produce hallucinogenic effects and to cause dilation of the pupils (used by ladies in Italy to give softness to eyes) as reported by Raghav (2003).The alkaloids responsible for these activities are 'atropine' and 'scopolamine' which vary considerably in roots and leaves. The use of the plant is also made in the treatments of toothache and for inducing sleep (narcotic and sedative in action). Atropine has been used for the effective treatment of people exposed to cholinesterase inhibiting nerve gases. Atropine from hyoscyamus is used as a premedication for anesthesia in gastro-intestinal disorders. The leaves and roots are strong painkillers, sedative and provide relief in rheumatism as well as neuralgic pains. Roots of belladonna are also narcotic, diuretic, antispasmodic, mydriatic and anodyne.

Belladonna is a most valuable plant in the treatment of eye diseases. Locally applied, it lessens irritability and pain, and is used as a lotion, plaster or liniment in cases of neuralgia, gout, rheumatism and sciatica. It is of value in acute sore throat, and relieves local inflammation and congestion.

Botanical Description

Belladona is an herbaceous, perennial often growing as a sub-shrub from fleshy root stock. Plant grows to 1.5 m with 18 cm long ovate leaves. Leaves are dull, darkish green in colour. The lower leaves solitary, the upper ones in pairs alternately from opposite sides of the stem, and one leaf of each pair much larger than the other, oval in shape, acute at the apex. The whole plant is glabrous, when crushed gives a disagreeable odour and the leaves have a bitter taste, when both fresh and dry. Well shaped flowers are purple and faintly scented. Flowers appear in June-July and continue to bloom till September. The fruits called berries are approximately 1.0 cm in diameter. They are sweet and consumed by animals and the seeds contain toxic alkaloids having poisonous properties. Cases of belladonna poisoning have been frequently reported. It contains several seeds. The root is thick, fleshy and whitish, about 6 inches long, or more, and highly branched.

Cultural Practices

Climatic requirements: Belladona thrives best under cool and humid climatic conditions. Frequent sunny conditions and winds have been found harmful for its growth.

Soil requirements: Belladona can be grown on variety of soils including gravelly and stony. However, its best growth takes place under well drained, sandy loam or loamy sand soils.

Propagation: The crop is propagated through seeds sown on the raised beds using 1.0 kg seed 100 m^{-2} during June-July. The so raised plants are then transplanted during August-September when seedlings attain the height of about 15-20 cm. Transplanting is done in the field at a space of 60x60 cm or 60x45 cm. Germination takes several weeks under alternative temperature condition but can be speeded up using gibrellic acid (GA).

Manures and fertilizers: Farm yard manure @ 2 tonnes ha^{-1} is added in the field at the time of transplanting. Recommended doses of N and P are 80, 40kg ha^{-1}. All of phosphatic fertilizer and one fourth of nitrogen is applied at the time of transplanting and the remaining nitrogen in 3 splits depending up on the growth condition of crop.

Flowering: Flowers appear in June-July singly and continue blooming up to September.

Yield: Average yield of fresh herb after 2nd and 3rd year varies from 4-6 tonnes and 12.5 to 15.0 tonnes ha^{-1}, respectively. Average yield of dry roots is upto 1 tonne per hectare in 3rd year.

References

Raghav, C. S. 2003. Medicinal plants on the touchstone of Science. *Science Reporter.* February Issue, pp 52-55.

Khajuria, G.L. 2005. Protecting medicinal and aromatic plants. *The Kashmir Times.* 62 (339): 13.

12

Desert Indian Wheat (*Plantago ovata* Forssk.)

Common name: Isbgol

Family: Plantaginaceae

Introduction

Isabgol is indigenous to the Mediterranean region and West Asia. It was introduced in India long ago, and is now mostly cultivated in Gujarat and parts of Rajasthan as well as some parts of Madhya Pradesh. Its other common names are Ishpaghula, Isabghil, Spogel seed and Ispaghal.

The word 'Ispaghal or Isapghol' has been derived from the Persian, Isap meaning horse and ghol meaning ear. Thus the name Isapghol means literally, the ear of a horse. In fact, the resemblance of the seed to a horse's ear is remarkable.

Medicinal Values

The dried seeds and husk of the plant are used as emollient, demulcent and laxative in the treatment of chronic constipation, amoebic and bacillary dysentery and diarrhoea. A decoction of the seeds is prescribed for coughs and colds. The crushed seeds are made into a poultice and applied to rheumatic glandular swellings. Extracts prepared from isabgol are useful for the food and pharmaceutical industries. Moreover, it is used as a basic stabilizer in the ice-cream industry. Isabgol is a common additive in many of the high fibre breakfast cereals and bulk forming laxatives.

Botanical Description

Isabgol is a 10-15 cm tall short stemmed annual herb. Leaves; 8-20 cm long are born alternately on the stem. Flowers are in terminal spikes, oval shaped. Fruit capsules oval in shape and seeds are translucent and concavo-convex, white in colour. The husk covers the seeds.

Cultural Practices

Climatic requirements: Isabgol grows best under warm temperature regions. It requires dry but cool weather and is sown during winter season. Sowing during first week of November provides good yields. Early sown crop makes it vulnerable to downy mildew disease, whereas, late sowing gives lesser period of growth in winter along with possibility of shattering of seeds due to rains during April-May. At maturity, if the weather becomes humid, its seeds shatter resulting into reduction in the yield. Heavy or even a light shower will proportionately decrease the yield and sometimes leading to even total loss of the crop. The temperature requirement for maximum seed germination has been reported to be 20 to 30 degree Celsius. It thrives well in low humidity in winter months.

Soil requirements: It is an irrigated crop which grows well on light textured soils. Soils with poor drainage are not conducive for good growth of this herb. A silty loam soil with pH range from slightly acidic to slightly alkaline (pH 7.2-7.9) is the best for growth of plants and high yield. However, Isabgol can be grown in soils having pH upto 4.7 (NMPB, 2005).

Varieties: Varieties recommended for cultivation in the semi-arid regions in India are; G_1, G_2 and Niharika.

Land preparation: The field must be free from weeds and clods. The number of ploughings, harrowing and hoeing vary depending upon the soil conditions, degree of weeds infestation and previous crop grown. Generally 2-3 ploughings are sufficient to get maximum tilth of the soil. Well decomposed organic manure @ 25.0 to 37.5 tonnes of FYM, compost or vermi-compost is required to be mixed into the soil at the time of last ploughing. The field should be divided into suitable plots of convenient size, usually 8 m x 3 m size for light soils.

Sowing time and seeds rate: Seeds at the rate of 5.0 to 7.5 kg ha^{-1} are sown. To obtain higher percentage of germination seeds must be selected from the crop harvested at the end of the preceding crop season. Old seed tend to lose viability under ordinary storage conditions.

Generally November to December but sowing in the first week of December is ideal. As the seeds are small and light, these must be mixed either with sand or sieved well with decomposed farm yard manure and then sown by broadcast method. After sowing germination begins within four to seven days

Irrigation: Immediately after sowing light irrigation is necessary. First irrigation must be given with light flow or shower of water otherwise with fast current of

water most of the seeds will be swept to one side of the field and the germination will not be uniform. If the germination is poor, second irrigation is given as and when required. Last irrigation is given when the maximum spikes are at milk stage. Usually 5-6 irrigations are sufficient for getting a good crop.

Weeding and hoeing: Periodical weeding and hoeing is required. No weedicides should be applied to control weeds, rather mechanical operations are required to be followed.

Manures and fertilizers: If well decomposed farm yard manure @ 25-37.5 tonnes ha^{-1} has been added, there is no need to apply any kind of fertilizers. 50 kg N + 25 kg P_2O_5 per hectare is recommended. Nitrogen is applied in two splits; 1st as basal and second 30 days after sowing.

Harvesting and yield: The crops mature within 110-120 days, generally in March-April. When matures, the spikes turns yellowish in colour and are pressed even slightly. At the time of harvest, the atmosphere must be dry and there should be no moisture on the plant, harvesting will lead to considerable seed shattering. Hence the crop should be harvested in the morning. The produce is threshed and seeds are separated. On an average crop produces 10- 12 q/ha seeds and net returns of Rs. 35,000/- per ha. During processing, seed is passed through series of grinding mills to separate out husk from the seed. The total recovery of husk is around 30 per cent.

References

National Medical Plant Board (NMPB). 2005. Isabgol (*Plantago ovata*). Department of AYUSH, Ministry of Health and Family Welfare, Government of India, New Delhi.

Jamwal, J. 2007. Medicinal/aromatic plants and value addition. *The Kashmir Times*, Jammu. 70 (338): 13 (December 6, Thursday).

13

Fenugreek (*Trigonella foenum* L.)

Common name: Methi

Family: Leguminosae

Introduction

Leafy greens such as spinach, kale, dandelion greens, turnip tops and fenugreek or methi are all amazing vegetables that provide iron plus lots of vitamin-C, good for preventing anaemia and scurvy disease of gums in teeth. Apart from this, green leafy vegetables also contain vitamin A, Vitamin B_2 and a number of minerals like calcium, magnesium, potassium and phosphorous, which have high protective food value especially for growing children and pregnant women. Leafy vegetables also possess a number of medicinal values and such values of methi are given as under.

Medicinal Values

Methi is used as a green vegetable vis-à-vis as a pulse. It can also be used for preparing different traditional and ayurvedic medicines. Seeds of methi are mucilaginous, demulscents and diuretic, tonic, carminative, astringent, emollient and aphrodisiac. Its leaves are cooling in effect. Leaves are also used as poultice for external and internal swelling as well as to improve eye sight (Mahajan, 2011).

Germinated seeds of methi have proved useful in diabetes and also in fever. In leucorrhoea, 3-6 g fenugreek powder and 6-12 g of sugar candy with 60-100 ml of fresh cow milk is recommended to be taken twice in a day. Methi is also used to stimulate metabolism and to help controlling blood sugar levels. Its use helps in stimulating and enhancing insulin secretion. Methi improves oxidation of glucose and reduces cholesterol level. Seeds of methi when are given in the range of 25 to 100g daily, assist in diminishing hyperglycaemia in diabetic patients.

Other Uses

Seeds of fenugreek are often used in the 'tadka' of Indian dishes or their powder is added to lend a warm aroma to a dish. Fenugreek has also been used for long time to improve the production of breast milk. Australian researchers showed that fenugreek could improve the male libido. It is believed that fenugreek contains a compound called 'saponin' that improves the production of sex hormones.

Botanical Description

Methi is a common vegetable crop especially in North India where its leaves are mostly cooked with potatoes as a great dish. It is an annual, leguminous, aromatic herb, 40-60 cm high. Stem is single or branched at the base, erect, long and tender. Leaves pinnately tri-foliolate; light green with 2-2.5 cm long, leaflets toothed. Flowers are solitary, yellow-white appearing in leaf axils sessile in summer. Fruit is a curved pod, 5-7.5 cm long with a long persistent beak and 10-20 seeds. Seeds are cuboid shaped, yellow to brown coloured.

Cultural Practices

Climatic requirements: Methi is a cool season crop. It can withstand frost and freezing temperature but for a short period. Areas with continuous high rainfall are not suitable for its cultivation and therefore, must be avoided for its cultivation.

Soil requirements: It can be grown on a wide variety of soils. However, heavy soils especially clay loams are best suited for its growth. Favourable soil pH is between 6.5 and 7.5.

Land preparation and sowing time: For successful cultivation of methi land should be ploughed twice or thrice followed by planking for breaking of clods and making it well levelled.

Sowing time varies for different agro-climatic zones. However, generally it is sown during July to November. August to September is the suitable time for sowing in areas of sub-tropical zone. Contrary to this, August to September and March-April or May-June is the best periods for sowing in sub-temperate and temperate areas, respectively.

Manures and fertilizers: Well rotten FYM @ 15 tonnes ha^{-1} is required to be applied at the time of sowing. If FYM is not available vermicompost or other kind of compost can be applied.

Recommended doses of N, P_2O_5 and K_2O in the form of nitrogenous, phosphatic and potassic fertilizer are 60, 20, 20 kg/ha respectively. Apply phosphatic and potassic fertilizers along with one fourth nitrogen at the time of sowing. The rest of the nitrogen (15 kg N ha^{-1}) should be applied after every cutting to obtain green, tender and succulent leaves.

Inter-culture and irrigation: One or two shallow hoeings are given. Sufficient soil moisture at the time of seed sowing is essential for quick and proper germination. Irrigation after each cutting is required to be applied.

Harvesting and yield: The young green shoots are nipped off in about one month after sowing. The picking of leaves is done by nipping off about 2 cm above ground level leaving the stubs. The stubs again produce new shoots.

The green leaf yield of 'Kasuri' type varies from 90-100 q ha^{-1} while the common type produces in the range of 70-80 q ha^{-1}. One crop of prickly seeded variety in a season gives approximately 3-4 cuttings with total yield of about 50-60 q ha^{-1} whereas, round seeded variety comparatively provides more number of cuttings which as a result give more yield ranging from 100-125 q ha^{-1}.

References

Mahajan, N. 2011. Eat vegetables for good eye sight. Daily Excelsior. 47(301): 4 (Magazine, October 30, 2011).

14

Garlic
(*Allium sativum* L.)

Common Name: Lasun, thome

Family: Liliaceae

Introduction

Garlic is a bulbous lilaceous plant possessing a pungent taste and very strong fragrance. Thus, garlic too was one of the precious nine gems bestowed upon the planet Earth to sustain life.

The Ayurvedic System of Medicine has very nicely ascribed qualities to garlic plant. These qualities compose both of prophylatic and curative values. Garlic has been used for treating indigestion, dysentery and flatulence and in curing nervous ailments. Furthermore, it has a prominent use in geriatrics to prevent aging and to promote the longevity of life.

In ancient Egypt and Greece, herbal physicians used to prescribe garlic against infections produced by different pathogens. The Chinese also had mentioned the name of garlic in their indigenous pharmacopoeia.

Garlic is an important spice crop of both vegetarian and non-vegetarian population. In spite of its spicy uses, garlic has medicinal values too. Madhya Pradesh leads in national production. Garlic is commercially cultivated in Malwa region of Punjab.

Chemical Composition

Food value/chemical composition of garlic presented in Tables 14.1 and 14.2 indicate that it is a good source of carbohydrates and proteins. It is also rich in phosphorus, potassium, calcium and magnesium. Besides these, it contains Vitamin

B_1, Vitamin B_2 and Vitamin B_6. Sulphur containing compounds, 'allicins' responsible for the characteristic fragrance of garlic are also present in it *vis-a-vis* some enzymes like alinase and pro-oxidase.

Table 14.1: Food Value of Garlic

Component	*Per 100 g*
Calories	35
Protein content	6.3 g
Calcium	30 mg
Iron	1.3 mg
Vitamin A (carotene)	Not detected
Vitamin B1	0.06 mg
Vitamin B2	0.23 mg
Vitamin C	13 mg

Source : KRIBCO, 1987.

Table 14.2: Chemical Composition of Garlic

Component	*In g 100 g^{-1}*
Moisture	62.0
Protein	8.3
Fat	0.1
Carbohydrates	29.8
Fibres	0.8
Eenergy K cal	145
Minerals	1.0

Source: Arya and Arya, 2007

Garlic contains a colourless, odourless, water soluble amino acid known as 'alliin'. Alliin changes to allicin, which serves as an anti-bacterial substance.

Medicinal Values

Garlic is believed to tone-up the nervous system and in curing gastro-enteritis problems and indigestion. Besides, it has also proved to be beneficial in curing heart troubles and different kinds of cancers. However, the use of garlic is prohibited in certain Indian communities on religious considerations. Medicinal values of garlic are detailed as under:

(i) **Prophylactic on curative values:** It is known to have antiseptic and antibacterial properties. It helps in the elimination of the noxious waste material from the body.

Garlic can be applied externally to the chest with the beneficial results, as it is rubefacient, resicant and disinfectant. It is also used as antihelminthic.

(ii) **Blood pressure and heart diseases:** A study conducted by the German general practitioners on heart patients, revealed that cholesterol associated with heart diseases is significantly reduced by regular use of garlic in the diet.

The other Western medical professionals admit that garlic reduces fats particularly cholesterol and even controls its formation in the body cells. They are of the opinion that garlic dilutes the blood and renders the blood plates less sticky and also inhibits the clotting of coagulant protein.

The substance which works against the cholesterol formation is the occurrence of alliin in the garlic.

(iii) **Mental tension and dizziness:** Apart from reducing blood pressure (hypertension), use of garlic also reduces mental tension because garlic has the power to ease spasm of small arteries. It slows the pulse, modifies the heart rhythm. It also relieves from symptoms of dizziness, short breath as well as formation of gases in the digestive tract.

(iv) **Breast, stomach, intestine and skin cancers:** Studies at Houston M.D. Anderson Cancer Institute under the Pennsylvania State University supported the evidence that certain ingredients in garlic block the cancer agents of breast, esophagus and colon, and skin.

(v) **Prostate cancer:** It has been found that men who used to take more than a third of an ounce (10 g) of garlic and onion were much less likely to be in the prostate cancer group.

(vi) **Other diseases:** Garlic's juice is used in the ear for the relief of earache. Garlic is a good carminate and an active appetizer. It is a good remedy for intestinal spasms and collics preventing indigestion, relieves flatulence and anti-helminthic.

Sir John Harrington recognized the medicinal property of garlic in the early 17th century *i.e.,* during 1609. He described the magical power of garlic as follows:-

"Garlic then have power to save from death,

Bear with it though it maketh unsavory breath,

And scorn not garlic some that think,

It only maketh men wink and drink and sink."

A new study suggested that garlic may be useful to treat high blood pressure (Anonymous, 2011). Garlic supplements have previously been shown to lower cholesterol and reduce high blood pressure in those with untreated hypertension. In another study, the researchers from the University of Adelaide (Australia) analyzed the effects of supplement of 4 capsules a day as aged garlic for 12 weeks and found systolic blood pressure to be around 10 mm Hg lower in the group given garlic compared with those given a placebo (Anonymous, 2011) and traditional methods of usage are given below:

Table 14.3: Traditional Methods of Garlic Consumption

Prescription	*Method*
Use of raw cloves in semi paste condition	Three to four small sized cloves are taken in semi paste condition with water in empty stomach as a preventive medicine.
Use of chopped cloves in soaked water during night hours	One to two big sized cloves are chopped and put in a glass of water during night. This water alongwith chopped garlic is then taken in empty stomach during morning. This will take care of all digestive disorders.
Use of crushed cloves with milk	One to two crushed cloves after mixing with milk are taken to reduce the bad smell of the breath in case of pulmonary gangrene.
Application of crushed cloves externally for controlling pulmonary gangrene	A paste after crushing two to three cloves is applied externally on chest to get relief from pulmonary gangrene, as it is rubefacient, resicant and disinfectant. It is also used as anthelmintics.
Use of cloves in mustard oil for joint pains (arthritis) and rheumatic pains	After frying or burning cloves in mustard oil, it is strained and applied as massage to the knee or joint pains. Massage of oil in which garlic has been fried is a useful liniment for rheumatic pains.
Use of garlic and clove	The oil in which garlic and cloves have been fried has proven very useful in arthritic pains.
Use of garlic oil extract	This is available in the market in garlic pearls. Daily intake of 2-3 pearls to get noticeable relief in hypertension (blood pressure).
Use of extracted juice	20-30 drops of juice diluted with water is injected into the rectum to treat ascariasis.
Use of juice with vinegar	Garlic juice mixed with vinegar is prescribed for soar throat and is a good cough expectorant.
Use of garlic as *"rasayan"* on *"lassanadi vati"*	Garlic is now available as a *"rasayan"* in the shape of *"oirasan pak"* which is prepared by boiling garlic in milk.
Odourless garlic tablets	These tablets have been prepared by the Germany's Lichtwer Corporation, and are being prescribed for curing heart diseases, high blood pressure and strokes. The tablet taken once a day is considered to be more effective than actual garlic.
Use of roasted cloves	Two roasted cloves of garlic swallowed every day with water keeps the blood cholesterol away.
Use of dry garlic	10 g of dry garlic swallowed every day with water maintains the blood pressure within normal limits.
Use of garlic oil or juice	Inhalation of garlic oil or juice has been recommended in case of pulmonary tuberculosis, cough and red eyes.
Repellent property	A formulation containing 1 per cent garlic extract gives protection to persons against mosquitoes for about 8 hours.

Vatsyayan (2011) reported that garlic is one of the most valued herbs since time immemorial, and has been in use by the old civilizations of the world. Whereas, the Indians, Chinese and Egyptian beliefs have many tales about its divine origin and curative qualities, the ancient Ayurvedic literature has gone to the extent of extolling it as 'Maha aushadhi', the great medicine.

Garlic is endowed with five different tastes as pungent, sweet, bitter, astringent and salty at the same time. Apart from garlic being digestive, carminative, cardiac stimulant, expectorant, analgesic, antioxidant and aphrodisiac, it is a natural scavenger of body toxins and a powerful digestive aid that helps resolve heavy and fried food by acting as an anti- flatulent and wind dispelling aid.

Recently, garlic has acquired a reputation as an effective immune-stimulant, anti-viral, anti-cholesterol, cardio-vascular tonic and also as a tumor inhibiting medicine (Vatsyayan, 2014).

Antimicrobial properties: The crude juice as well as various solvent extracts of garlic has been shown to be highly inhibitory to the growth of *Salmonella typhi, Bacillus subtilis* and *Pseudomonas pyocyaneus* (Abdou *et al.*, 1972).

***Antioxidant and food preservative properties*:** Garlic, besides its role as flavourings and seasonings, strong medicinal, preservative and antioxidant properties and, thus, contribute to overall safety and preservative of food products (Sudha and Sathyan, 2006).

Botanical Description

Garlic (*Allium sativum*) belonging to the family Liliaceae, is a bulbous plant. It grows all over the country including Kashmir valley.

The garlic bulb is short, compressed breaking-up into 10-12 bulbils which are also known as cloves. The cloves of the garlic are, infact, used to season the foods. These cloves when dried can easily enumerate the bulbils, each of them covered by a very thin paper like layer.

Garlic possesses a pungent taste and a strong smell, due to the presence of sulphur containing compounds like allyle sulphide and vinyl sulphide etc. (Gupta and Mahajan, 2005). The inflorescences produce both seeds and bulblets. The latter alongwith cloves and leaves have been hitherto used for flavouring salads, soups and sauces. Garlic is largely used in Indian cookery for flavouring curries, chutneys, *pakoras* and other preparations.

Cultural Practices

Climatic requirements: Like onion, garlic is also a cool season crop. Although sub-tropical and sub-temperate climatic conditions are most suitable for its cultivation yet it can tolerate frost to some extent. It can also be grown under temperate climatic conditions. Cultivation of garlic in cold arid region of Ladakh (Jammu and Kashmir) has recently been started (Singh *et al.*, 2006). At the time of bulb maturity, long days and dry weathers, are favourable, whereas, short days and low temperature favour its vegetative growth.

Soil requirements: Garlic can be grown on a variety of soils. But well drained light soils *i.e.*, loamy sand, sandy loam and loam are the most suited. Heavy soils *i.e.*, clay loam and clayey soils are altogether harmful because they hinder the proper development of bulbs in their true shape and size. Such soils, however, become suitable if large amount of organic manures are added in them.

Propagation: The crop is propagated by segments (cloves) of bulbs which are dibbled during October to November in flat beds at a spacing of 15 cm between rows and 5-10 cm between plants. In mid-hills, it is propagated during September-October and in colder climates during April-May. Planting depth should not be more than1.5 to 2.0 cm.

Seed rate: 5–7 quintals of cloves are sufficient to plant an area of one hectare.

Manures and fertilizers: Well decomposed organic manure should be applied at the time of field preparation during last ploughing at the rate of 25-30 q ha^{-1}. Fertilizer dose (N, P_2O_5 and K_2O) varies with locality. However, an application of 100 kg N, 60 kg P_2O_5 and 40 kg K_2O is the general recommendation.

All FYM, phosphate (P_2O_5) and potassic (K_2O) fertilizers and ½ N *i.e.*, nitrogenous fertilizer should be applied at the time of sowing. Remaining nitrogenous fertilizer is applied at the time of bulb formation.

Irrigation: After sowing, the fields are immediately irrigated and subsequent irrigations are given after every 10-15 days. Irrigation is stopped about 15-20 days before harvesting.

Harvesting and yield: The well developed bulbs ripen in end of April or 1st to 2nd week of May and can be harvested by pulling out the plants. On an average, yield varies from 100 to 120 q ha^{-1} or 60 to 85 q ha^{-1} (Saxena *et al.*, 1999).

References

Abdou, I.A., Abdou-Zeid, A.A., El-Sherbeeny, M.R. and Abou-El-Gheat, Z.H. 1972. Antimicrobial activity of *Allium sativum, Allium cepa, Raphanus sativus, Capsicum frutescens, Eruca sativa* and *Allium kurrat* on bacteria. *Quality of Plant Materials and Vegetation*. 22 : 29-35.

Altaf, T. 2002. Garlic. The Greater Kashmir. Sunday Supplements – Sensor. Nov. 24, 2002.

Anonymous. 2011. Garlic could help treat hypertension. *The Kashmir Times*. 68 (166): 12 (Thursday, June 16).

Arya, L. and Arya, K. 2007. Garlic – Natural medicine. *Agrobios Newsletter*. 5 (8) : 44-45.

Gupta, R.D. and Mahajan, N. 2005. Medicinal value of garlic, from cooking to curing. *The Kashmir Times*, Jammu. 62 (221) : 13 [Thursday, August 11, 2005].

Raghav, C.S. 2003. Medicinal plants on the touchstone of science. *Science Reporter*. February 2003 Issue, pp. 55.

Saxena, A.K., Tiwari, K.B. and Thakur, P. 1999. Adoption of improved garlic production technology. *Agricultural Extension Review* 11 (1) : 17-18.

Singh, N., Kumar, N. and Ahmed, Z. 2006. Cultivation of garlic in cold arid region. *Indian Farmers Digest*. 39 (8) : 34-35.

Sudha, S. and Sathyan, N. 2006. Antibacterial property of garlic. *Agrobios Newsletter*. 5 (4) : 43-44.

Vatsyayan, G. 2014. Garlic from God to man. *Ayurveda*. 1(24): 26.

Vatsyayan, R. 2011. Who should be more careful in using garlic. *The Tribune*. 130 (304): 12 (Wednesday, November, 3).

15

Ginger
(*Gingiber officinale* Rosc.)

Common name: Adrak

Family: Zingiberaceae

Introduction

In Sanskrit, ginger is known as *"Vishwa Bheshaja"* which means Universal Medicine. Ginger is a plant which has been used since time immemorial as an important spice crop of the world (Kushwaha *et al.*, 2004) and grows best in the warmer climate of China, India and Jamaica, which has been discovered more than 2,500 years ago. Rhizomes (underground stems) of ginger are medicinally and culinary useful parts of the plant (Sivakumari and Rajendran, 2006). Although ginger is being used since time immemorial as spice crop of the world yet it has been listed as neglected and under-utilized crops in Asia (www.ipgri.cgiar.org).

Indian ginger is very popular for its flavour, texture and taste. More than a spice, ginger is considered to be a taste maker, an appetizer and a drug. Its medicinal value such as carminative, stimulant of the gastro-intestinal tract, chest diseases etc. is of worth mentioning.

Distribution

Ginger is believed to have been originated in India, and thereafter, it was introduced in China at early date. It seems to have been utilized as a spice by the Indians and the Chinese since time immemorial. A number of references with regard to its use are available in Sanskrit literature and in Chinese medical treatises.

In Europe, ginger was known during first century A.D., which has been mentioned by Dioscorides and Pliny. As rhizomes of ginger are very easy to transport, the plant soon spread to all tropical and sub-tropical countries. These days, the main ginger producers are China, India and Taiwan. Ginger is, indeed, a major spice crop cultivated in India and marketed as fresh and dried. India is the largest producer and exporter accounting for more than 70 per cent of world production. It is cultivated in almost all tropical and sub-tropical parts of the country, especially in Kerala, Karnataka, Tamil Nadu, West Bengal, Himachal Pradesh, Uttar Pradesh, Uttrakhand, Maharashtra, Bihar, Jammu region of Jammu and Kashmir. Ginger is cultivated all over India, but that grown in Kerala has been found to be the most superior to other places, in aroma and taste.

Chemical Composition

Ginger is available in two forms, fresh and dried. Both of these forms have effective food value. The proximate composition of the ginger as elucidated in Table 15.1 indicate that fresh ginger is a good source of carbohydrate, protein and fibre. Fat is also present and is a source of energy. The dried ginger scraped or peeled before drying constitutes the spice. It is esteemed for its flavour, pungency, aroma and medicinal value. The pH value of the ginger varies from 6.0 to 6.3.

Table 15.1: Chemical Composition of Ginger (100 g^{-1} of Edible Portion)

Type of Ginger	*Moisture (per cent)*	*Protein*	*Fat*	*Fibre*	*Ash*	*Carbo-hydrates*	*Energy (K Cal)*
Fresh	80.9	2.3	0.9	2.4	1.2	12.3	67
Raw	ND	8.1 to 11.6	1.2 to 3.9	1.4 to 9.5	6.1 to 9.6	ND	ND

Ginger contains shogaols, gingerone and gingerols which give it its flavour and have analgesic and antibacterial properties (Anonymous, 2009). It also maintains the production of saliva and is known to have arthritis related pain relieving. The ginger yields an essential oil with pungent yellowish in colour and viscous principle and is used in manufacturing of flavouring essence and in perfumery. The volatile oil content in ginger varies from 1.25 to 2.81 per cent. It is sparingly soluble in alcohol and insoluble in water.

The main constituent of ginger oil is a sesquiterpene hydrocarbon predominantly zingeberene (35 per cent), curcumin (18 per cent) and farnesene (10 per cent) with lesser amounts of bisabolene and B-sesquiphellandrene. A smaller percentage of atleast 40 different monoterpene hydrocarbons is present with 1,8-cineole, linalool, borneol, citral, neral and geraniol being the most abundant. Other organic compounds present in the ginger oil include camphene, methyl heptenone, n-decylaldehyde, n-nonylaldehyde, sesquiterpene alcohol and zingiberol. Ginger rhizomes also contain a potent proteolytic enzyme called zingibain.

Physio-chemical characteristics of the oil vary considerably, depending upon the variety, period of storage of ginger, method of extraction etc. Major pungent principles along with their empirical formulae, have been given in Table 15.2 which are confined to ginger oleo-resin.

Table 15.2: Major Pungent Principles in Ginger

Pungent Principles	*Empirical Formula*
Zingiberene	$C_{15}H_{24}$
Zingiberol	$C_{15}H_{26}O$
Zingirone, 3-methoxy-4-hydroxy-phenyl-ethylmethyl ketone	$C_{11}H_{14}O_3$
Shogaol	$C_{17}H_{24}O_3$

Source: Kushwaha *et al.*, 2004.

Medicinal Values

In India, ginger has been used as medicine since Vedic period and is called *"Maha-aushidi"* which means the greatest medicine. Ancient physicians used ginger as a carminative and anti-fermenting medicine. Galen, the Greek physician used ginger as medicine to rectify the defective humours of the body. He used ginger in the treatment of paralysis caused by phlegmatic imbalance in the body. Avicenna, another Greek physician, used ginger as an aphrodisiac. Pomose, yet another Greek physician also used ginger in the treatment of gout centuries ago.

Ginger is now widely used in local formulations in India. It has digestive and carminative properties and is widely used for dyspepsia and bowel colic. It is a general stimulant and is one of the spices which are regarded as aphrodisiacs. Many traditional Chinese medicines contain ginger, and its use generally seems to be safe.

It is a valuable drug for disorders colic bowel, vomiting, flatulence and painful infections in the stomach. Ginger being an expectorant is an excellent remedy for coughs and colds, bronchitis, asthma, whooping cough and tuberculosis in lungs. Ginger is an excellent pain killer in curing all kinds of pains like toothache, headache and earache. It is highly beneficial in treating sexual weakness. It tones up the sex centers and cures impotency, premature ejaculation, spermatorrhoea besides being useful in menstrual disorders. Traditional and simple methods commonly used for curing different diseases/disorders, have been mentioned in Table 15.3.

Table 15.3: Traditional Methods for Using Ginger against Different Diseases/Disorders

Sl.No.	*Disease*	*Method*
1.	Menstrual disorders like dysmenorrhoea and amenorrhoea	A piece of fresh ginger should be pounded and boiled in a cupful of water for few minutes. The infusion sweetened with sugar should be taken thrice daily after meals for relief of these disorders.
2.	Sexual debility like impotency, premature ejaculation and spermatorrhoea	Half a teaspoon of ginger juice should be taken with a half-boiled egg and honey once daily at night for a month. It tones up the sex centers and cures the said sexual disorders.
3.	Respiratory disorders (i) Influenza fever (ii) Bronchitis asthma, whooping cough and lungs tuberculosis	A teaspoonful of the fresh ginger juice mixed with a cupful of fenugreek decoction and honey to taste is an excellent diaphoretic mixture for increasing sweating to reduce fever in influenza. Fresh half-teaspoonful juice should be taken daily after mixing with honey in the morning. Alternatively, a piece of fresh ginger after peeling and washing is boiled in water. This boiled water (3-4 spoonful) is taken with a teaspoonful of honey early in the morning (empty stomach).
4.	Aches and pains (i) Headache (ii) Earache	In headache, ginger ointment made by rubbing dry ginger with a little water and applied to forehead affords relief. In case of earache, a few drops of ginger juice will give relief.
5.	Colds and coughs	Extract juice of ginger and take it with honey in case of coughs. In case of colds, ginger should be cut into pieces and boiled. It is then strained and half-a-teaspoonful sugar added to it and taken while hot. Ginger tea prepared by adding few pieces of ginger into boiled water before adding tea leaves is also an effective remedy for colds and fever resulting from cold.
6.	Digestive disorders – dyspepsia, flatulence, bowel colic, vomiting	Chewing a piece of fresh ginger after meals regularly as insurance against these ailments. This protective action is attributed to excessive secretion of saliva, diastase enzyme and volatile oil.
7.	Nausea and vomiting due to biliousness, indigestion caused by intake of heavy non-vegetarian and fried fatty food	Half-a-teaspoonful of fresh ginger juice, mixed with one teaspoonful each of fresh lime juice and fresh mint juice and a teaspoonful of honey constitutes an effective measure of these orders.
8.	Morning sickness, jaundice and piles	The mixture prescribed above should be swallowed thrice daily in the treatment of these conditions.
9.	Fractures, rheumatism, arthritis	Ginger oil is used in the treatment of these disorders. Infact, massage of the oil is done on the affected parts.
10.	Motion sickness	Powdered ginger (about 1 g) is found to be effective in reducing the symptoms of this sickness.

Contd...

Table 15.3–*Contd...*

Sl.No.	Disease	Method
11.	Abdomen pain	When there is pain in abdomen due to flatulence or acidity, a mixture of half-tablespoonful of fresh juice of ginger and asafetida be given two to three times a day.
12.	Common cold and headache due to common cold and sinusitis	For common cold, a decoction of dry ginger, cinnamon and sugar should be given twice at four tablespoons. Paste of dried ginger and *barian*-sweet flag (*Acorus calamus*) should be applied to relieve the said disorders.
13.	Inflammation of joints in arthritis and rheumatism	A decoction of dry ginger with one tablespoonful of castor oil if taken daily at night gives relief of joint pains.
14.	To improve blood circulation	In collapse or shock, dried ginger powder should be rubbed over the palms and soles of feet to improve the blood circulation.
15.	Flatulence and toothache	Dried ginger burnt to charcoal is best remedy for flatulence. This powder mixed with common salt is rubbed over the teeth to cure the ache and sourness.
16.	Pharyngits and hoarseness of voice	Chewing of fresh ginger with clove and crystal of common salt is useful for curing the said disorders. One tablespoonful ginger juice mixed with one tablespoonful fresh mint juice and one tablespoonful of honey, if taken thrice a day will not only cure pharyngits and hoarseness of voice but also improves digestion, bronchitis.
17.	Joint swellings (arthritis)	Fresh ginger fried in coconut oil can be applied as liniment in case of joint swelling.
18.	Diarrhoea and dysentery	Fresh ginger and buttermilk fried in *ghee*, is used in diarrhoea and dysentery.

Ginger is used in countless classic Ayurvedic preparations (Vatsyana, 2006). As single drug, the dose of fresh ginger varies from 2-5 ml as its extracts and dried powder from 1-9g in a divided amount.

Recent studies have revealed that mixture of oil and resin, *i.e.*, oleo-resin occurs in ginger used as an anti-emetic. In large amounts, it seems to inhibit platelet aggregation. Curcumin in ginger is known to stimulate the activity of glutamine S-transferase which inhibits cancer.

Botanical Description

Ginger is a perennial herb, growing 60-90 cm in height with underground branching stems called rhizomes which are swollen and tough. They are white or light yellow outside and become grey brown with age, upto 2.5 cm in diameter. The plant has leafy shoots of 0.50 to 0.75 m height, leaves sheathy, alternatively arranged linear with 15 cm long. Flowers are sessile, borne on a spike, condensed, oblong and cylindrical with numerous scar bracts. Flowers are numerous, yellow in colour with dark purplish spots, bisexual, epigynous, stamens only one, ovary inferior, three carpelled. Fruit is oblong capsule, seeds glabrous and fairly large. Leaves and rhizomes have characteristic fragrant odour when cut or bruised. Rhizomes are dugout after the leafy parts are dried.

Cultural Practices

Climatic requirements: Ginger grows best in warmer climate. The crop can be grown upto an altitude of 1500 m. In north-western Himalayan states, it is cultivated in Himachal Pradesh, Uttarakhand and Jammu and Kashmir. Dadhau and Kangra areas are the best producers in Himachal and Arnas and parts of Udhamupr in Jammu and Kashmir. It is also cultivated in almost all tropical areas of country especially in Kerala, Karnataka, Tamil Nadu, West Bengal and Bihar as well as Maharashtra, Madhya Pradesh, Punjab and Haryana.

Soil requirements: Ginger thrives best on well-drained soils with well-distributed rainfall. Among soil types loamy sand and sandy loam are considered the best. It grows on other soils also except heavy textured soils.

Sowing time: In tropical or sub-tropical areas it is sown during March or April, while in intermediate or sub-temperate climatic zone, the ginger is cultivated during mid-April to mid-May or sometimes during June to July.

Seed rate: For sowing, the rhizomes are cut into pieces and each piece or bit having a bud weighing 20-30 g, is sown 5 cm deep on about 15 cm high ridges. About 15-17 q of seed (rhizomes) is required ha^{-1}. The required row-to-row and plant-to-plant spacing is 20 cm 15 cm.

Manures and fertilizers: The recommended doses of fertilizers @75 kg N, 50 kg P_2O_5 and 70 kg K_2O ha^{-1} have been found to be optimum.

FYM, P_2O_5, K_2O and ½N are applied at the time of planting. The remaining N should be applied in two equal split doses, first one month after planting and the second one month after the first dressing.

Irrigation: The crop is immediately watered after planting or sowing. Thereafter, irrigation is given at 4-10 days interval as and when required.

Intercultural operations: Shoots of ginger emerge within 10-20 days after sowing. Weeding is done 3 to 4 times and plants are earthed up once or twice.

Pests and diseases: Rhizome rot is the major disease. Owing to this, the rhizomes become soft and on pressing collapse easily. The leaves of the affected plants become yellow, and rot at the basal portion. The shoots finally wilt and collapse.

Before sowing, the rhizomes should be immersed in 0.1 per cent mercuric chloride for 3 minutes. However, use of mercuric chloride nowadays is discouraged, so neem based products should be used.

Harvesting and yield: The crop is ready for harvesting in about 8 months when leaves turn yellow. Rhizomes are collected in December or January when plants wither after flowering period. Rhizomes are carefully dugout with the help of digging forks or spades. Aerial stems, fibrous roots and buds are removed. They are washed to remove mould, clay particles attached to them.

Rhizomes are peeled on flat surface as well as between the fingers thoroughly, washed in running water, then dried completely by keeping them in the sun on mats for 3-4 days and hand-rubbed. Excessive sun-drying should be avoided as it will bleach the produce. Yield varies from area-to-area upto 200-250 q ha^{-1} as fresh weight.

References

Anonymous. 2009. Ayurveda reinterpreted. *Infotainment: The State Times*. 14 (219): 5. (Saturday, August 8).

Kushwaha, A., Kulshrestha, K. and Singh, P. 2004. Food that heal : Ginger. *Indian Farmers Digest*. 37 (5-6) : 61-64.

Sivakumari, V. and Rajendran, A. 2006. Ginger – A beneficial medicinal plant. *Agrobios Newsletter*. 5 (1) : 41-42.

Vatsyana, R. 2006. Ginger-More than a digestive aid. *Ayurved*. December Issue, pp. 22-23.

www.ipgri.cgiar.org. Sept. 9, 2003.

16

Green Chirayta (*Andrographis paniculata* Nees.)

Common name: Kalmegh

Family: Acanthaceae

Introduction

Kalmegh is an annual herb found throughout India, especially in dense forests. It is under cultivation in many states of India. Kalmegh is an herb, whose complete plant, leaves and roots are used as a drug. It can be grown easily in tropical and sub-tropical areas and would provide a good income for the peasants or orchardists. By growing this herb like others a farmer can ensure that even a small piece of land can fetch for him the required income.

Medicinal Values

In Ayurveda the plant is described as bitter, acrid, cooling, laxative, antipyretic, astringent, antiperiodic, anti-inflammatory, expectorant. Kalmegh can be used for curing acidity and liver inflammation. Its use is also very common in curing the malaria fever and skin diseases. Not only this, it is also used by the diabetic patients as well as those who are afflicted with dysentery, diarrhoea, cholera and abdominal pains. It is used in de-worming. The decoction of roots is utilized against liver problems.

'Andrographolide' is the key medicinal compound present throughout the plant with leaves containing its maximum content while in the stem it is in lesser amount. Leaf juice along with cardamom, cloves and cinnamon in the form of globules is prescribed to infants for relief from bowl problems, irregular stools and loss of appetite. The plant is also very effective for treating snake bites and jaundice (Kaur *et al.,* 2011). It is used as bitter tonic, febrifuge, stomachic and anthelmintic. Decoction of its roots is given in dyspepsia and liver disorder. Recent experimental findings have indicated that kalmegh is having antityphoid, immunosuppressive and antibiotic properties. It has been proved to be hepatoprotective drug and useful in influenza, cough, sore throat, tonsillitis, osteodynia, bronchitis, itches, arthralgia, mensural and gonorrhoea.

Botanical Description

An erect, branched annual herb growing straight with alternate leaves up to 0.3-0.9 meters height; leaves - lanceolate, acute, undulate, pale beneath; flowers small, solitary distant, in axillary or terminal racemes or panicles,pink coloured with white spots. Flowering time in India is November-December. Three strains of *Andographis paniculata* are considered to be the most superior namely; AP-3, AP-10 and AP-89 which were evaluated for their growth, biomass production and alkaloid content and yield (Anonymous, 2008). Strain AP-3 was found to excel over others in all these parameters.

Cultural Practices

Climatic requirements: Kalmegh can be grown under tropical to sub-tropical climatic conditions but sensitive to severe winter. It is worthwhile to mention that kalmegh grows well both under irrigated and un-irrigated condition. In India, it is cultivated as rainy season (*Kharif*) crop.

Soil requirements: Kalmegh can be grown over a wide variety of soils in tropical and sub-tropical regions having low rainfall. However, the plants prefer to grow in sandy loam to laterite, well drained soils having good fertility with fair amount of organic matter. It can be cultivated on shady wastelands.

Time of sowing and seed rate: May - June is the best time for seed sowing in nursery. Sowing is done in well prepared field using seed @ of 2.0 -2.5 kg ha^{-1} for nursery sowing.

Transplanting: Transplanting is done during June -July and transplanting at 45 cm row to row and 30 cm plant to plant spacing. Well grown seedlings of 30-35 days are transplanted in well prepared fields.

Intercultural operations: One weeding/hoeing is given to the crop after 35-40 days of transplanting.

Manures and fertilizers: FYM or compost @ 10 tonnes ha^{-1} should be added before transplanting. Although chemical fertilizers @ of175, 85 and 50 kg through urea, DAP and potash, respectively have been recommended yet their use is restricted. It is because with the increasing awareness of healthy living and the harmful effects of chemicals, there is good demand of organically produced crop. For this purpose, use of vermi-compost, bio-fertilizers, green manuring is being encouraged to meet the fertilizer requirements.

Irrigation: At least two irrigations are essential to obtain good yield of Kalmegh; one at the time of transplanting and the other before flowering. Low soil moisture content at flowering stage results in dropping of flowers. Timely irrigation promotes good vegetative growth and production.

Weed control: Need based weeding/hoeing is essential to kill the weeds.

Harvesting and yield: To obtain maximum biomass, duration of the crop varies from 90-100 days. Accordingly, August - September is the period of its harvesting.

Approximately 8.0 q dry matter is obtained ha^{-1}. The whole plant is used for medicinal purposes. The roots, leaves, branches, buds/flowers are collected separately. The average rate in Indian market is Rs. 1800-2500 per quintal and net returns per hectare are Rs.65,000/-.

References

Anonymous. 2008. Medicinal plant varieties developed by Central Institute of Medicinal and Aromatic Plants, Lucknow. *Gram Vikas Jyoti*. April-June Issue, pp 52-53.

Kaur, N, Kaur, B and Singh, K. 2011. Medicinal plants of India. *Agrobios News Letter*. 10(1): 26-27.

17

Holy Basil

(*Ocimum sanctum* Linn.)

Common name: Tulsi

Family: Lamiacreae

Introduction

Basil has been found to occupy important place in the traditional folklore medicine in general and Ayurvedic System of Medicine in particular throughout the world. It is an ubiquitous and a delicate little plant with which Hindu mythology is replete with legends. If space is not available, it can even be found growing in pots.

It is also known as *Vishnu Priya* (Love of Lord Vishnu). It is looked upon with great reverence and is considered to be very dear to Lord Krishna (incarnation of Lord Vishnu) also. The name of Tulsi denotes the embodiment of wealth and wisdom, and is worshipped as Lakshmi and Saraswati. It is also believed that salvation is assured to anyone who tends the Tulsi plant with great care and emotion. The plant is regarded to be so sacred that even during solar or lunar eclipses; its leaves are kept in stored food articles to protect them from the bad effects of eclipses.

Because of medicinal properties, Tulsi has been popularized as plant of devotion and worship. No other herb is as popular as Tulsi-a common name in many Indian languages. Although tulsi is native to Asia yet it has been grown throughout the world as culinary and medicinal herb. English name – Holy Basil, refers to a Greek word, 'Basilikon' meaning royal, perhaps because it was crowned as a king among all herbs (Patwari, 2005). Accordingly, the word Tulsi literally means unmatched or unparallel. A mention of Tulsi has been made in most religious books that a house with tulsi plant becomes sacred and dwellers of that house will not face sudden death and fatal diseases. Even the air around it is purified (Gupta, 2005; Krishnamurthy, 2005).

Types

There are two types of tulsi plants *viz.*, black or purple type and white or green type. The green type Tulsi is also called as *"Sri Tulsi"* or *"Rama Tulsi"* which is the most common whereas, second type bearing purple leaves is not so common and is known as *"Krishna Tulsi"* or *"Shyama Tulsi"*. Black or purple coloured Tulsi plants are, however, considered to be more superior to those of white coloured.

It is because black coloured tulsi plants are slightly better, pungent, fragrant, light and appetizing than white coloured plants. Black Tulsi has more medicinal values like curing gas trouble, cough, intestinal worms, skin diseases, kidney disorders, regulating flow of urine, and restoring the balance in the system by cleansing the body of toxin and strengthening the tone of every organ.

Chemical Composition

The plants contain alkaloids, glycosides, saponins and tannins. The leaves are rich in ascrobic acid and carotene. The leaves on steam distillation yield a bright yellow volatile oil which possesses antibacterial and insecticidal properties. It checks the *in vitro* growth of *Mycobacterium tubercle* and *Micrococcus pyogenes*. The oil is also an excellent mosquito repellent and is very effective against *Salmonella typhosa*. Ether and alcohol extracts of leaves are active against *Escherchia coli*.

Seeds of are rich in mucilaginous and demulcent substance and hence useful in gentio-urinary disorders. It also finds use in cosmetics, toiletries, pharmaceuticals, perfumes and aromatherapy.

Medicinal Values

Tulsi gives a pungent, bitter taste and every part of the plant has its medicinal use in one or the other form. In the first place, its presence in the precinct is believed to purify the air around the house. Persons can get relief from ailments like stomach upsets, gastric troubles, cough, fever, tuberculosis and snake bites (Kumar, 2012). Indeed, the leaves have expectorant properties and its juice is usefully given to the patients with bronchitis and throat troubles apart from tuberculosis.

Tulsi can cure a wide range of diseases like blood pressure, heart ailments, respiratory trouble, bronchitis, asthma, kidney and urinary disorders, piles and even cancer (Hussain, 2013). It is believed to lower cholesterol in blood. Details of its medicinal properties have been enumerated in ancient and medicinal literature namely "Charak Samhita" and "Encyclopaedia of Edible Plants of the World".

The plant has also been found useful in leucoderma, vomiting, foul smell, lumbago pain, eye soar, hiccup, purulent ear discharge. Also, it is reported to have abortificient, anti-fertility and adoptogenic, antibiotic and anti-stress activity.

Tulsi in Agriculture

Tulsi has the power and capacity to keep insects and mosquitoes at bay and is, therefore, called "Mosquito Plant" in South Africa.

The extract of the leaves have also been used as an antidote for poisonous reptiles, as is believed they do not dare to enter the house where the fragrance of tulsi pervades. It had been found to be useful and effective plant to combat pollution including soil contamination known to be because of a particular kind of gas which purifies the atmosphere including soil, air is emitted by the plant around its ambiance. The plant possesses antimicrobial activity and hence protects the environment from microbial pollution.

Various Methods of Consumption of Tulsi

There are several methods in which parts of tulsi plant like leaves, stems, seeds and fresh roots can be used or consumed and described in Table 17.1.

Botanical Description

The plant forms a very good species as an understorey of the forest vegetation and thus increases organic status of the soil. Under favourable conditions, the plant grows to a considerable size and furnishes with a woody stem large enough to make beads from rosaries used by the Hindus on which they count the number of recitations of their deity name. The plant is an erect, much branched annual herb growing to a height of 40 to 60 cm. It has hairy stems. Leaves are simple, opposite, ovate-round, green or purple, up to 5 cm long, with entire or toothed margins, highly aromatic. Flowers are small, purple to reddish, in compact clusters in spikes up to 5 mm. Fruit is small with yellow to reddish seeds.

Table 17.1: Traditional Methods of Consumption of *Tulsi*

Sl.No.	*Method of Consumption*
1.	Fresh leaves can be swallowed every day with water after washing them. It would be much better if these leaves are taken empty stomach immediately after brushing teeth.
2.	Juice extracted from the leaves diluted with water and taken empty stomach in the morning.
3.	Fresh leaves after washing are boiled in glass of water for few minutes. Two to three spoonful of the boiled but cooled extract so obtained can be taken early in the morning after mixing with a glass of water.
4.	Its leaves can also be mixed with leaves of either *bael* (*bilwa*), coriander, curry mint or all and their juices extracted together.
5.	Juice of tulsi can also be mixed with vegetable juices like carrot, beet root, spinach, tomato and cabbage.
6.	A teaspoonful of tulsi and ginger juice mixed with diluted honey is excellent for cough, cold, bronchitis, cataract, fever and asthma.
7.	It is believed that powder prepared from 50 g of leaves with half of the quantity of black pepper prepared into pea-sized pellets and dried in the shade and its regularly use can cure all kinds of fever.
8.	Fresh leaves with ginger can cure stomach ache. For nausea and vomiting mix some powdered seeds with milk and take it slightly hot, the relief will be instant.
9.	Used for de-worming in children, tulsi leaves juice with salt. For skin infections including eczema, mix leaves juice and lemon.
10.	To cure toothache, brush teeth with dried powdered leaves, this will strengthen gums, prevent bad breath and cures pyorrhea too.
11.	Its decoction prepared from seeds and roots is given for malarial fever to bring about sweating.
12.	Tulsi tea is a popular preparation. Wash leaves and boil them with some jaggery powder for 5 to 10 minutes. After boiling, strain it and add a few teaspoonfuls of milk. Add some pounded raisins to neutralize the bitterness and make the drink better taste.
13.	Pounded leaves, mixed with sandalwood paste, is applied on the forehead for relief from heat, headache and to retain coolness in general.
14.	The fresh roots ground with water can be applied to the stings of wasps, bees and bites of worms and leaches.
15.	The bruised fresh roots, stems and leaves are applied to the bites of mosquitoes.
16.	Daily use of leaves along with water kept overnight in a copper pot definitely helps in gaining general health and vitality.
17.	Fresh leaves are made into paste with curd and taken 2-3 times a day with honey on empty stomach, checks cancer spread and improves memory.
18.	About 2 g of seeds taken with luke warm water removes constipation and in cold water checks diarrhoea.

Cultural Practices

Climatic requirements: Tulsi plant is found throughout India upto an altitude of 1800 m in Himalayas, cultivated and grown in garden and temples, and is frequently found as an escape. The plant even can be grown under arid, semi-arid, moist conditions of cold sub-tropical regions of the world but sensitive to frost.

Soil requirements: It can be grown on a variety of soils, yet it grows more luxuriously in sandy loam or loam, well drained, levelled soils.

Propagation: Plant is generally propagated by seed which is easy to grow without much care. Within a few days sowing its seeds germinate. However, flooding the patch of land or pot with water will cause the plants to die. 1 kg seed is required for 1 ha area. Transplanting is done in the month of March - April at a spacing of 45 x30 cm. It is 85-90 days crop.

Manures and fertilizers: 10-15 tonnes of FYM or 5 tonnes of vermicompost/ ha at the time of field preparation is recommended. P and K @ 40:40: kg/ha, prior to transplanting and 80 kg/ha N in there splits after the establishment of the plants.

Harvesting and yield: The flowering stems are harvested when the flowers first appear in summer right through to autumn. The leaves and flower heads are then dried and blanched for use in infusions or decoctions. The first harvest is taken 90-95 days after planting and subsequently harvests after 65-75 days interval. The crop planted in February-March and May-June, are ready for harvest in April-May and August-September, respectively.

It is estimated that on an average 15-20 herbage/ha can be obtained on distillation 100 kg oil/ha can be obtained and net returns of Rs. 50,000/- per ha can be realized.

References

Gupta, R.D. 2005. Why *Tulsi* worship? *The Daily Excelsior*. 41 (308): 3 [Sunday Magazine, November 6, 2005].

Husain, S. 2013. Tulsi treatment. *The Daily Excelsion*. 49 (322): 4.

Krishnamurthy, R. 2005. 'Vishnupriya' *Tulsi*. *Bhavan's Journal*. 52 (7): 89-93.

Kumar, P. 2010. Sacred and medicinal herb Tulsi. *Ayurveda*. 1 (24): 4-5.

Patwari, H.N. 2005. Treatment from *Tulsi*. *The Daily Excelsior*. 41 (2): 1 [Sunday Magazine, January 2, 2005].

18

Indian Aloe
(*Aloe barbadensis* Mill.)

Common name: Quar gandal

Family: Liliaceae

Introduction

A variety of Aloes varying in size ranging from less than 3 cm to massive 60 cm plants have been reported. Although many of the Aloes have some medicinal or commercial values yet the most commonly known is the *Aloe barbadensis*, which is commonly known as *Aloe vera*.

Aloe vera is widely known because of its use in many cosmetic products. Over the years, this plant has been known by a number of names such as "the wand of heaven", "heaven's blessing" and "silent healer". It is also called "first aid plant", "medicinal Aloe", "burn plant" and "unguentine cactus", "medicinal plant", "plant of immortality", "lily of the desert". The name 'Aloe' has been derived from Arabic word *'Alloeh'* which means bitter because of the bitter constituent present in the leaves and *'Vera'* from Latin word which maens true.

Indian Aloe *i.e., Aloe barbadensis* or *Aloe vera* Mill. is a drought hardy succulent and perennial plant species. It is found as wild in warmer climates and is known by various names in different languages (Table 18.1). The plant is so named as it furnishes with youthful energy and boasts of feminity.

It is a native of North Africa, Canary Islands and Spain and from there it has spread to the rest of the world. In Indian market it is known by the name 'Mussabar'. The species is also found in wild in Jammu region of J&K state and commonly known as Quar Gandal. Aloe is believed to have been used to preserve the body of Jesus Christ. Reference to its use as healing agent can be found in early Egyptian, Chinese, Greek, Christian and Indian literatures. Legend indicates that it was the desire for Aloe plants that caused Alexander the Great to conquer the Island of Socotra, where Aloe used to be cultivated in the fourth century B.C. Leaves of *Aloe vera* are also thought to have been one of secrets of Cleopatra's beauty. It has number of medicinal, ornamental and cosmetic uses.

Table 18.1: Names of *Aloe vera* in different Languages

Language	*Name*
Botanical	*Aloe vera*
English	*Barbolin* and *Barbados Aloe*
Sanskrit	*Kumari* (meaning young virgin girl)
Hindi	*Ghee Kunwar*
Dogri	*Kunwar Gandal* or *Quar Gandal* or *Kawaar Gandal*

Historical Background

There are conflicting reports about the number of Aloe species. According to Pareek *et al.* (1999), the genus Aloe is represented by 275 species, of which only four are reported from India. They further added that Aloe is indigenous to the Mediterranean countries like Greece and Italy.

There are over 250 species of Aloe in the world mostly native to Africa (Elanchezhyan, 2006). Jamwal (2006) has reported that there are about 500 species of Aloe which mostly belong to African continent. Ancient Greeks, Arabs and Spanish and of course Chinese, West Indies and Indians used to utilize this plant throughout the millennia. African hunters still rub the gel of Aloe on their bodies to reduce perspiration and their smell.

In many of the Indian villages, it is used for heeling of wound or burn/injury due to acid or a sharp weapon or knife. The Aloes are sometimes confused with agaves, but the latter have fibrous leaves whereas the leaves of the former are succulent and not fibrous.

Aloe vera in India

Two strains of the Indian Aloe are known to occur (Pareek *et al.*, 1999). These are: (i) vegetative type; and (ii) non-vegetative type. The vegetative type has lower content of bitter component and the leaves-pads are somewhat sweet. Its pads are, therefore, used as vegetable. They can also be utilized for preparing pickle and for salad making.

The non-vegetative type owing to much higher content of bitter component, its pads are generally used for medicinal purposes. The newly emerging immature inflorescence stalks in both vegetative (sweet) and non-vegetative (bitter) types are almost free from bitter components and make a delicious vegetable. In industry, the mature pads are extensively used to prepare shampoos, face creams and moisturizing agents.

This species possesses the ability to close its stomata completely to avoid loss of water, thereby enabling it to survive long periods of drought. Thus this species has potential to be grown on degraded lands under rainfed conditions. This species is promoted for commercial cultivation under "National Ayush Mission". 20 per cent subsidy is provided to the growers for cultivation of this species under this mission.

Chemical Composition

The *Aloe vera* plant being succulent in nature contains about 96 to 99 per cent water. The remaining part contains 20 minerals, 18 amino acids and 12 vitamins. The other compounds include essential oils, carbohydrates, glycoproteins, enzymes, phenolic compounds or anthraquinones, lignins, saponins, sterols and salicylic acid. Different kinds of organic and inorganic compounds like monosaccharides and polysaccharides, minerals, vitamins and enzymes found in the gel have been given in Table 18.2.

Table 18.2: Chemical Composition of *Aloe Vera*

Sl.No.	*Name of Component*	*Type*
1.	Monosaccharides	Glucose, L-rhamnose, mannose, galactose
2.	Polysaccharides	Cellulose
3.	Minerals	Phosphorus, potassium, sodium, calcium, magnesium, iron, zinc, manganese, copper, chlorine
4.	Vitamins	Carotene-precursor of vitamin A, vitamin B, vitamin C, vitamin E
5.	Enzymes	Amylase, phosphatase, kinase, catalase, cellulase, lipase, nucleotidase, phosphokinase, protease
6.	Anthraquinoes	Aloin, barbaloin, isobarbaloin, anthranol, anthracene
7.	Organic acids	Cinnamic acid, aloetic acid, chrisophanic acid

Aloe vera gel is a thin jelly like substance obtained from the parenchymatous tissues that makes up the inner portion of the leaves. The gel contains carbohydrates polymers such as glucomannans or pectic acid as well as other organic and inorganic compounds.

The latex, commonly referred to as "Aloe juice", is a bitter yellow exudate from the pericyclic tubules just beneath the outer skin of the leaves. It also contains number of inorganic and organic components.

Recently, it has been found that aloe contains vitamin B_{12}, folic acid and choline besides vitamins A, C and E which are anti-oxidents (Khandale and Dadhich, 2013). Similarly, enzymes like bradykinase, alliase and carboxypeptidase are also present in the aloe apart from the enzymes which have been presented in Table 18.2. Bradykinase helps to reduce excessive inflammation when applied to the skin topically, while others help in the breakdown of sugars and fats. Not only this, a glycoprotein with anti-allergic properties called 'alprogen' and a novel anti-inflammatory compound, C-glucosylchromone, have been isolated from its gel (Khandale and Dadhich 2013).

Medicinal Values

Aloe vera is considered to be one of the most valuable medicinal herbs which rejuvenates the entire system and can be consumed to have benefit at any age and during any season. The juice of leaves contains an important component of variety of Ayurvedic formulations. *Quar Gandal* or *Guarpatha* leaves contain 1.6 to 68.7 mg 100 ml^{-1} "aloin", an active ingredient having laxative and healing properties. For pharmaceutical use as a laxative, the juice is often dried to produce "Aloe" granules which become dark brown owing to exposure of air. These have shown to have analgesic and inflammatory effects also. The pharmacological active ingredients are, indeed, concentrated both in gel and rind of the *Aloe vera* leaves.

Aloe vera gel has been used for treatment of wounds, minor burns and skin irritations. It can also be used as a beverage. Various products of *Aloe vera* have been promoted for internal use against constipation, coughs, wounds, ulcers, cancer, diabetes, headaches, arthritis, immune system, deficiencies and have a modulating effect on Human Immunodeficiency Virus (HIV). It possesses antimicrobial properties also. It is a favourite remedy for intestinal worms in children.

Aloe vera has long been in use to cure several diseases, particularly connected with the digestive system. In fact, the juice is soothing to digestive tract irritations such as colitis and peptic ulcers. As a food supplement, *Aloe vera* is known to facilitate digestion, aid in blood and lymphatic circulations as well as help in kidney, liver and gall bladder functions. *Aloe vera* contains at least three anti-inflammatory fatty acids which are helpful for the stomach, small intestines and colon. Naturally, the juice of *Aloe vera* is alkaline to neutral in reaction and this prevents acidity in the digestive juice.

A newly discovered compound in *Aloe vera* called acemannan, is being studied for its ability to strengthen the body's natural resistance. Acemannan also boosts T-lymphocyte cells that aid the immune system.

The clear gel is used commercially in variety of preparations like first aid creams, shampoo and soaps. The bitter Aloe in dried sap is very potent laxative.

Aloe vera can be used as a beverage and products for internal use *i.e.*, for controlling constipation, coughs, ulcers, diabetes, cancer, headaches, arthritis, immune system, deficiencies etc. However, only substantiated internal use is as a laxative. It also acts as a general liver tonic and its regular use tones up the system. It reduces laxity of large intestines and helps to invigorate the abdominal walls. *Aloe vera* has been found to be stimulant to the uterus and vagina. And as such, it has long been used as household remedy to promote menstruation and to relieve muscle spasms. The leaf pulp of this plant has been found very effective when is applied externally on piles.

The juice of Aloe is taken as snuff which is useful to get relieve from jaundice and given with turmeric powder in spleen enlargement (Khandale and Dadhich, 2013).

When the leaf is broken, its gel can be placed to relieve pain and prevent blisters. It may reduce inflammation, decrease swelling and redness and accelerate wound healing.

Aloe can help in keeping the skins supple (flexible), and has been used in the control of acne and eczema. Different methods of using *Aloe vera* pulp, juice, leaf etc., have been shown in Table 18.3.

Table 18.3: Traditional Methods of Using *Aloe vera* Pulp, Juice, Leaf, etc.

Prescription	*Method*
Use of pulp extract	The simplest method is to extract the pulp after grinding in the mixer and diluting with water. Take about 10 ml of the diluted extract after mixing with 5 g of pure honey daily in the morning and evening.
Use of pulp	The pulp can also be mixed with wheat flour or gram flour prior to making the bread. One such bread is sufficient for an adult. It should be taken early in the morning. About 50 g pulp is sufficient to prepare 5 breads.
Use of juice with milk	The juice of the *Aloe vera* (2 to 5 ml depending upon the age) can be taken. If a glass or cup of milk is taken after eating the juice, it proves very useful and is considered the best technique.
Use of cooked juice	Another best method to make use of *Aloe vera* juice is to cook in the milk and take in the breakfast.
Use of juice with glycerine	Its juice after mixing with glycerene can be rubbed on the face, hands and feet.
Use of juice	If the hair is weak or grey prior to aging the juice of *Aloe vera* should be applied under the hair and after a few hours, wash the hair. This practice should be continued for about a month. After a month, the hair appears to become black again.
Use of gel	To tone the inner muscle and to cure inflammation of vagina and other local infections, prepare a douche of two tablespoons of the gel in cup of warm water and add two pinches of turmeric powder to the solution. The douche should be used every other day for about 4 days.
Use of leaf exudates	The leaf exudates – a bitter yellow liquid can be heated, concentrated and dried to a black powder. This becomes the source of drug *Aloe*, which is used as purgative.
Use of whole leaf	The whole leaf of *Aloe vera* or its products can be used directly on radiation burns, thermal burns, wounds and diabetic ulcers.
Use of juice with honey	Crush leaves; take six tea spoonful of extracted juice thrice daily with honey for 10 days. This will help in checking menstrual disorders.

Botanical Description

Although *Aloe vera* is a succulent perennial herb having stolons just like cactus yet unlike the latter, the former's juice (gel) is not harmful to the eyes of the humans. White coloured latex of the cactus is harmful to the eyes of human beings due to its acidic nature.

Aloe vera grows as a clump up to 60-90 cm. The plant has well developed fibrous root system. The 5-7 cm long thick stem is densely crowded with long leaves having broad base. The fleshy grey-green leaves are arranged in vase shaped like rosette condition. The leaves, in fact, are dagger-shaped with jagged thorny edges, and remain about 45 cm to 60 cm long and 5 cm wide at the base. They are slightly grooved on top with a sharp point in the end. The leaves contain gelitinous material (semi-solid). This pulp on drying becomes solidified and is known as *Quar* or *Kumari-sar*. During winter and spring, the plant bears small tubular orange, yellow or red flowers. The flowering shoot arises from the centre of leaves. Older species may even bloom producing a tall stock covered with bright coloured coral flowers. *Aloe vera's* flower nectar is a favourite of humming birds.

The inflorescence is simple, dense raceme and bears actinomorphic flowers, having lanceolate-acute bracts, yellow cylindrical perianth tube, superior trilocular ovary with axile placentation.

All Aloes are semi-tropical succulent plants and may only be grown outdoors in areas where there is no chance of freezing temperature. However, they make excellent house plants when they are having sufficient light.

Cultural Practices

Climatic requirements: Climatically, *Aloe vera* is a drought hardy plant and can be grown best under hot and arid region with annual rainfall of 100-300 mm. Its plants being susceptible to low temperatures or frost, are required to be protected. Water logged conditions are not congenial for its cultivation. In India, *Aloe vera* is found wild in arid and semi-arid regions of Rajasthan, Gujarat, Andhra Pradesh, Madhya Pradesh, Maharashtra, Tamil Nadu and Karnataka. Indian Aloe is also grown extensively in Malta, Cyprus, Canary Cape etc.

Soil requirements: The plant prefers well drained, gravely sandy loam, loamy sand and/or sandy or clay soil. It can even be cultivated in sandy soils having low soil fertility as well as on rocky and sandy wastelands under rainfed conditions. The soils of *Kandi* belt of Jammu *vis-a-vis* those of Punjab, Haryana, Himachal, Uttrakhand and Chandigarh are also well-suited for the cultivation of *Aloe vera*.

Propagation and planting: The seed is not formed in this species due to male sterility and thus it is propagated vegetatively through suckers. A healthy mother plant may produce as many as 4-10 suckers during one year. The plantlets are removed from mother plants when they are about 2 months old and about 2.0-5.0 cm tall at 4-5 leaf stage. Propagation by stem cuttings is also possible but on a limited scale. Healthy suckers with 4 to 6 leaves are considered to be the best for planting.

The land should be ploughed twice, harrowed and planked. Arrangement for proper drainage should be made. Planting should be done at a spacing of 60 60 cm in well prepared beds in which sufficient quantity of FYM, compost or vermicompost has been incorporated. Application of 10-15 tonnes ha^{-1} FYM has been found beneficial for good production of leaves in desert soils of Bikaner (Rajasthan) (Pareek *et al.*, 1999). Application of the wood ash at the time of planting helps in establishment of plants and their subsequent growth.

The planting can be done throughout the year except November, December and January. Planting is done during monsoon season *i.e.* from July to September in small holes in which about 100 g of ash has to be added to protect the plants from an assault of termites. However, where irrigation facilities are available planting of *Aloe vera* can be done at any time during the year, but preferably during January-February. When planting is done during this period, application of irrigation becomes necessary every day upto 7 to 10 days after its plantation. This ensures establishment of plants cent per cent. Rooted plantlets should be planted at a distance of 30- 45 cm in rows which are 60 cm apart. About 37000-55,500 plantlets are required to plant a hectare area. Coarse sand, pebbles and gravels should be spread near the base of the plantets so as to avoid rottening of the leaves.

Manures and fertilizers: In general, neither irrigation nor fertilization practices are adopted. Although still systematic experimental studies have not been done so far with regard to their nutrients and water requirements yet the *Aloe vera* plants have shown positive response to fertigation.

Application of 25 g diammonium phosphate (DAP) plant^{-1} year^{-1} has given good results (Pareek, *et al.*, 1999). This dose works out to be 350 to 400 kg of DAP ha^{-1}, with 14000 to 16000 plants ha^{-1} equivalent to 63 to 72 kg of N and 161 to 184 kg P_2O_5 ha^{-1}. Sharma *et al.* (2006) recommended the application of potash also in addition to N and P_2O_5 @ 300, 150 and 300 kg as N, P_2O_5 and K_2O ha^{-1} respectively.

Irrigation: For establishment of plants, weekly irrigation is essential up to one month after planting. When established, the plants can tolerate drought but irrigation would give higher yield. In all, 4-6 irrigations are required per year according to the moisture content of the soil. However, water logging should be avoided.

Intercultural Operations: Hoeing is essential every month to control weeds and break the crust of the surface soil. 2-3 weedings per year should be done. In case of more weed infestation, the frequency of weeding should be increased. Use of mulch with local material throughout the year has been found useful in conserving moisture and adding organic matter.

Pinching to remove upper 1-2 cm tip of leaves has been found to increase thickness and mucilage content in the pads.

Inter-cropping: *Aloe vera* plants can also be grown as intercrop within the rows of Ratanjot (*Jatropha curcas*), a bio-diesel plant.

Pests and diseases: The plant is quite hardy and is not affected much by pests and diseases of serious nature. Only leaf rot and anthracnose diseases caused by *Alternaria alternata and Fusarium solani* have been reported in *Aloe vera*. The disease can be controlled by spraying Dithane M-45 or Dithane Z-78 @0.2 per cent at monthly interval. For controlling mealy bug, spraying *neem* oil (10 ml L^{-1} of water) at 20 days interval during early stages should be done. For control of diseases integrated disease management practices like sanitation, removal of infested leaves, irrigation at optimum time should be adopted (Sharma *et al.*, 2006).

Harvesting and yield: *Aloe vera* starts giving yield from second year normally after its 8-9 months of planting. The matured leaves are harvested for vegetable purpose at monthly interval. On an average, 3-5 kg leaves per plant can be harvested per year in 2-3 pickings. Fully mature leaves should be harvested from outer margins of the plants. Aloin content has positive correlation with short, broad and thick leaves. Leaves are carefully harvested without injuring the stem. After harvesting, the leaves are washed, packed and transported to the market for sale.

For industrial use, harvesting is done in bulk for transportation to distant places and economical yield can be obtained after 3-4 years of its planting. On an average annual yield in terms of green leaves from 3-4 cuttings is 10-20 tonnes ha^{-1}. About

20 tonnes leaves may be obtained ha^{-1} in 14-16 months after planting (Ramamoorthy *et al.*, 2012). For cultivation of one hectare of land it costs about Rs. 1, 25,000 while the profit comes around Rs. 50,000 per hectare.

Sap prepared from the fresh leaves should be kept at low temperature for some time otherwise it may spoil.

References

Elanchezhyan, K. 2006. *Aloe vera* – A wonder plant. *Agrobios Newsletter*. 5 (5) : 41-42.

Jamwal, J. 2006. *Aloe vera* : The plant of immortality. Personnel Communication. 189 Sector-1, Trikuta Nagar, Jammu, J&K State.

Khandale, S. and Dadhich, O. P. 2013 Gire friendly-aloe vera *Ayurveda*. March-April Issue, pp 8-9.

Pareek, O.P., Nath, V., Singh, R.S. and Bhargava, R. 1999. Grow Indian Aloe in arid region. *Intensive Agriculture*. 37 (5-6) : 21-23.

Ramamoorthy, K., Shingare, R.P. and Kalaichelvi, K. 2012. Aloe vera- A mineral plant. *Agrobios Newsletter*. 10(11): 41-42.

Sharma, K.K., Sambyal, K.S., Sharma, N.K., and Gandotra. S. 2006. Agronomical Practices for the cultivation of High Demand Plant Species of Medicinal and Economic Importance. Directorate of Command Area Development, Jammu, J&K State.

Sharma, O.P. 2000. 151 Healing plants of Jammu and Kashmir. Directorate of Social Forestry Project, Jammu, Government of Jammu and Kashmir.

19

Indian Atees (*Aconitum heterophyllum* Wall.)

Common name: Patees, atees, vatsnabh

Family: Ranunculaceae

Introduction

Aconitum genus includes number of important medicinal species. On the basis of the type of aconitine present in them, they have been classified into poisonous and non-poisonous aconitum. *Aconitum heterophyllum* is one of the non-poisonous species, which is mainly found in alpine and sub-alpine Himalayan belt of Uttrakhand, Himachal Pradesh and Jammu and Kashmir (Punetha *et al.*2013; Chauhan; 2006 and Vidyarthi, 2014). It usually grows between 3000 to 4300 m altitudes. Like other species of *Aconitum*, *Aconitum heterophyllum* is a medicinal herb.

Medicinal Values

Vatsnabh is a valuable febrifuge and a bitter tonic. It is used against diarrhoea and dysentery. Powdered root mixed with honey is effective against cough, fever and vomiting. Root is also used against toothache. Tubers and roots are bitter in taste and are claimed to be useful in acute inflammation, loss of memory, piles and throat diseases.

Aconitum heterophyllum has been previously reported to have anti-diarrhoeal activity along with astringent and tonic properties. It is now used as a key ingredient of anti-diarrheal medicine 'Diarex vet'. The activity of 'Diarex vet' against lactose induced diarrhoea in Wistar rats has found it as a potent anti-diarrhoeal agent.

Botanical Description

Aconitum heterophyllum is a perennial, erect herb of 30-90 cm in height and occurs at high elevations in wet alpine zone. Leaves are alternate, ovate, lanceolate, toothed. Flowers are helmet shaped, bright blue to yellow greenish blue with purple veins in long. There are many flowered peduncles. Fruit is a 5-celled capsule. Seeds are dark brown.

Cultural Practices

Climatic requirements: The plant prefers temperate to alpine/sub-alpine type of climate. This is why it is mostly found in the sub-alpine and alpine zones of the Himalayas from Indus to Kumaon at altitudes between 3000-5000 m (Chauhan, 2006).

Soil requirements: The herb flourishes well in humus rich soil with good retention of moisture. Sun shade or partial shade is required for its good growth.

Propagation: The herb can be cultivated/propagated through seeds. Seeds have to be sown during March-April after snow melts and the seedlings are transplanted at a distance of 60x60 cm after about 1.5 to 2.0 months of sowing.

Roots are collected during September - November from the wild pockets and at the same time from cultivated fields after three years of planting. Roots are dried, freed from dust, root fibres and then graded before packing and storing.

Harvesting: Usually flowering occurs during July to September and seeds can be collected from the fruits during September to October.

Previously in North-West Himalayan states, various species of *Aconitum* including *A. heterophyllum* used to occur abundantly but now have become rare due to their over exploitation. Infact, their niche is facing the risk of survival. So it is essential that state governments must take suitable steps to conserve them. On an average 3-4 quintals dry roots per hectare are obtained after 3 years.

References

Chauhan, N.S.2006. Medicinal and Aromatic Plants of Himachal Pradesh. Indus Publishing Company, FS-5 Tagore Garden, New Delhi-110027.

Punetha, H., Khokhar, D and Gaur, AK.2013. *Aconitum* - A genus of high medicinal value. *Indian Farmers Digest*. 46(05): 26-28.

Vidyarthi, O.P.S. 2014. 151 Medicine Plants of Jammu and Kashmir Directorate of Social Forestry, Jammu and Kashmir Government.

20

Indian Pennywort (*Bacopa monnieri* Linn.) Syn. *Herpestic monnieri*

Common name: Brahmi buti

Family: Scrophulariaceae

Introduction

Brahmi and Brahmi *Buti* (*Centella asiatica*) or *Ghor-sumbi* (Jammu) of family Apiaceae, are often called the "thinking person's herbs" or "Indian water hyssop". Both these herbs are greatly revered in the ancient Ayurveda medical systems and used from school children to scholars. In Ayurveda, Brahmi has been described as *"Medhya rasayana"* or brain tonic with the ability to promote mental functioning along with general rejuvenative characteristics.

In fact, two different herbs *viz.*, *Centella asiatica* and *Bacopa monnieri* are the classical brain and nerve tonics in the Brahmi category of herbs. These give benefits both to the mind and spirit and improve the intellect and consciousness. These are, regarded to be the main rejuvenating herbs for nerve and brain tissues/cells, and have, therefore, played a very important role in Ayurvedic therapies for the treatment of cognitive disorders of aging. The indigenous system of medicines and their entire plant constitute drugs, used as potent diuretic, cardio-tonic as well as a valuable nerve tonic for curing insanity, epilepsy, mental disorders etc.

A. Brahmi Buti (*Centella asiatica* Syn. *Hydrocotyl asiatica*)

It is one of the species of Brahmi which is locally known as "Brahmi *Buti*" or *Ghor-Sumbi*. It is also called *Gotu Kola* (Anonymous, 2003). Its Sanskrit name is *"Manduk Parni"*. Like *Bacopa monnieri*, it has pharmaceutical effects of naturally revitalizing the brain cells and enhancing mental capacities like memory concentration and verbal articulation. According to Singh (2005) Brahami buti is a prostrate herb which is found along the canals, streams and other wild moist habitats.

Chemical Composition

All parts of the plant *viz.*, fresh and dried leaves, stems, roots and seeds are used in preparing medicines. The herb contains saponins. Saponins are glycosidic triterpenoids, which are soluble in water giving stable forms. These compounds protect against fungal attack. Some of the substances present in saponins like aristolochic acid-1/aristolochic acid-A, medica genic acid-3 and medicogenic acid-3, triglucoside (triterpenoid saponin) act as antifertility, antifungal and haemolytic factors, respectively.

Nutitional Value

Both the herbs are nutritionally very rich as they also contain vitamins A, B, E, G and K, and magnesium (Anonymous, 2003). They have high content of Vitamin B_1 (thiamin), Vitamin B_2 (riboflavin), Vitamin B_6 (pyridoxin) and carotene or provitamine A. Other nutrients include numerous free aminoacids especially serine, threonine, alanine, lysine, histidine. Aspartate, glutamate and amnobutyrate are also present. In the traditional Chinese medicine *Gotu Kola*, particularly is believed to provide longevity and thus called "foundation of youth" herb.

Gotu Kola contins brahmoside, brahminoside, brahmic acid, isobrahmic acid, asiaticoside and centelloside. The primary active ingredients include triterpenoid compounds that are basically "saponins". Saponins are known to have a beneficial effect on collagen, the material that makes up the connective tissue or the basic glue that holds the body cells together. The centella oil obtained from brahmi is a green, strongly volatile oil composed of unidentified terpene acetate, camphor, cineole and other essential oils. It also contains fatty acids, various plant sterols, such as campesterol, stigmasterol and sitosterol and various polyacetylene compounds.

The protein kinase activity and new protein synthesis occurs in the brain cells which are particularly concerned with learning and memory.

Medicinal Values

In India, the formulation has great reputation as brain tonic. It has tranquilizing, sedative (tending to soothe the nerves) and spasmolytic (convulsive tightening of muscles) and antiamoebic properties. It has also anti-anxiety and anti-stress action. Being anabolic in action, it enhances the retarded growth of the children besides improving their behavioural pattern.

Brahmi builds up nerve cells and detoxify them from buildup of heavy metals and toxins. Brahmi *Buti* or *Gota Kola* is a rejuvenating brain tonic that is even known to provide relief in epilepsy. Although Brahmi has been shown to be helpful in treating Attention Deficit Disorder (ADD), age related mental deterioration and concentration difficulties due to stress, it may be incredibly beneficial in improving cognitive function of normal individuals. It induces a sense of calm and peace for those who suffer from stress and nervous anxiety, while at the same time invigorates the mental processes (Rashmi *et al.*, 2012). Brahami herb acts as a natural remedy which is reputed to stimulate hair, skin and nail growth.

Used extensively in Asia for the treatment of leprosy, healing of skin disorders like dermatitis and cellulites. It basically enhances protein growth or keratinization in the skin and stimulates formation of lipids and proteins necessary for healthy skin. Traditional usage has been described in Table 20.1 below:

Table 20.1: Methods of Using *Brahmi Buti*

Parts Used	*Use and Advantages*
Leaves	The leaves are used to prepare elixir (*Sharbat*) named as PEP-UP syrup. It increases appetite, improves digestion, renders the mind cool and refreshes the brain and sharpens the memory too.
Fresh leaves with sugar or *misri*	After mixing with small amount of sugar or *misri* (sugar crystals) 4-5 leaves are chewed early in the morning, empty stomach. Its continuous use daily for about a month not only increases the vigour of the brain but also enhances memory.
Fresh leaves with sugar or *misri* and black pepper	Due to cooling effect of the leaves, its use must be made during summer season *i.e.*, in the May-June. Sometimes its use with sugar or misri daily during summer months does not suit children, in such cases 2 to 3 grains of black pepper should be mixed prior to chewing the leaves.
Fresh leaves with sugar and milk	Eight to ten leaves of the herb are mixed with 8-10 kernels of almond (which have already been soaked in water overnight). These are then ground in a mixer or mortar with pastel. Thereafter, 200-250 ml of milk is added in the ground paste followed by sugar according to the taste, taken empty stomach in the morning. It should be used continuously for about a month. By this act, there is an improvement in digestibility, strengthens the limbs and muscles of the body, sharpens memory and revives the vitality of whole of the body including brain.

Botanical Description

It is commonly found in plains particularly on the banks of the canals and their tributaries. In Jammu region of the Jammu and Kashmir, it mostly occurs naturally on the banks of the irrigation canals and along sides of the *Khuls*. It prefers to grow on moist or damp places with sandy loam to loam or silty loam to clay loam soils, (Gupta, 2002).

The Sanskrit name *"Mnduk Parni"* indicates very closely resemblance of the leaves of this plant with the claws of the frogs. It is a prostrate mat forming herb and stems bear roots on their nodes.

Leaves of the plant are small in size with mostly 1 to 2 cm in diameter which sometimes may extend upto 4 cm. They are reniform in shape and stalked. Occasionally, the leaves become more rounded or broad kidney shaped. Leaves are somewhat bitter in taste. Flowers are very minute in size. They are of varying colours but mostly either pink in colour or red or sometimes green (Sharma *et al.*, 2006). Flowers are usually 7 to 9 in number like a ridge.

Cultural Practices

As already stated, "Brahmi *Buti*" grows naturally so not much work has been done so far on its various cultural practices. However, the information with regard to its climate and soil requirement is as under:

Climatic requirements: Brahmi *Buti* grows well under varying climatic conditions, ranging from sub-tropical (300 m) to sub-temperate or intermediate climatic zones upto (1300 m). It also grows in Himalayan range from 600 -2800m (Singh, 2005).

Soil requirements: "Brahmi *Buti*" grows on variety of soils, ranging from loamy sand/sandy loam to loam or silty loam to clay loam. The pH of the lies between 7.5 and 8.5. However, it can also grow well naturally in acidic soils having pH range of 5.5 to 6.5 in high rainfall humid temperate localities as in Palampur, district Kangra, Himachal Pradesh.

B. Brahmi (*Bacopa monnieri*) Linn. Peneil

It is also called *Neer*-Brahmi, *Jal*-Brahmi, *Lonkia, Jalneem* and estimated that about 10,000 tonnes of fresh herb is collected from wild habitat every year for various herbal formulations available in the market. Retail price of dry herb in different parts of India varied from Rs. 20 to Rs. 50 kg^{-1}. Annual requirement of Brahmi is around 2700 tonnes of dry biomass which is solely being met from natural habitats leading to their gradual depletion.

Studies have shown that single oral dose of 20-200 mg or 100-300 mg daily for about 4 weeks were safe and did not produce any reaction or side effects. A significant improvement in logical memory has been observed from 4th week onwards in the group treated with the standardized extract without any side effects.

Chemical Composition

The systemtic chemical examination of the plant has revealed the memory enhancing effects of Brahmi as a mixture of triterpenoid saponins designated as bacosides A and B. Brahmi contains 1.7 per cent N, 0.4 per cent P and 1.8 per cent K. Apart from them, it also contains hersaponin which resembles with resperine and chloropromazine in the central action and is reported to be a promising tranquilizer.

Medicinal Values

Brahmi (*Bacopa monnieri*) is unique in its ability to invigorate mental process, reducing the effects of stress and nervous anxiety. It is well known for its salutary effect in anxiety, depression, hypertension, sleeplessness, mental retardation, insanity and hysteria.

It is also used for treating bronchitis, chronic cough and asthma, hoarseness, arthritis, rheumatism and backache. Brahmi has proved boon to inflammatory conditions, fluid retention, blood cleanser, chronic skin conditions, constipation and hair loss.

Use of Brahmi controls digestive problems, fevers, post-natal depression, improves blood circulation and strengthens capillaries. Consumption of Brahmi removes impotence, premature ejaculation and irregular menstruation and physical fatigue. Also Improves mental clarity, confidence, intelligence and memory retention.

Brahmi has been esteemed as a cardiac tonic nerve or nervine and brain tonic, insomnia and overactive mind, forgetfulness, confused and loudy thoughts. No side effects have been reported. Brahmi can be taken practically by everyone, regardless of age and gender. It also possesses laxative properties. Brahmi has even been used to treat rheumatism, blood diseases, hepatitis and high blood pressure and usage of different plant parts has been described in Table 20.2.

Table 20.2: Use of *Brahmi* (*Bacopa monnieri*)

Parts Used	*Use and Advantages*
Leaves	These are placed on the chest in acute bronchitis and coughs in children.
Leaves and stems	These are placed around the urinary tract to cure urinary problems.
Juice of plant	About 8 to 10 ml of juice once a day has been traditionally used to increase the speed of learning and memory power.
Dry powder	Instead of juice, 3-4g of dry powder (approximately 6 sprigs of stem 10 cm long with leaves) is also effective in increasing learning and memory power.
Whole of the herb as tea infusion	The herb can be prepared as a tea infusion, tincture or decoction, it can also be sweetened with honey or *Stevia* leaves
Whole plant as salad, soup, dish	In India, the plant is used in salads, soups, as cooked vegetable and pickle.

It has complimentary effects on the functioning of liver, kidneys and adrenal glands. It also strengthens the circulatory system as it improves blood flow. Many studies conducted on Brahami have shown its potential role in cancer treatment, cardio-vascular disorders, Alzheimer's and Parkinson diseases etc. (Rashmi *et al.*, 2012). Brahami is one of the simple herbs which has a soothing effect on the irritated nerves (Vatsyayan, 2010).

Botanical Description

Bacopa monnieri is a perennial herb commonly found in damp and marshy areas near streams/canals and on the periphery of ponds almost throughout India (Sharma *et al.*, 2006). The plant grows exceptionally well in waterlogged areas.

Brahmi is a prostrate, succulent, glabrous herb rooting at the nodes with numerous branches which form a dense mat of vegetation covering the soil surface. The leaves are oval-shaped, 1.5 cm long, in pairs along the stems. Flowers white (with a pale hue of blue on the outside of the petals), tubular 5 petals formed in the leaf forks and can blossom over many months of the year. Since its roots are shallow, so if grown in the garden it becomes important to water well during times of dry weather.

Cultural Practices

Climatic requirements: *Bacopa monnieri* grows well under varying climatic conditions but ideally suited for sub-tropical conditions with temperature (33-40° Celsius) and humidity (65-80 per cent) (Gautam and Chauhan, 2004). The herb naturally grows in the mid hills of Jammu (Sharma *et al.*, 2006) and is traded as a medicinal plant locally.

Soil requirements: *Bacopa monnieri* can grow on all soil types except sandy. Loamy sand/sandy loam, loam and silty loam are considered to be the best provided they are rich in organic carbon and available N, P and K contents.

Propagation: Although Brahmi is a small succulent herb with slow initial growth and creeping on soil but is easy to propagate by stem cuttings. It can also be multiplied through seeds. 100 kg suckers or stem cuttings are required for an hectare.

Method of sowing/planting: Deep ploughing followed by planking is done. Stem cuttings are sown directly in the field. Suckers having length, of 5 to 7.5 cm are sown in the nursery and then about 2 months old plantslets are transplanted in the field.

Irrigation: Brahmi is a long duration crop with high water requirements. Its growth period, however, coincides with the monsoon season and as such during rainy season it does not require much water. After planting of the cuttings/suckers irrigation is essential. In winter season, irrigate after 20 days interval and in summer after 15 days interval.

Time of sowing: In irrigated areas, it is sown during February-March and in un-irrigated condition, best time of sowing/transplanting is rainy season (July-August).

Manures and fertilizers: 15-20 tonnes of FYM ha^{-1} is the recommended dose to get good crop. Usually no inorganic fertilizers are applied if sufficient FYM has been applied. However, in case of shortage of FYM, 50, 30, 30 kg of N, P_2O_5 and K_2O are added ha^{-1}.

Weeding and hoeing: In the initial stage of crop growth weeding and hoeing can be done at an interval of 15-20 days. This will check the growth of weeds.

Harvesting and yield: The plant foliage is cut about 4 to 5 cm above ground level, dried in oven at 70-80° Celsius for half an hour and then dried in shade. A yield of 100 kg dry herb ha^{-1} is obtained.

References

Anonymous. 2003. Brahmi-a brain tonic. *Science Reporter*. February issue, pp. 33.

Gupta, R. D. 2002. Beneficial butti. *The Daily Excelsior*. 38 (222): 1 (The Sunday Magazine).

Rashmi, I., Neenu, S., Radha, T. K, Mandal, A. and Shivangi, K. C. 2012. Brahmi-Herb of grace. *Agrobios News letter*. 10(1): 39-41.

Sharma, K.K., Sambyal, K.S., Sharma, N.K., and Gandotra. S. 2006. Agronomical Practices for the cultivation of High Demand Plant Species of Medicinal and Economic Importance. Directorate of Command Area Development, Jammu, J&K State.

Singh, S. S. 2005. Fate of the Medicinal Plant Resources of Jammu and Kashmir. Directorate of Environemnt and Remote Sensing, Jammu.

Vatsyayan, R. 2010. When winter makes you feel depressed, your diet can help. *The Tribune*. 130 (338): 12.

21

Indian Snakeroot (*Rauvolfia serpentina* L. Benth.ex Kurz)

Common name: Sarpgandha

Family: Apocynaceae

Introduction

Serpentine plant was named 'Rauwolfia' after name of German Scientist Dr. Leonard Rauwolf of sixteenth century. This plant is mostly cultivated in Pakistan, Burma, Java, Nepal, Sri Lanka and India. In India, the plant occurs in Himalayan region of Uttrakhand, Sikkim and Assam. It is also cultivated in West Bengal, Bihar and South India.

Serpentine is widely used medicinally in all the systems, in the Modern Western Medicinal system and in the Ayurvedic System of Medicine as well as in the Unani and in folk medicine (Jamwal, 2012).

It is a perennial, under-shrub. The species is widely distributed in tropical and sub-tropical regions of North, North-West and South India. The plant has also been reported in Jammu region of J&K State.

Chemical Composition

The major alkaloid present in root, stem and leaves of the plant is 'proserpine', the content of which varies from 1.7 to 3.0 per cent. The root and bark has more than

90 per cent of the total alkaloid. Ejatine, ejmaline, sarpagine, serpentinine are the other alkaloids present.

Medicinal Values

In Ayurvedic System of Medicine, Serpentine has been used for centuries for the relief of various central nervous system disorders, both psychic and motor neurotic including anxiety states vis-à-vis excitement, insanity, insomnia, asthma, acute stomach ache and painful delivery of child besides controlling high blood pressure and epilepsy. It is useful for treating hysteria.

Extract of the roots are valuable for treatment of intestinal disorders, particularly diarrhoea and dysentery, and to kill intestinal worms. Mixed with other plant extracts, it has been used in the treatment of cholera. The extract of the leaves is used for eye disorders. Serpentine herb is the best remedy for high blood pressure and this has been accepted by the medical fraternity in most of the countries. Those alkaloids which have a direct effect on hypertension have been found in it and are widely used by the practioners of modern medicine. But they have certain unpleasant side effects when the drug is taken in the natural form (Jamwal, 2012). Half a teaspoon of serpentine's root powder taken thrice a day is effective in relieving hypertension.

The roots contain pharmacologically important alkaloids *viz.* reserpine, ajmalicine, ajmaline, neoajmaline and serpentine which are used as sedative, and in certain forms of insanity. Due to large demand for roots, the species has been destructed from its wild habitat leading to its endangered status.

Other Uses

The drug has a tranquilizing effect and is used as an antidote in the stings of insects and the bites of poisonous reptiles. The root of serpentine also possesses anti-bacterial properties. Infusion, decoction and extract of the root are employed to increase uterine contraction for expulsion of the foetus to treat painful affections of bowels, diarrhoea cholera, and fever.

The root of the herb is believed to stimulate uterine contractions and is, therefore, recommended for use in child birth.

Botanical Description

Serpentine is an evergreen perennial, glabrous and erect under shrub. The maximum height of the plant is up to 60cm. The leaves are lanceolate shaped, glabrious and whorled. Inflorescence is cymose with gamopetalous flowers of white or pink colours. Flowering occurs from March to May (Jamwal, 2012) or upto November (Morsing *et al.*, 2009). The roots of this plant are snake like so it is commonly known as serpentine.

Cultural Practices

Climatic requirements: Serpentine grows well in frost free tropical to sub-tropical conditions with high annual rainfall (>150cm.). Where annual rainfall is low, its cultivation can be taken with assured irrigation facility only. It prefers hot humid climate. A climate with a temperature range of 10 – 30 C seems to be well suited for this plant. It performs well under partial shade (50 per cent shade).

Soil requirements: The plant requires slightly acidic to neutral soils. pH in the range of 6.5 to 7.5 is for good growth with medium to deep well drained fertile soil having clay loam to silty loam texture rich in organic matter.

Propagation: The crop can be propagated by seed, stem cutting and root cutting/root stumps. However, propagation done by seeds is the best method for raising commercial plantation. Hardwood stem cuttings of 15-22cm length are planted in the nursery during June. The sprouted cuttings are transplanted in the field after the formation of roots (in August). Sprouting percentage in stem cuttings varies from 40-65 per cent.

Root cuttings of 5cm length are planted in the nursery beds during February-March which sprout within three weeks of planting and can be transplanted in the field during rainy season. About 100 kg of roots are sufficient for planting one hectare area. Root cutting alonwith portion of stem above the collar is directly planted in the field during rainy season. This method gives about 90-95 per cent success, but requires large number of plants.

Seed Sowing: Seed is the best method for raising the commercial plantation, however, seed germination is highly variable (10-45 per cent). Two types of seeds (light and heavy) are obtained which can be easily separated by simple water flotation. Light seed should be discarded as they consist of atrophied embryo and do not germinate. Germination of heavy seed can be enhanced by using pre-sowing technique of alternate wetting and drying for two days. About 6 kg seed is sufficient for planting one hectare area.

Raised beds of size 10 x 10m^2 are made under partial shade. Seed bed should have equal quantity of sand, soil and FYM. About 500 m^2 area of nursery is sufficient to produce seedling for one hectare area. The seed should be treated with Bavistin @ 2g/kg and then sown in rows which are 2 to 3 cm apart, during third week of May. The seeds are to be covered with fine mixture of soil and FYM and the seed beds should be kept moist. The furrows are them covered with fine mixture of soil and FYM. Keep the beds just moist by light irrigation. Germination starts after 25-30 days of sowing and continue upto 45 days. Seedlings are ready for transplanting by mid-July.

The seedlings are transplanted at 30 cm distance within the rows spaced at 45cm. If rains do not occur during or immediately after transplanting, irrigation is necessary for better crop stand. Serpentine is long duration and slow growing crop especially in the initial stage, thus different intercrops should be tried.

Varieties/Selections

(i) **R S1** selection released by Jawahar Lal Nehru Krishi Vishwa Vidyalaya, Jabalpur.

(ii) **CIM-Sheel** variety released by Central Institute of Medicinal and Aromatic Plants, Lucknow.

Manures and fertilizers: The medicinal plants have to be grown without chemical fertilizers and pesticides. Organic manures like FYM, compost and vermicompost should be used. Wherever possible, green manuring is required to be done. The crop responds to inorganic fertilizers also. A dose of 10 kg Nitrogen, 60kg Phosphorus and 30 kg Potassium is applied as basal, whereas, 20kg of nitrogen is applied into two splits *i.e.* after 50 days and 170 days of transplanting, respectively.

Irrigation: It is grown as a rainfed crop in sub-tropical areas receiving well distributed mean annual rainfall of 150cm or above. Low rainfall areas with high temperature need regular irrigation. About 15-16 irrigations are required per year with an irrigation interval of 20 days in summer and 30 days in winter.

Intercultural operations: The field should be kept weed free in the initial period of growth. Two or three weeding and two hoeing are sufficient, where sole crop is to be taken. When intercropping is practised, 5-6 weeding are sufficient. In the second year, one or two weeding cum hoeing would be sufficient.

Intercropping: This is a long duration (18-36 months) and slow growing crop in the initial stages. Hence, to increase the net returns, intercropping is to be practised. Being a shade loving crop, it can be grown as intercrop under orchards of Peach, Amla, Citrus, Harad etc. However, yield of roots may be higher under monoculture but net returns would be more in intercropping.

Pest control: To prevent diseases, bio-pesticides prepared from neem (seeds, leaves) and drek can be used. The seedling should be treated with 0.1 per cent bavistin solution before transplanting to protect them from soil born diseases.

Harvesting and yield: The harvesting period coincides with the shedding of the serpentine leaves during early autumn next year (18 months after planting). At this stage, the roots harvested in winter contain maximum concentration of total alkaloids. As at harvest, the roots may go up to 40 cm deep in the soil so harvesting of the roots is done by digging. The root bark should be kept intact while digging out the roots, since it constitutes 40-45 per cent of total weight of the root and contain 90 per cent of the total alkaloid present. A light irrigation is to be given before harvesting to facilitate easy digging of roots. After digging roots are cleaned, washed and cut into 12 to 15 cm pieces for convenience in drying and storage. The dried roots are stored in gunny bags in cool dry places to protect from mould.

On an average root yield varies from 15 to 20 q ha^{-1} of dry weight under irrigated and optimum soil fertility and crop stand conditions. Net returns to the tune of Rs.1,25,000/- per ha can be realized under optimum conditions.

References

Anonymous, 2012. Sarpgandha. Bulletin published by Division of Agroforestry, SKUAST-Jammu.

Jamwal, J. 2012. *Rauvolfia serpentina* or Sarpagandha. *The Kashmir Times* 69(76): 7 (Saturday, 17 March).

Morsing, M.R, Kharde, A.V., Jadhav, R.M. and Nale, R.S. 2009. Medicinal value of serpentine. *Agrobios News Letter.* 7 (8): 57.

22

Kuth
(*Saussurea lapa* C.B. Clarke)

Common name: Kuth, kut, kustha

Family: Asteraceae

Introduction

Kuth is usually found at a height of 2,500 to 3,900m in South Asia including the Himalayas, Western Ghats and Kishan Ganga valley. Besides Jammu and Kashmir, it also grows in Sikkim, Arunachal Pradesh, Himachal Pradesh and Uttrakhand. No account of herbal and medicinal plants is complete without mentioning kuth. It is because of its importance that before the 2nd world war, its roots were used to be collected to the extent of 100 tonnes per year from the forest of Jammu and Kashmir also (Anonymous, 2013).

Medicinal Values

Kuth is reported to have antispasmodic, anodyne, aphrodisiac, astringent, bronchodilator, carminative, stimulant, stomachic and as hair tonic. The roots of kuth are aromatic called 'chobe-kot, are scented like iris and many are stimulant, useful against cough, asthma, dyspepsia, tooth ache and many skin diseases. It has proved beneficial to those who are suffering from chronic odontaliga, gout, headache, herpes and hiccough. It is used to lower the blood pressure (Khajuria, 2005).

Use of kuth helps to normalize and strength digestion, cleanse the body of toxic accumulation, enhances fertility and reduces pains, curing fever and cholera. Its dried powder is a principal ingredient in an ointment for ulcers. It is also used

for hair wash and relieving from ring worms. Pregnant mothers should seek medical advice if to be consumed. People suffering from high blood pressure should take extra precautions as this herb can lower the blood pressure.

Botanical Description

Kuth is a tall, perennial herbaceous herb characterized by the presence of leaves in the form of dense basal rosette. Stem is up to 2 m tall. Leaves are long-stalked pinnate of 20-40 cm size, with a triangular terminal leaflet of 25-30 cm size. Upper leaves are smaller in size, clasping the stem. Flower heads are purple with multiple florets with dense white to purple woolly hairs and look like balls. Flowering takes place in July–August. Its root is bitter in taste.

Cultural Practices

Climatic requirements: Kuth requires temperate to sub-temperate and alpine type of climate for its growth with an elevation ranging from 2500 to 3900m.

Soil requirements: Soils having sandy loam to loam texture, rich in soil organic matter and high water holding capacity are considered the most suitable for its growth.

Sowing time: The seeds are sown @ 32 kg ha^{-1} in nursery during April-May. The seedlings get ready when reach at a height of 15 cm. These are then transplanted in the well prepared pulverized fields.

Manures and fertilizers: Well decomposed organic manure or compost should be added @ 6.0 to 7.5 tonnes ha^{-1} while preparing the fields prior to transplanting and mixed in the soil thoroughly.

Irrigation: Five to six irrigations are required to get good yield. First irrigation is done when the seeds are sprouting. The remaining irrigations are given at 15 days interval subsequently before harvesting the crop.

Harvesting and yield: Flowering starts in August-September and seed collection is done in the months of September-October. Harvesting usually starts 2-3 years after planting when root tubers become fully matured/developed. However, good yield is obtained only from 3 years onwards. Roots are harvested in early September to October or in early spring *i.e.* March-April. About 500-625 kg of dried root tubers ha^{-1} are obtained.

References

Anonymous. 2013. Conservation and sustainable utilization of Medicinal Plants in Jammu and Kashmir State. Plan Proposal for Support under the PDF Scheme of the Global Environment.

Khajuria, G. L. 2005. Protecting medicinal and aromatic plants. The Kashmir Times. 62 (339): 13.

23
Lamb's Quarter (*Chenopodium album* Linn.)

Common name: Bathu

Family: Chenopodiaceae

Introduction

'Bathu' or 'Bathua' is a weed of cultivated crops in rabi season in plains and rainy season in hilly areas. It is well distributed in Europe, North America and Central and West Asia. In India, it occurs in plains and hills of Northern India as a weed.

Despite a seasonal weed in the rabi crops, it is relished by the cattle as fodder particularly during the acute scarcity. In addition to this, it is also cooked as a vegetable for human consumption with or without mustard (sarson) stalks or leaves. Leaves are rich source of minerals like calcium and iron as well as Vitamins especially Vitamin A and Vitamin C, and as such it possesses lot of medicinal uses.

Nutritional Value

Bathu is a good source of carbohydrates and proteins vis-à-vis crude fibre (Table 23.1) and total ash content (Pandey and Singh, 2007) as well as protein, minerals and Vitamins.

Medicinal Values

It is used in special diseases of spleen and bile disorders, and is also used as a laxative. Due to the presence of ascorbic acid (Vitamin C) it has proved useful in scurvy disease. The occurrence of beta-carotene serves as precursor of Vitamin A which eventually helps in curing night blindness in human beings.

Table 23.1: Nutritional Contents of *Chenopodium album*

Content	*Per 100g Edible Part (g)*
Moisture	83.92
Total ash	3.02
Crude fat	0.31
Crude fibre	1.13
Crude protein	4.35
Carbohydrates	7.27
Energy Kcal	53.00
Ascorbic acid	33.64
β-carotene	1838.0
Iron	5.25
Calcium	265

Source: Pandey and Singh, 2007.

Botanical Description

An annual or perennial halophytic herb which occurs (cultivated occasionally) in the hills up to 1200m. Plant is 20-30 cm tall and bears alternate, simple, exstipulate irregularly loved leaves. Lower surface of the leaves is silvery, inflorescence a branched spike. Flowers are minute, small greenish to dull white perfect or sometimes unisexual. Fruit is one seeded achene. It is entirely a rainy crop in the hills and attains a height of 1.8m in plains.

Cultural Practices

Climatic requirements: Bathu can be grown from sub-tropical to sub-temperate type of climate *i.e.* from 300m to 1200 m amsl.

Soil requirements: Bathu requires moist, rich soil. For such condition, the soil must of clay loam or silty clay loam in texture with higher content of organic carbon.

Propagation: The plant is considered as weed and rarely grown. Even then, if needed it should be cultivated through seed in moist, rich soil in the month of October. The leaves of the plant are ready for harvest after about two months. Harvest, can be done at 20 days interval up to March ending. The seeds formed at the top ripen and shed on the soil. It is not commercially cultivated but yield of 0.2-0.6 tonnes per hectare has been reported from farmers fields in India.

References

Pandey, D.T. and Singh, D.K. 2007. Minor vegetables of Uttaranchal for nutritional security. *Indian Farmers Digest*. 40(1): 5-8 and 24.

24

Lawn Grass (*Cynodon dactylon* Linn.)

Common name: Dub, drub grass

Family: Gramineae

Introduction

Lawn grass also known as 'Dog grass' is very hardy and creeping grass, rooting at nodes and forms a thick mat. Although this grass grows in almost all the kharif crops like sugarcane, cotton and maize, yet is a very popular grass for the lawns, play grounds, grassy plots and turfs. It is also a very useful fodder particularly for milch cattle and horses. Besides this, grass possesses a lot of medicinal uses (Singh, 2005).

Medicinal Values

Its decoction is diuretic and is useful in dropsy as well as in hysteria, epilepsy and insanity. Chewing of grass gives immediate relief in toothache. Juice of the grass can be applied on bleeding cuts and wounds. Roots and rhizomes are specific in urino-genital disorder. Lawn grass is a rich source of vitamins and minerals. Regular drinking of raw juice in moderate quantities makes up all nutritional deficiencies. It is also employed in natural cure of diseases and solves intricate health problems.

B-sitosterol and B-carotene have been isolated from the entire plant of drub grass. Also contains vitamin C, palmitic acid, triterpenoids, alkaloids, ferulic, vanillic and p-coumaric acid. Glucose, fructose, sucrose and starch are the main components of carbohydrates. Also contains lignin, hydrocarbons and esters (Chauhan, 2006).

Botanical Description

It is a creeping herb, rooting at the joints and sending smooth upward stems. The roots are whitish, tough, creeping and almost woody with smooth fibres. Leaves tapering to a sharp pointed end, ribbed with smooth sheath and hairy stipules. Flowers are purplish in colour arranged in two close alternate rows in equally crowded 4 or 5 terminal linear spikes.

Cultural Practices

Climatic requirements: It can grow under variety of climatic conditions. Being very hardy this grass can be grown in turfs up to 2,600 m amsl.

Soil requirements: Lawn or drub grass can be grown on wide variety of soils ranging from sandy loamy to loamy sand, silty loam to silty clay and clay loam to clayey.

Propagation: Its method of propagation consists of the stem blades which throw roots in the soil from nodes as the grass spreads on the land surface. This results in a dense and compact growth.

Drub grass acts as a weed which grows in almost all the kharif crops and if it is not eradicated, it causes damage to these crops. The crops do not make good progress and they neither tiller well and not they are vigorous. The extent of damage may be as high as 40 to 60 per cent reduction in crop yield. The method of eradication consists of deep hoeing with 'Khurpas' in early stages. In the later stages inter-culture may be given by means of cultivators. On an average dry matter yield of 5-15 tonnes per hectare in summer season is obtainable.

Flowering and fruiting: Flowering occurs during the month of August and fruiting takes place in the month of September.

References

Chauhan, N.S. 2006. Medicinal and Aromatic Plants of Himachal Pradesh. Indus Publishing Company, FS-5, Tagore Garden, New Delhi – 110027

Singh, S.S. 2005. Fate of the Medicinal Plant Resources of Jammu and Kashmir. Directorate of Environment and Remote Sensing Jammu, J&K Government.

25

Lemon Grass (*Cymbopogon citratus* DC. ex Nees)

Common name: Ginger grass, agiaghas

Family: Gramineae

Introduction

Lemon grass is a native herb from India. But now it is cultivated in number of other tropical and sub-tropical countries, such as Australia, China, Africa etc. (Verma and Verma, 2014). In India, it grows in 3,000 ha area, mostly in Kerala, Karnataka, U.P., Assam and foothills of Arunachal Pradesh and Sikkim. The annual production ranges between 300-350 tonne/annum. It is used as a basis of a popular drink in the tropics known as "Takral" in Thailand and used in many Thai cuisines. In the western world, it is often used in curries, marinades and sea foods soups. Lemon grass possesses number of medicinal uses such as carminative and insect repellent, perfumes, cosmetics, herbal teas and non-alcoholic beverages.

Medicinal Values

Although there are inadequate scientific data on the medical claims of lemon grass yet, it has been used in traditional medicine for treatment of several diseases (Atal and Kapur,1982). These diseases consist of fevers, digestive disorders, menstrual disorder, rheumatism and joint pains. The infusion of decoction of aerial parts of lemon grass has been widely used in folk medicine. This plant has been recommended to treat inflammation, nervous disorder, headaches, fever, nervous exhaustion and stress-related conditions. Its oil has proved to be beneficial to cure ailments like cough, cold, rheumatism, lumbago, digestive problems, bladder affliction, leprosy and as mouth wash for the toothache and swollen gums. It has also been claimed to stimulate diuretic, anti-purgative and sudorific to reduce fever (Chopra, 1985, Luthra, 2006).

The 'Citral' rich oil has germicidal, medicinal and flavouring properties. 'Citral' a volatile constituent of oil (0.4 per cent), is the starting material for the preparation of important ionones. Some of these are used in flavouring soft drinks, cosmetics and perfumes while others are used in the manufacture of synthetic Vitamin-A. Besides these, oil is also used to make bactericides, insect repellents and medicines.

Botanical Description

Lemon grass is an aromatic plant which is a tall, clumped, perennial grass growing to a height of 1-2 m depending on the climate. Several species of lemon grass such as *Cymbopogon refractus, Cymbopogon ambiguus, Cymbopogon nardus, Cymbopogon citratus,* are found in countries like Australia, China, India, Africa and others. Leaves are long, green, linear, tapering upwards and along the margins, form the top portion (90 cm in length and 0.5 cm in breadth) and the solid portion consists of solid, fibrous stem. The culm is stout and erect. The rhizome produces new suckers that extend vertically as tillers to form dense clumps. It is a short day plant and produce profuse flowering in South India. The inflorescence is a long spike (1m in length) and flowers borne on decompound spatheate with panicles 30 - 60 cm long.

Cultural Practices

Climatic requirements: Lemon grass requires tropical to sub-tropical climatic conditions. However, ideal conditions for growth are warm and humid climate with sufficient sun-shine and about 250-330 cm rainfall annually, evenly distributed over most part of the year. It can also be cultivated in semi-arid regions with low to moderate rainfall.
Soil requirements: Lemon grass grows well on medium fertile soils. Well drained sandy loam soil is the most suitable for optimum growth. It can be grown on a variety of soils ranging from loam to sandy loam or sandy loam to silty loam. Even it can be grown in laterite type of soils mostly found in South India.

Varieties: The commonly cultivated varieties of lemon grass in India include 'Sugandhi', 'Pragati', 'Praman', 'Jama Rosa','RRL -16' and 'CKP 25' which yield oil to the extent of 70-80 kg per hectare.

Propagation: It can be best propagated both through seed and vegetative means. 2.5 kg seed is required to produce seedling for planting one hectare area. Through vegetative means, the rooted slips are planted at a spacing of 60x80 cm with about 55,000 slips per hectare. Lemon grass is generally planted during July- August *i.e.* during rainy season. Maximum age of lemon grass is 18 to 24 months, while it is necessary to renew its plantation after every 6 to 8 years (Anonymous, 1950; Atal and Kapur, 1982).

Nursery raising and planting: Nursery raised seedlings are transplanted instead of direct sowing. The seed rate is 4-5kg/ha. Germination starts in 5-6 days and the seedlings are ready for transplanting after 60 days.

Seedlings are planted at a distance of 40x40 cm, 40x30 cm, 40x60 cm apart depending upon fertility of land. Preferably planting is done on the ridges in areas receiving high rainfall and generally 1-2 rooted slips per hole (15 cm deep) are planted.

Irrigation: In North India, 4-6 irrigations are given during summer months (February- June) and require to be irrigated at an interval of 3 days up to one month if rains are erratic and 7-10 days subsequently. Once the plants are established, irrigation is done depending on water holding capacity of the soil and local weather conditions.

Manures and fertilizers: Application of 30 kg N, 30 kg P_2O_5 and 30 kg K_2O per ha as basal application is recommended. Remaining 60-90 kg N is applied as top-dressing in 3 to 4 splits during the growing season. In soils having low fertility levels, the dose of nitrogen should be increased.

Intercultural operations: The field must be kept weed free up to 3 - 4 months after planting and 2-3 weedings are necessary in a year.

Harvesting and yield: Lemon grass is generally cultivated as a ratoon crop. It is first harvested at 4 to 6 months after planting followed by subsequent harvests at 2 to 3 months interval. Harvesting is done by cutting at 20 cm above the ground level, allowed to wilt and then transported to the distillation plant.

Green leaf biomass production was found in the order of 471.8, 1988.0 and 2917.3 q ha^{-1} during the first, second and third year of planting, respectively. The corresponding oil yield extractable was estimated to the order of 2.36, 9.94 and 14.59 kg ha^{-1} (Dass, 2008). Oil is obtained through steam distillation. The oil is yellowish in colour (conatin 75-85 per cent citral and traces of other aromatic constituents). The recovery of oil from the grass ranges from 0.5 - 0.8 per cent. It takes about 4 hours for complete recovery of the oil.

References

Anonymous. 1950. The wealth of India (Raw material). Vol. 11, CSIR. India, pp. 411-415.

Atal, C.K., and B.M. Kapur. 1982. Cultivation and Utilization of aromatic plants. Regional Res. Labs. CSIR Jammu, pp. 314-317.

Chopra, R.N.1985. Indigenous drugs of India. Dhur and Sons. Pvt. Ltd. Calcutta, India.

Dass, A.2008. Conservation of arable lands through vegetative barrier of lemon grass 2008. *Indian Farmers Digest*. 41(5): 5-7.

Luthra, P.K. 2006. Medicinal properties of 50 trees of India. *Gram Vikas Jyoti*. July to September Issue, pp. 63-68.

Verma, K.C and Verma, S.K. 2014. Lemon grass (*Cymbopogon citratus*): Plant with multiple uses. *Indian Farmers Digest*. 47(12): 26.

26

Lineseed
(*Linum usitatissimum* Linn.)

Common name: Alsi

Family: Linaceae

Introduction

Linseed/flaxseed, is found in the areas between eastern Mediterranean to India. It was domesticated in the Fertile Crescent and extensively cultivated in ancient Ethopia and ancient Egypt. A discovery reported in 2009 of spun, dyed and knotted wild flax fibres in cave in Republic of Georgia, the plant was already in use by human beings at the surprisingly early date of 30,000 BC.

Linseed is the top ranking crop for the production of technical oil in India. It is cultivated over 0.82 m ha in the country, of which more than three-fourths of the area lies in Madhya Pradesh, Uttar Pradesh and Maharashtra (Gupta, *et al.*, 2012).

The linseed is mainly grown for seed and oil, whereas, flax is grown for fibre and edible oil (Anonymous, 2010). In spite of its various nutritional benefits, linseed is still a mystery to many of us. It is only in recent times, that linseed/ flaxseed has started to regain its popularity in the world of nutrition. A number of nutritionists and scientists believe that promoting flaxseed is the most important health promoting supplement next to a multivitamin. It is considered to be one of the Nature's richest natural resource of omega-3 fatty acids, which are required for overall health and the general well being of a person.

Chemical Composition

The chemical composition of flax seeds consists of 20-40 percent of mucilaginous material soluble and insoluble fibre and fixed oil besides other nutrients like vitamins, carbohydrates, proteins and minerals like potassium, calcium, magnesium and iron (Vatsyayan, 2010).

Medicinal Values

Flax seeds are sweet and bitter in taste, heavy, unctuous and hot in effect (Vatsyayan, 2010). With a biter post digestive outcome, they allay the 'vata' but aggravate the 'pitta' and the 'kapha'. They have been described as expectorant, anti-inflammatory, wound healer, absorbent diuretic and aphrodisiac. Flax seeds and their oil have been used in numerous ways for their medicinal properties for generations. According to modern concept, a high content of alpha linolic acid present in flax seeds is responsible for many of its salutary effects on human body.

Several studies have revealed that linseed oil as well as ground linseeds, can lower cholesterol, thereby significantly reducing the risk of heart disease. Taking linseeds oil may also protect against angina (chest pain) and high blood pressure. It can help to prevent elevated blood pressure by inhibiting inflammatory reactions that cause artery hardening plaque and poor blood circulation.

Linseed controls constipation, hemorrhoids and gallstones. Because they are high in dietary fibre, ground linseeds can help easy passage of stools and thus relieve constipation and hemorrhoids. If taken for inflammatory bowel disease, linseeds oil can help to calm inflammation and repair any intestinal tract damage. In addition, the oil may prevent painful gallstones from developing and even dissolve existing stones. It can treat acne, eczema, psoriasis and sunburn. The essential fatty acids in linseeds oil are largely responsible for its skin-healing powers. Sunburn may heal faster when treated with the oil as well. It also promotes healthy hair and nails. Linseed may lessen the severity of diabetes by stabilizing blood sugar levels.

Not only this, linseed increases the production of energy and improves stamina. It can also shorten recovery time for fatigued muscles after exertion and accelerate the healing of sprains and bruises. The lignans in linseed oil play role in protecting against breast, colon, prostate and perhaps skin cancer. It also protects against various effects of aging.

Use of linseed in treating menopausal symptoms, menstrual cramps and female infertility is common. The hormone balancing lignans in linseed oil help stabilize a women's estrogen - progesterone ratio, which can have beneficial effects on the menstrual cycle and relive the hot flashes of pre-menopause and menopause. Linseed oil also improves uterine function and thus, its use can be done in treating fertility problems. Its use is made to fight prostate problems and male infertility.

The linseed oil is a potent source of essential fatty acids that are critical to health which the body cannot make its own. Apart from having omega-6 fatty acids in the form of linolic acid, the flaxseed oil boasts of having omega -3 fatty acids in the form of alpha – linoleic acid which help in treating many other ailments like Parkinson's and Alzheimer's disease (Wadhwa, 2008).

Botanical Description

Linseed grows 1.2 m tall, with slender stems. The leaves are bright green, slender and 20-40 mm long and 3 mm broad. The flowers are pure, pale blue,14-25 mm diameter with five petals. Sometimes the flowers are bright red in colour. The fruit is a round, dry capsule 5-9 mm diameter, contains many glossy brown apple shaped, 4-7 mm long fruits. The oil type linseed plants are dwarf and multi-branched with higher seed yield but poor quality of fibre production. On the other hand flax plants are characterized by tall, smooth and scarcely branched stem with low capsule bearing ability (Baily and Soper, 1985).

Cultural Practices

Climatic requirements: The seed crop of linseed does well under moderately cold but the fibre grows best in cool but moist climate. Although areas having annual rainfall ranging from 45-75 cm are the best suited for its cultivation (David and Toms, 2006) yet it can be grown in high rainfall areas *e.g.* linseed grows very well in certain areas of Kangra district in Himachal Pradesh where rainfall varies from 100 to 200 cm or even more.

Soil requirements: Linseed crop can be grown on different kinds of soil except very light (sandy) and badly drained heavy (clay or clay loams). The soil reaction must be normal *i.e.* in the pH range of 6.5 to 7.5 or slightly acidic to moderately acidic pH 6.4 to 5.5 (Singh *et al.*1999).

Seed sowing and seed rate: The crop is either broadcasted or seeds are drilled in lines 20-30 cm apart. In the case of broadcasting, the seed rate is 40 kg ha^{-1} while for line sowing the seed rate varies from 20-30 kg ha^{-1} depending upon size of the seed.

Sowing time: In many parts of India, linseed is grown during September to October, however, in North Western Himalayan States (Himachal Pradesh, Uttrakhand etc.) linseed being a Rabi crop, is sown during October to November or even first week of December.

Manures and fertilizers: To produce 16 q of linseed ha^{-1}, it removes 62 kg N, 13kg P and 72 kg K (Gupta *et al.*, 2012) hence application of manures and fertilizers is essential to obtain higher yield of this crop. Generally linseed is grown in dryland/ rainfed conditions, where the application of potassic fertilizer is not recommended. However, the recommended doses of N and P for the soils of Madhya Pradesh vary from 10 to 30 and 30 to 40 $kgha^{-1}$respectively. Contrary to this, the recommended doses range between 20 to 40 kg ha^{-1} and 20 to kg ha^{-1} for linseed in case of Uttar

Pradesh soils. In Himachal Pradesh, the result of influence of nitrogen levels on linseed grown at Palampur, have indicated that its levels from 30-90 kg ha^{-1} resulted in significantly higher fibre yield, fibre length and fibre per cent (Singh *et al.*1999). The available quantity of organic manure must be applied before sowing *i.e.* at the time of final field preparation along with one third of nitrogenous fertilizer and full phosphorus. Second dose of nitrogenous fertilizer should be applied at the time of first irrigation and third at the flowering stage.

Harvesting and yield: The crop is harvested by hand pulling and cutting through sickles in ending March or first to second week of April, when it is fully matured. After harvesting, the crop is sun dried and threshed with wooden sticks. Yield of linseed varies from variety to variety and place to place. However, on an average, the yield in Himachal Pradesh, Punjab, Haryana and Jammu region of Jammu and Kashmir varied from 11.5 to 12.0 q ha^{-1}.

References

Anonymous 2010. Get flexible. *The Tribune*. 130(323): 4 [In *Jammu Tribune*, November 17]

Baily, L.D.and Soper, R.J. 1985. Potassium in Agriculture, pp 754-98, Munson, R.D (Ed). American Society of Agronomy, Wisconsin.

David, L.A. and Toms, B. 2006. Plant flora future: Linseed. *Agrobios News Letter*. 4(12): 38-39.

Gupta, R.D, Arora, S. and Gupta, S.K. 2012. Withering Yellow Revolution in the Indian context. In: Technological Innovations in Major World Oil Crops, Volume 2, Springer Science + Business, Media, LLC, 2012, Spring Street New York.

Jamwal, J.2012. Alsi as a much better medicine. *The Kashmir Times*. 69 (285): 9[October 13].

Singh, R, Bassi, K. and Badiyala, S.D. 1999. Influence of nitrogen levels, harvesting and retting methods on fibre yield and quality of dual purpose linseed (*Linum ustiatissimum*) varieties. *Indian Journal of Agricultural Sciences*.69 (9): 668-669.

Vatsyayan, R. 2010. Fight diseases with flax seeds. *The Tribune*. 130 (102): 16 (April 14).

Wadhwa, K.2008.Healing with flaxseeds. *Ayurveda*. 1(10): 14-15.

27

Liquorice
(*Glycyrrhiza glabra* Linn.)

Common name: Mulathi, malathi

Family: Leguminosaceae (sub-family Papilionaceae)

Introduction

It is a medicinal plant and a well known perennial branched shrub. It is also called sweet wood or *Meethi Lakari*. It has a number of medicinal values which are detailed hereunder along with description of the shrub.

Economic Importance

The mulathi has a large demand in flavouring and blending of tobacco products (Sharma *et al.*2002).It can also be utilized by brewers, being added to porter and stout to give thickness and blackness. Mulathi is also used in confectionary. The spent roots after extraction are used in production of mushroom compost, insulating mill board and fire extinguishing boards.

Medicinal Values

Mulathi is one of the most popular healing herb in Ayurvedic System of Medicine. It is as popular today as it was 3,000 years ago and is freely available today even at the grocer's shop. It is a good alternative *i.e.*, it alters the body from a diseased negative state to a healthy state. Mulathi is supposed to bestow lasting youth and is a soothing rejuvenator (Rasayan). It is demulcent, expectorant and germicide with laxative properties. Mulathi has been used with muscle problems

because of anti-inflammatory, anti-arthritic properties. Due to its strong sweet taste, the herb is sometimes used in recipes to mask the unpleasant taste of another herb.

Mulathi can also be used to treat skin inflammation and infections. Chewing a piece of mulathi, refreshers the mouth and cleanses the teeth arresting tooth decay. It heals ulcers in the mouth. When mulathi is boiled in water, it releases its sweet taste in it. So, one can drink a warm infusion or decoction quite akin to tea. An extract hardened into various shapes becomes even more effective.

In China, mulathi is used for strengthening muscle and bone, for increasing physical strength and for curing wounds. The herb stands next to Ginseng in importance in China (Patwari, 2012) pharmacy being great corrective adjunct and harmonizing ingredient in a large number of recipes. It is called, "great detoxifier and is thought to drive poisons (toxins) from the system". Many clinical studies have validated its usefulness in liver disorders, lungs and urinary bladder problems. Its decoction is given in inflammatory affections of respiratory system, bowls, and catarrh of urine-genital system, asthma and dysuria. It is a tonic and laxative and is in combination with many other drugs. Mulathi is added to various laxative powders, cough syrup, confections, lozenges etc.

The roots of the mulathi contain glycyrrhizin and glycyrrhizinic acid as well as asparagine, mucilage, phosphoric, malic and sulphuric acids. Besides, the large proportions of the roots are composed of polysaccharides like starch and others. Presence of both glycyrrhizin and glycyrrhizinic acid in mulathi impart number of medicinal properties. As such mulathi can be used to treat gastric ulcers and inflammation. It is also beneficial in the treatment of problems caused by the excessive acidity in the digestive system. Taking 'mulathi tea', after meals aids digestion and relives from constipation since it has laxative action.

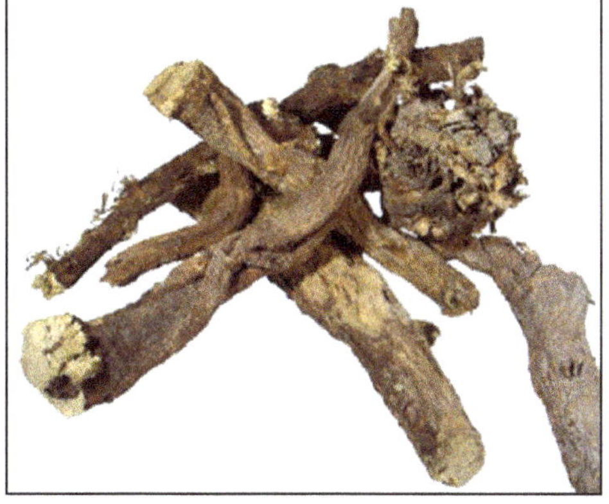

Mulathi releases phlegm and soothes cough, eases asthma, stimulates the immune system. Mulathi can also be used to treat skin inflammation and infections. Research carried out and results published during 1980 indicated the inflammatory effects of glycyrrhizinic acid against erythema (reddening of the skin). The powder of mulathi mixed with ghee or oil can be used externally to heal wounds both septic and non septic.

The mulathi should not be taken during pregnancy. Its use is also restricted for the patients afflicted with high blood pressure. As mulathi is known to cause fluid retention and so should be sparingly used by people who have this tendency.

Botanical Description

It is called 'Yashtimadu' in Sanskrit. It is a perennial shrub with upright branched stem to about 1m in height bearing alternate leaves made up of 7-17 bright green leaflets. Its small pale blue to lilac flowers grow in spikes. It has thick woody rhizomatous roots and creeping stems. The flowers are generally borne in spring season on older bushes in axially racemes, 1cm long, blue purple in colour. The flower possesses 2-3 grey coloured and oval shaped seeds.

Cultural Practices

Climatic requirements: Mulathi plant grows well under warm temperate climate regions having long day conditions and longer winter season.

Soil requirements: The plant grows best in well drained light soils with loamy sand to sandy loam texture. The pH of these soils should be in the range of 5.5 to 8.5. According to Sharma *et al.* (2006), mulathi grows well in undulating lands along periodically river flooded areas during rainy season and consequently receiving annual silt deposits.

Land preparation: The land is well ploughed deeply and exposed to the sun for 10-15 days prior to cross harrowing and leveling followed by laying out small fields. Well decomposed farm yard manure (FYM) to the tune of 10 tonnes ha^{-1} should be added and mixed well with the soil at the time of land preparation.

Time and method of Sowing: February-March and July-August are the best times for growing of mulathi plants. The mulathi is generally raised from 15-20 cm long semi-woody underground stem cuttings (stolons) having 2 or more eye buds. About 3 quintals of such cuttings are sufficient for planting in an hectare of land.

Intercultural operations: The crop requires 2-3 weeding cum harrowing during cultivation period. Use of weedcides is discouraged owing to the organic food production regime.

Irrigation: As mulathi is a hardy crop and it requires less irrigation. However, during dry season irrigations are required to get good growth and yield of the crop. In accordance with Sharma *et al.* (2006), 7-10 irrigations are necessary for obtaining a very good yield of mulathi.

Harvesting and yield: The roots of the crop are dug in dry autumn season. These are cleaned and cut into 15-20cm long pieces and split if more than 2cm thick. The roots are then dried in open sun for 7-10 days. Drying of the roots through hot blast at 40 degree Celsius is an ideal procedure. The produce having 12-15 per cent moisture content can be stored in dry cool places. It can even be stored for a long time without loss in quality.

The yield of the crop varies according to the age of the crop. It is nearly 2.5-4.0 tonnes ha^{-1} from 20 months old crop. On the other hand, the yield is 6.0 to 7.0 tonnes ha^{-1} from 30-36 months crop.

References

Patwari, H.N. 2012. Multipurpose- Mulathi. Personnel Communication, *Daily Excelsior*.

Sharma, K.K., Sambyal, K.S, Sharma, N.K. and Gandotra, S. 2006. Agronomical practices for the cultivation of high demand plant species of medicinal and economic importance. Directorate of Command Area Development, Jammu, J&K.

28

Marigold (*Tagetes erecta* L.)

Common name: Gutta, gainda

Family: Composite

Introduction

Like rose, the marigold is also one of the traditional flowers traded in the market (Nazir, 2011). Marigold is infact, one of the most popular and commercial flower (annuals), cultivated by the small, marginal and landless farmers in different parts of the country. In Bihar most of the small and marginal farmers are engaged in the cultivation of marigold as cash crop throughout the year (Singh *et al.,* 2011). Even the landless farmers grow marigold in their kitchen gardens and earn good returns. Marigold cultivation in Jammu district was started in 1989 (Gupta, 2007) for local market which was further extended in many areas of districts of Udhampur, Samba and Kathua.

Medicinal Uses

Marigold has great demand for garland, cut flowers and decorative purposes, at religious offerings, decoration of *mandaps* and social functions.

Marigold flowers are mostly in demand for children ailments such as fever, bladder and kidney problems as reported by Kalita, (2013). If the marigold flower is rubbed on the affected swelling caused by a wasp or bee, it gives relief.

During both the American Civil War and the First World War, marigold was used to treat wounds and their infections with toxins and bacteria (Bevinakatti, 2013). The marigold flowers were either made into a poultice or infused oil for application on the wound. Besides this industrial uses of carotenoids extracted from marigold flowers in pharmaceuticals, food supplements, animal feed additives and colourants in food and cosmetics has been exploited commercially.

Cultural Practices

Climatic requirements: Marigold requires mild climate for luxuriant growth and the flowering. The climatic conditions of some states like Bihar, Jammu region of J&K, parts of Punjab, Haryana and Himachal Pradesh etc. are favourable to grow marigold throughout the year. The optimum temperature range for its profuse growth is 18-20° Celsius. Temperatures above 35° Celsius restrict the growth of the plants which leads to reduction in the size and number of the flowers. In severe winter, plants and flowers are susceptible to frost.

Soil requirements: Marigold can be grown on wide range of soils, as it is adapted to different soil types. Dwarf (French type) marigolds are best cultivated in light textured soil whereas; rich, well drained and moist soils are considered the best for tall marigold (African type). However, sandy loam soil having pH 7.0 to 7.5 with good aeration and drainage are the ideal for its cultivation. It can be planted in any type of well drained soil (pH 6.5 to 7.5) with sufficient irrigation (Wazir, 2013).

Varieties: There are many varieties of marigold and can be divided into four basic species.

i) African Marigold – *Tagets erecta*

ii) French Marigolds – *Tagets patula*

iii) Triploids – A hybrid of *Tagets erecta* and *Tagets patula*

iv) Signet Marigold – *Tagets tenuifolia* (signata) *pumila*

'Atlantis mixed F1', 'Galore mixed F1' and 'Gold Coin Mixed' are the main varieties of African Marigold, while 'Disco Mix Large', 'Heromix and Safri Mix' are those of French Marigolds.

Planting season: Marigold can be planted in all the seasons *i.e.* summer season, (February –March), rainy season, (May to July), winter season (August to September) and spring season (November). Their transplanting period varies from April-July, June-October, November and February-April, respectively.

Propagation: Marigold is propagated by seeds as it germinates easily and produces vigorous plants. The marigold seeds count about 300-350 per gram and remain viable for about 1-2 years. Their germination of fresh seeds is about 90-95 per cent.

Manures and fertilizers: To improve the fertility of the soil, FYM (25 tonnes ha^{-1}) should be applied 15 days prior to transplanting. A fertilizer dose of 90 kg N, 90 kg P and 75 kg K ha^{-1} is the optimum for crop growth and yield.

Irrigation: Frequency of irrigation varies according to the season and soil type. Generally in winter, irrigation is given at an interval of 8-10 days, while during summer frequent irrigation is required at interval of 4-5 days.

Harvesting and yield: Marigold flowers are plucked when they have attained the full size. Harvesting of flowers is done in the evening, just before the flowers are fully opened.

The yield of African Marigold and French Marigold varies from 8-12 tonnes ha^{-1} and 11-18 tonnes ha^{-1}, respectively.

References

Bevinakatti, S.K. 2013. The medicinal uses of marigold flower. *Agrobios Newsletter*. 12(7): 28.

Gupta, R.D. 2007. Crop diversification pays more: Success stories. *Agrobios Newsletter*. 5(10): 2-24.

Kalita, N. 2013. Medicinal flowers and uses. *Agrobios Newsletter*. 11(12): 57-60.

Nazir, M. 2011. Floriculture-Domestic scene. *The Daily Excelsior*. 42 (53): 6 (Tuesday, February 22).

Singh, V.N, Shahi, B.P., Shahi, A and Singh, J. 2011. Marigold cultivation: A profitable venture. *Indian farmers Digest*. 44 (1): 27-30.

Wazir. T.S. 2013. Marigold: Poor man's saffron (Personnel communication).

29

Mint
(*Mentha* spp.)

Common name: Pudina

Family: Lamiaceae

Introduction

Mint is a perennial herb found in various forms *viz.*, Jangli Pudina (*Mentha arvensis*), Peppermint (*Mentha piperata*). The former is common herb which grows along water channels and wet places up to 1000m above mean sea level. Plant is strongly aromatic and finds many uses in home remedies and ethnic needs. It is very effective in stomach disorders and coughs. The latter, on the other hand, occurs up to 1300m altitude in the same habitat but in lesser quantity and is more valuable medicinally (Singh, 2005). This species *i.e., Mentha piperata* is also found in other forms known as Spearmint, Lemon mint and Japanese mint.

Medicinal Values

Mint is a popular herb used for its distinctive aroma and medicinal value. The essential oil 'menthol' present in it acts on cold sensitive receptors in the skin, mouth and throat, responsible for evolving sensation when inhaled, eaten or applied to the skin. The leaves contain many imperative B-Complex vitamins like folates, riboflavin and vitamin B6. The herb is also an excellent source of vitamin C (Behera, 2012).

Mint has been used for a long time to settle a nervous stomach. It is because mint is well known for its ability to soothen the digestive tract as well as to reduce the severity and length of stomach aches. It also eases the discomfort associated with

irritable bowl syndrome. Mint slows down the growth of many harmful bacteria and fungi. The antifungal properties of mint play a vital role in the treatment of asthma and allergy conditions.

Mint also reduces the nervous tension and anxiety. It works effectively for the diseases like colic and vomiting. On the whole mint works well to refresh the body and has an amazing power to cure cancer. Mint contains a phyto-nutrient called perillyl alcohol. This phyto-nutrient prevents the occurrence of colon, skin and lung cancer. The use of mint for cancer treatment has fewer side effects than conventional treatments. Mint works by attacking the tumor's blood vessels, starving cancer cellls to death.

The leaves are acrid, aromatic, thermogenic, stimulant, anodyne, deodorant, antiseptic, anthelmintic, carminative, sudorific, dentifrice, antispasmodic, febrifuge and contraceptive. They are beneficial in vitiated conditions of vata, dyspepsia, flatulence, colic-peptic ulcer, diarrhoea, cardiac debility, cough, asthma and bronchitis.

Usage of Mint

Mint can be used in a variety of ways. Since leaves are used for medical treatment, so many prefer to eat them as a part of diet. The most popular way of consuming the mint is through brewing a mint tea. The mint tea can be prepared by pouring hot, but not boiling water over the mint leaves. The preparation has to be covered while preparing the mint tea as it prevents valuable oil from evaporating. There are a number of preparations of mint in pills form. Mint is present in many food items as well as breath freshener.

Botanical Description

The plant is found throughout India including Western Himalayan region. It is an erect aromatic herb up to 60cm in height with opposite leaves. Leaves of mint plant are simple, opposite, shortly petioled, oblong, ovate or lanceolate, crenate, serrate, cuneate at the base. Flowers are lilac in auxillary distant whorls but are irregular with two lipped corolla, four stamens, four lobed ovary which ripens into nutlets each with one seed. Fruit nutlets are very smooth (Kumar and Shalla, 2010).

Mint Types

It is small spreading herb about 50cm tall. Its aromatic leaves yield a fragrant oil used to flavour foods and also in pharmaceuticals. Two commercial varieties of peppermint are widely recognized *viz.* Black mint and White mint. Black mint is hardy plant containing high percentage of oil, while white mint is less hardy and low oil yielder but the oil has superior aroma.

Other types of peppermint are: Spearmint, Lemon mint and Japanese mint. Spearmint is a perennial spreading herb about 50 cm tall. Its aromatic leaves yield oil resembling "Caraway" like flavor. It is similar to peppermint in its cultural requirements. It is used in confectionary. The oil possesses 55 per cent carovane which gives it a characteristic aroma.

Lemon mint is a robust branching perennial herb and grows 60 cm tall. Japanese mint is a perennial bushy herb, native of China and yields aromatic oil rich in menthol.

Cultural Practices

Climatic requirements: Mint including peppermint requires wide range of climate ranging from tropical/sub-tropical to sub-temperate and temperate. For example, mint is cultivated in Western Himalayas with mainly sub- temperate and temperate climate. Similarly peppermint is grown presently in the valley of Kashmir only. Whereas, lemon mint is cultivated under tropical to sub-temperate climate conditions, and Japanese mint in Tarai area of Uttar Pradesh (Now in Uttrakhand). It is point to mention that mint grows throughout India having sub-tropical to tropical climate.

Soil requirements: Both mint and peppermint prefer to grow under well drained loamy to silty loam, and loamy to clay loam soils having good soil fertility. The pH of the soils should be in the range of 6.5 to 7.5. It is worthwhile to point out that mint crop does well on plots that support good crops of potatoes.

Planting time: Mint is propagated by cuttings. There are generally four planting time *i.e.*, early spring, early June, late September, and October-November. Planting of fresh stolons during early spring, however, provides good yield. Propagation is done by 8-10 cm long stolons.

Irrigation: Light irrigation or light showers at planting and ample sunshine at the time of harvesting very well suits for its cultivation. However, stagnating water damages the crop and frost cause a temporary setback at planting time.

Harvesting and yield: The first harvest is taken before the onset of rains after about 100-105 days of planting at the time when the flower buds begin to open. The harvested crop is left in the field for about 4-6 hours for wilting to reduce its bulk prior to distillation.

On an average yield is nearly 8-10 tons ha^{-1} as fresh herbage per harvest producing 75-100 kg oil in a year.

References

Behera, B.N.2012. Fitness Mantra- Healing herbs. *The Tribune*. 132(132): 1 [Spectrum 13th May, Sunday].

Kumar, A. and Shalla, A. 2010. Mint: A priceless herb. Personal Communication, Division of Agronomy. SKUAST, Jammu.

30

Natural Sweetener (*Stevia rebaudiana* Bertoni.) Syn. *Eupatorium rebaudianum*

Common name: Stevia, sweet leaf, sugar leaf

Family: Asteraceae

Introduction

Stevia (*Stevia rebaudiana*) is a small perennial natural herb with low calories and is more than 300 times sweeter than cane sugar. Steviocide obtained from this species has been used in Paraguay and Brazil from time immemorial as sweetener and to lower blood sugar. Indeed, in 1931, a French chemist extracted a white pure crystalline compound from the leaves of Stevia plant and named it as steviocide. Steviocide extracts were found to be 100 to 300 times sweeter than table sugar and has been used for many years in the treatment of diabetes among the Indians in Paraguay and Brazil (Sharma *et al.*, 2006 a).

The other names by which Stevia is known, are 'sweet leaf', 'honey leaf', 'sweet herb', 'honey yerba', 'honey grass', 'sweet herb of Paraguay', 'sugar leaf', 'khaa jee' and 'caa-he-e'. Stevia is extensively grown and cultivated in places like Brazil, Paraguay, Uruguay, Central America, the United States, Israel, Thailand and China but is native to Paraguay and Brazil. Because of its medicinal values, now *Stevia rebaudiana* is being cultivated in many places in India also. Till date no other plant has been discovered having more sweetening power than Stevia (Singh *et al.*, 2010) and thus it is the future crop of India (Autade and Choudhary, 2015). It is remarkable to note that now the use of stevia as a sweetener has been approved by USFDA (US Food and Drug Administration), WHO (World Health Organization) and EFSA (European Food Safety Authority). As a result it is now being cultivated in Korea, Japan, Canada, Indonesia and Tanzania besides Israel, Thailand, China, Brazil and United States (Autade and Choudhary, 2015).

Chemical Composition

Fresh Stevia leaves possess water content ranging from 80-85 per cent. The main active constituents present are glycosides, organic acids, pigments, growth hormones, vitamins and a number of metals which are presented in Table 30.1. Apart from these, isoquercitrin, isosteviol, lupeol, kaempferol, stigmasterol, quercetin, quercitrin, laurene, umbelliferone and scooletin, are also present.

Table 30.1: Chemical Composition of *Stevia rebaudiana*

Sl.No.	Components	Kinds
1.	Glycosides	Steviocide, steviol and rebaudiocide A and B as well as diterpene glycosides.
2.	Pigments	Chlorophyll and beta-carotene, xanthophylls.
3.	Acids	Ascorbic acid (vitamin C), caffeic acid, formic acid, chlorogenic acid, gibberellic acid.
4.	Hormones	Gibberellic acid, gibberellin.
5.	Metals	Chromium, cobalt, iron, zinc, magnesium, potassium, phosphorus, tin.
6.	Vitamins	Riboflavin, thiamin.
7.	Other compounds	Apigenin, austroinulin, avicularian, beta-sitosterol, compesterol, caryophyllene, centaureidin, cynaroside, dulcosides A-B, foeniculin, indole-3-acetonitrile.

Medicinal and Other Uses

Stevia rebaudiana has hypoglycemic, hypotensive, vasodilator, taste improving sweetening, antifungal and anti-viral, anti-inflammatory, anti-bacterial and urination increasing actions. It was used as sweetener for bitter drinks such as mate for centuries in South America. It is an ideal substitute for persons suffering from diabetes and/or obesity linked with cane sugar. It is also used for the treatment of obesity, hypertension, fatigue, depression and infections like yeast infection. It is also used to stimulate alertness and counter fatigue. It has proved useful for improving digestion.

Owing to the presence of anti-bacterial properties, it prevents tooth decay and promotes oral healing. It is used in skincare, healing cuts, and without scars optimizes blood pressure and soothes gastric ulcers. It is effective in treatment of several infections including common cold. As the extracts of Stevia are fund to be 300 times sweeter than cane sugar, the extract is used in the daily diet of obese persons to reduce weight (Behera, 2012).

Steviocide is used in making a number of sweetened dietaries like chewing gums, tooth pastes, mouth washes and even anti-smoking lozenges. Stevia leaves have been used in herbal tea by diabetic patients in the Asian countries. No side effects were observed in these patients after many years of its continuous consumption (Dewan, 2006). The Indian tribes of South America have used it as digestive aid, and have also applied it locally for years to help wound healing.

Botanical Description

It is a small perennial shrub which grows upto 2 m height in wild, and the cultivated varieties grow up to about 1 m. It is a spindly, much branched fibrous root system plant. Fine root spread out on the surface of the soil, while a thicker part of the root grows deep into the soil. The stem is hairy, wand-like and covered with leaves. Leaves are opposite and toothed, fibrous and dark green in colour. Flowers are white, tubular and bisexual. The fresh and dry leaves are sweet in taste. Out of the 240 species of the genus *Stevia, Stevia rebaudiana* is the only one which is cultivated and used in food in East Asia, including China, Korea, Tiawan, Thailand and Malaysia (Verma and Verma, 2011).

Stevia is a short-day plant. The concentration of steviocides in the leaves increases when it is grown under long day conditions.

Cultural Practices

Climatic requirements: Stevia grows very well in sub-tropical regions of India, especially in the sub-mountain or foothills of Siwaliks. However, it also grows in the sub-temperate region of the country below snowline. Average minimum and maximum temperature should not exceed 10° and 37° Celsius during the active growth period (Jamwal, 2006; Sharma *et al.*, 2006 b).

Soil requirements: Stevia grows on a wide variety of soils. However, well-drained loamy sand, sandy loam and loam-textured soils are best suited. The soil should be well pulverized prior to sowing of the crop with good structure (granular to crumb) to promote plant growth. Availability of optimum soil moisture is essential for proper growth of plants. Soil pH should be in the range of 5.5 to 7.5. Stevia plants do not tolerate salt accumulation in soils. It can be grown as pot plant.

Propagation: *Stevia rebaudiana* can be propagated both from cuttings and through seeds.

(*i*) **Nursery raising:** Seeds for raising nursery are sown in the month of February to March in raised beds of 1.25 m 10 m size. After sowing the seeds, irrigation should be done daily upto 5 days in the morning. Germination of the seeds occurs within 10 to 15 days after sowing. Seeds can also be grown in plastic trays, pots and wooden boxes having good mixture of soil, organic manure and sand in the ratio of 4:2:1.

Two months old seedlings, having 5-6 leaves and with height of 8-10 cm, get ready for transplanting. Transplanting is done in furrows during April to May at a spacing of 45 cm 45 cm and 45 cm 30 cm in soils having high and medium fertility, respectively. For this purpose, land should be well prepared by repeated ploughing and harrowing.

(*ii*) **Vegetative method:** Seedlings can also be raised successfully by vegetative means using terminal cuttings of 10 to 15 cm having 4 to 6 nodes. Stem cuttings should be taken from plants in the active growth during either in February/March or during July to August. Extreme cold or hot months should be shunned. After trimming the lower leaves, cuttings are raised in pots or polysleeves. In nursery beds, the cuttings are planted at 15 cm 15 cm spacing, keeping one node inside the soil.

Cuttings raised under partial shade and under high humidity develop roots at faster rate within 8 to 10 days. Watering of plants should be done daily through sprinkler for about 10 days. Later on, irrigation should be done once or twice a week depending upon the soil condition.

Manures and fertilizers: *Stevia rebaudiana* is mostly fertilized with organics (well decomposed FYM, compost, poultry manure, vermicompost). Although exact doses of these manures have not been worked out so far under different situations yet 25-30 tonnes FYM ha^{-1} or 20 tonnes vermicompost ha^{-1} has been found useful to grow a successful crop (Sharma *et al.*, 2006b). These manures are required to be applied during field preparation.

If chemical fertilizers have to be applied, they can be added by maintaining the ratio of 4:2:1 in the form of N, P_2O_5 and K_2O fertilizers.

Irrigation: *Stevia rebaudiana* cannot tolerate drought situations and, therefore, demands irrigation, especially after transplanting. Other irrigation is required after about 3 to 4 days of transplanting. Subsequent watering is done after weekly interval till the onset of monsoon. However, actual number of irrigations depends upon soil moisture holding capacity and occurrence of rainfall.

Weeding and hoeing: Two manual weedings at 15-20 days interval are sufficient to destroy the weeds. Mulching of any local organic material helps in checking the weeds and hastening the germination of *Stevia rebaudiana* crop.

Harvesting and yield: Harvesting is done leaving 8-10 cm height of the stems from the ground level. First harvesting is done about 75-90 days after transplanting the crop. Subsequent second harvest is taken after 60-75 days of the first harvest at the flower bud initiation.

In case of late transplanting, the crop can be grown only for single cut, harvesting is done after 3-4 months of transplanting and continues till flowering begins. Once, the crop planted in the field, perennial crop may continue upto 4 years. It is point to mention that maximum number of leaves is produced during 3rd and 4th year. Flowering of the plants should be avoided and pinching of apical buds should be carried out to enhance bushy growth of the plants with side branches. An average dried leaf yield of 17, 20, 23 and 15 quintals ha^{-1} can be harvested as total biomass yield in 1st, 2nd, 3rd and 4th years, respectively.

On an average, total fresh biomass yield from two harvests has been recorded to be 20 to 30 tonnes ha^{-1} $year^{-1}$ which becomes about 4 to 5 tonnes on dry weight basis.

References

Autade, R.H., and Chouhary, R. S. 2015. Stevia-Future crop for India. *Agrobios News Letter*. 13 (8): 60-62.

Behera, B. N. 2012. Healing herbs under Fitness Mantra columns. *The Tribune* 132(132): (Sunday Spectrum, May 13).

Dewan, M. 2006. Stevia-A harmless sweetner. *Ayurveda*. April Issue, pp 24-25.

Jamwal, J. 2006. Medicinal and herbal plants - Stevia. In : Rural Round Up columns. The Kashmir Times, Jammu. Thursday, September 14, 2006, pp 13.

Sharma, K.K., Sambyal, K.S., Sharma, N.K. and Gandotra, S. 2006a. Agronomical practices for the cultivation of high demand plant species of medicinal and economic importance. Directorate of Command Area Development, Jammu.

Sharma, N., Kaushal, N., Chawla, A., Mohan M., Sethi, A. and Sharma, Y. 2006b. *Stevia rebaudiana* – A review. *Agrobios Newsletter*. 5 (7): 46-48.

Singh, A. K., Kumar, P and Singh, D. 2010. King of natural sweetner-Stevia. *Indian Farmer's Digest*. 43 (4) : 22-24.

Verma, K. S. and Verma, S. K. 2011. Stevia: Medicinal Properties and cultivation techniques. *Indian Farmer's Digest*. 44(7): 27-29.

31

Nut Sedge
(*Cyperus rotundus* Linn.)

Common name: Deela, motha grass

Family: Cyperaceae

Introduction

Like Dub grass, 'Deela' grass is also a weed which is mostly found during kharif season. It causes damage not only to crops but also to vegetables. The extent of damage is really great often varying from 50 to 75 per cent of the crop yield. Unchecked growth of the weed might totally ruin the crop or vegetable. Its eradication becomes difficult unless, the root tuber is dugout as it continues to sprout after each weeding. Therefore, the weeding requires deep hoeing by means of 'baguri' or spade. Deep inter culture to crops and a vegetable sown in rows is also beneficial.

Despite being a weed, it is also fed to the cattle besides possessing number of medicinal properties.

Chemical Composition

Tubers contain terpenes, oleanolic acid and glycosides, beta-sitosterol and essential oil. Essential oil contains pinine traces of cineole, monoterpenes, sesquiterpenes and patchoulenone. Also possesses fatty oil to the extent of 2.7 per cent.

Medicinal Values

The roots of the grass are stomachic, tonic, diuretic, demulcent, stimulant and galactagouge. They are given in fevers, dyspepsia, diarrhoea and cholera. Root paste is applied for healing wounds and sores. The extract is anti-inflammatory, antipyretic and analgesic. Its tea is taken during colds and against jaundice (Chouhan, 2006).

Botanical Description

Dila grass is a weed of kharif season crops, especially in rice fields. Culms are erect, triangular, with rounded angle. Leaves are radical, sheathing, shorter than the culms, roots tuberous, sweet smelling, rounded black and hard like a knot. Tubers are black externally, white internally and smell like acorus.

Cultural Practices

Climatic requirements: It requires sub-tropical to sub-temperate type of climate and grows from 300m to 1500m elevation. It is distributed in moist areas, rice-fields and along water courses.

Soil requirements: It grows on variety of soil ranging from sandy loam or loam to clay loam soils having good amount of organic matter.

Propagation: Propagation takes place both by root tubers and seeds. It grows as a gregarious weed in moist places during rainy season (mid June to mid August). Flowering and fruiting season varies from July to September. Crop may produce 4 tonnes of underground tubers per hectare.

References

Chauhan, N.S.2006 Medicinal and Aromatic Plants of Himachal Pradesh. India's Publishing Company FS-5, Tagore Garden, New Delhi – 110027.

32

Onion

(*Allium cepa* Linn.)

Common name: Payaz

Family: Liliaceae

Introduction

Onion is one of the most important bulb crop. It is a native of Central Asia, but a rich diversity also exists in Greece, North-West India, Pakistan, Turkey and the then USSR (Dixit, 1999).

Onion is a common condiment vegetable grown from sea level to over 1800m amsl. It is a cool season crop and can be grown in all climatic and soil conditions. Onion is an annual plant, sometimes bi-annual, bulbous producing linear leaves upto 50 cm. The importance of the onion cannot be over emphasized as no dish is complete without onion. Even many of the rural people take their bread (chapatis) with pickle and peeled onion pieces. In addition to this, onion has lot of medicinal value, which are described below.

Food Value

Onion is a very good source of calcium and B group of vitamins. However, its total food value varies according to its stage of maturity. Ripe onion contains large amount of moisture along with other minerals like phosphorus and iron besides, sufficient amount of fibre and nutrients such as carbohydrates and proteins. Its typical odour is attributed to the presence of organic sulphur compounds (Vatsyayan, 2010). Apart from this, onions are also rich source of chromium and Vitamin K (Mohindra 2014).

Medicinal Uses

Onion is a common remedy of homes and possesses lot of medicinal uses. Roasted poultice is applied to boils, abscesses. Fresh juice is applied in opthalamia and ear-ache. Liberal doses of onions fresh juice are given in cholera, vomiting, diarrhoea, dysentery, jaundice, biliousness as well as diabetes, respiratory troubles and sexual debility (Sood, 2011).

Onion is also very effective in disorders of spleen and liver. It is prescribed in gout, rheumatism, arthritis, diabetes, reducing the need of artificial insulin to one half. Fresh juice of onion has antibacterial properties, reduces blood sugar, and controls cholesterol. Seed of the onion is effective in caries of teeth and in tooth ache. Juice of the onion is also fed in hysteric infections. It is eaten raw as salad.

Charak, Sushruta and leading lights of Ayurveda have described onion to be carminative, digestive, liver stimulant, anti-inflammatory, expectorant, diuretic, emmenagogue (that enhances the menstrual flow) and aphrodisiac in effect (Vatsyayan, 2010). Scientific studies have found that onion is an antioxidant, anti-tumour, lipid lowering and bronchodilator agent. Many of its therapeutic effects match to those of garlic.

The use of onion has also been found to be effective in conditions like anorexia, indigestion, flatulence, osteoporosis and skin blemishes. Ancient references indicate that regular use of onion enhances glow on one's face and keeps premature changes of age at way. Researchers have found that under controlled studies onion inhibited platelet aggregation and increased the bleeding time (Vatsyayan, 2010).

Botanical Description

Its Sanskrit name is 'Palandu', whereas in Hindi it is called 'Piyaz' and in Dogri as 'Ghanda'. Onion is a condiment. Plant is annual, growing from a sub-terranean bulb up to 70 cm in height. Stem is an underground bulb having shallow roots. with inner fleshy scaly leaves and outer membranous dry scales. Its leaves are sessile, simple cylindrical, hollow green with fleshy sheathing base arising from the underground stem. Leaves grow up to 40 cm in height. Immature green leaves and bulbs are also eaten as raw. Bulbs are uniform in shape (spherical to nearly cylindrical), size and skin color (white, yellow, brown, red or purple). The inflorescence is terminal, erect, developing from the ring-like apical meristem. It bears white to pink flowers in a spherical umbel at the top, consisting of 200-300 small individual flowers. Fruit is a capsule of about 5 mm size.

Cultural Practices

Climatic requirements: Onion thrives well under mild cool climate. But at maturity dry weather is favourable. Drought conditions are un-favourable for plant growth and bulb formation. Generally it requires a temperature range of about 12 to 24 degree Celsius before bulb formation and about 16-21 degree Celsius during bulb formation for about 8-10 hours a day with relative humidity of about 70 per cent. Onion is also being grown in cold arid zone of Lahual Spiti and Ladakh (Dixit, 1999)

Soil requirements: Onion can be grown on a variety of soils. However, the best soils for the production of onion are considered to be light ones. *i.e.* loamy sand/sandy loam to silty loam or loam. These soils must be rich in organic matter.

Field preparation, seed rate and sowing: Onion seedlings are produced in raised nursery beds after ploughing twice or thrice. Seeds @ 10-15 kg ha^{-1} are sown in shallow furrows, 1-1.5 cm deep and 5- 7 cm apart.

After about 8-10 weeks, seedlings become ready. These are then transplanted in well prepared fields which are ploughed 3-4 times to get good tilth. The plants are required to be transplanted at 15 cm spacing each way.

Sowing time depends on the agro-climatic conditions. For tropical to sub-tropical zones, its best time is middle of October to middle of November and transplanting time is December to January (mid). The sowing period for temperate and sub-temperate regions is August and September respectively and transplanting time is November.

Manures and fertilizers: Onion crop is a heavy feeder, especially the nitrogen and phosphorous. Well decomposed manure in the form of FYM, compost or vermicompost @ 20-30 tonnes ha^{-1} should be applied before transplanting depending upon its availability. Its quantity may be increased up to 50 tonnes ha^{-1}.

Although 100 kg N, 50 kg P_2O_5 and 50 kg K_2O ha^{-1} is the recommended dose to be applied through fertilizers. However, use of inorganic fertilizers is discouraged due to more popularity of "Organic produce" which has been found good for nature as well as health.

Intercultural operations: Being a shallow rooted crop, shallow hoeing twice and thrice will assist in good plant growth with almost free from the weeds.

Irrigation: Irrigation is given as and when required. Fields should not be allowed to dry. A dry spell followed by irrigation may cause splitting of the outer scales. Irrigation should be stopped when the tops mature and start falling down.

Harvesting and yield: Generally, the bulbs of onion crop get ready after 120 days of its transplanting. The maturity of the bulbs can be judged by the cessation of the plant growth and dropping of the leaves. Average weight of the harvested bulb is 50-70 g. Bulbs are quite firm with good keeping quality good for dehydration. Yield varies from 175-200 q ha^{-1}.

References

Dixit, S.P. 1999. Summer onion in Spiti Valley. *Intensive Agriculture*. 37 (12): 5 – 9.

Mohindra, V. 2014. 7 Vegetables you should eat to boost your disease fighting potential. *Ayurveda*. 1 (26): 18-19.

Sood, S. 2011. Onion – King of vegetables. *Ayurveda*. 1(20) : 14-15.

Vatsyayan, R. 2010. Use onions, be healthy. *The Tribune*. 130 (359): 12 (Wednesday, December 29).

33

Opium
(*Papaver somniferum* Linn.)

Common name: Afim

Family: Papaveraceae

Introduction

Opium poppy is cultivated in Eastern and Southern Asia, and South Eastern Europe mainly in Egypt, Japan, Yugoslavia, Russia, Argentina, Spain, Bulgaria, Hungary, Czech, Slovakia, Poland, Germany, Netherland, Portugal, China and India. It is believed that it is originated in the Mediterranean region. India is one of the largest producer and supplier of opium poppy in the world. The *'Ain-e-Akbari'*, a historical record of Mughal Emperor - *Akbar* (1556 to 1605AD) states that opium was cultivated in all the provinces of North India scattered over an area of more than one million sq. km. Although poppy comprises of about hundred species and is one among them, the opium poppy is more important. It is because of its being the source of priceless drug 'opium' which possess variety of medicinal uses. In India, cultivation of opium poppy is mainly done in Madhya Pradesh, Uttar Pradesh and Rajasthan, largely in Chittorgarh district (Sharma and Rajput, 1999).

This species, along with numerous other *Papaver* species, is also planted as an ornamental for its attractive flowers and foliage. Habitat areas include fields, clearings, stream banks, railroads, roadsides and other disturbed sites usually between 0-1300 m. An estimate shows that about 2,000 tonnes of opium is being produced annually for use in medicinal products.

Medicinal Values

Opium is used as an ingredient of many pharmaceutical preparations in the form of tincture, powders and tablets. The medicinal property of opium is due to the presence of alkaloids/analgesic opiates including morphine, codeine, narcotine, papaverine and baine.

Poppy oil is used for culinary purposes and for the manufacture of paints and artists ink. Though serious attempts have been made to replace the various alkaloids of opium with synthetic substances, no real substitute has been developed as yet. Poppy seeds are rich in oil, carbohydrates, calcium, and protein. Poppy oil can also be added to spices for cakes, or breads. Poppy products are also used in different paints, varnishes, and some cosmetics (Krzymanski and Jonsson, 1999).

Botanical Description

Papaver somniferum is an herbaceous, annual/biennial herb, greyish-green in appearance growing to 1.2-1.5 m. Leaves are lobed, alternate, sparsely covered with coarse hairs. The flowers are attractive with 4-6 white, pink, mauve or red petals, normally 12 cm in diameter, the petals lie flat before falling. Flowering season is late spring to early summer. The fruit is a hairless, rounded capsule, topped with 12–18 radiating stigmatic rays and turning brown at maturity. Seeds are numerous either black or deep grey or white seeded in India.

Cultural Practices

Climatic requirements: Poppy prefers moderately cool weather and open sunny locations but can be grown successfully during winter in sub-tropical regions. Severe cold spell, frost, dull cloudy weather, high winds and high rainfall during the lancing period adversely affects the yield and quality whereas, dew during lancing increases yield. Dull, cloudy or rainy weather tends to reduce, not only the quantity, but also the quality of the drug exudations.

Soil requirements: The opium poppy thrives best in a well drained soil having optimum fertility, light black or loam soil with an optimum pH around 7.0. It does not tolerate water logging at any stage of growth.

Field preparation: The field should be thoroughly ploughed 3 or 4 times to prepare well pulverized soil. The field is then prepared into beds of convenient size (Verma *et al.,* 2010).

Method of sowing and sowing time: The seed is either sown by broadcasting or in lines during late October to early November. Line sowing is preferred over broadcasting to ensure uniform growth besides advantages like lower seed rate, good crop stand and ease in carrying out intercultural operations.

Seed rate is 7-8 kg ha^{-1} for broadcasting method and 4-5 kg ha^{-1} for line sowing. Seed treatment with fungicide before sowing improves germination. A spacing of 30 cm between lines and 30 cm between plants is normally adopted.

Varieties: A large number of races under different local names as varieties are being cultivated in different agro-climatic regions such as; 'Telia', 'Haryali', 'Safed Dhanti' or 'Kanth Bhabutia', 'Katila', 'Kali Dandi' or 'Kali Danthi', 'Choura Karia', 'Damia', 'Mororia', 'Variegated poppy' etc. They vary in leaf and floral morphology and capsule characters.

Irrigation: The crop is given about 15 irrigations in all depending upon the soil type and variety sown. Moisture deficiency at the time of capsule formation adversely affects the yield (Verma *et al.,* 2010). Three irrigations at 12-15 days interval upto flowering stage and subsequently at 8-10 days at flowering and capsule formation stage give good yield.

Manure and fertilizers: Farm yard manure @ 20-30 tonnes ha^{-1} is applied by broadcasting while the field is being prepared for sowing. Besides this, 60-80 kg of N and 40 kg P_2O_5 ha^{-1} is recommended. Half of N and entire P_2O_5 are applied at sowing time through placement and remaining half of N is applied at the rosette stage.

Pests and diseases: The crop is attacked by the leaf minor which can be controlled by spraying with 0.2 per cent Metasystox or Rogor.

Downy mildew, which usually occurs at the time of capsule formation, can be controlled by the application of 0.2 per cent dithane Z-78.

Lancing and latex collection: Opium starts flowering in about 95-115 days after sowing. The petals start shedding after 3-4 days of flowering. The capsules mature after 15-20 days of flowering. Lancing is done at a stage when capsule colour changes from greenish to light green.

Lancing is done longitudinally in bright sunlight and the latex (crude opium) which oozes from the wounds dries during the following night and is scraped with the sharp knife in the early hours of the next morning.

Harvesting, threshing and yield: The crop is left for drying for about 20-25 days after the last lancing, the capsules are then picked up and the plant is removed with sickles. Harvested capsules are dried in open and seeds are collected by beating with a stick.

The average yield of raw opium varies from 13 to 33 kg ha^{-1} though much higher yields have been recorded with good agronomic practices (Verma *et al.*, 2010).

The average yield in India was recorded to be about 35-38 kg latex ha^{-1}, whereas, in Chittorgarh district of Rajasthan, the average yield of latex was 46-50 kg ha^{-1}.

References

Kryzmanski, J. and Jonsson, R. (1999) Poppy. In: Robbelon, G., Downey, R.K., Ashri, A.(eds.), Oil Crops of the World. Their Breeding and Utilization. McGraw Hill, New York, p. 388-393.

Sharma, F.L and Rajput, V.S. 1999. Opium production Technology in Rajasthan. *Agricultural Extension Review*.11(6): 29-31.

Verma, S., Singh, Y., Anandhi, P. and Srivastava, D.S. 2010. Opium poppy: A useful Medicinal flat. *Agrobios Newsletter*. 9(7): 45-46.

34

Periwinkle (*Catharanthus roseus* L.G. Don) Syn. *Vinca rosea*

Common name: Sadabahar

Family: Apocynaceae

Introduction

Periwinkle also known as 'Old Maid', 'Sadaphuli', 'Sadabahar', 'Nayantra', is one of the few plants which has generated interest both in scientific, horticultural and medical communities around the globe. It was originated from the tropical forests of Madagascar and is currently cultivated in China, India, Israel, Madagascar, United States and in the countries of Central and Southern Europe. In India it is cultivated as a pot herb and produces pink or white flowers, but its commercial cultivation is done in about 3,000 ha in Tamil Nadu, Maharashtra and to some extent in Gujarat (Halder and Rai, 2012).

In ancient days, the plant was recognized as symbol of love and charms. But now, its future looks quite bright as it possesses tremendous commercial as well as pharmaceuticals potential to growers in different agro-climatic zones of the world.

Medicinal Values

The anti-diabetic action of the periwinkle was made use in folk medicine of the Philippine Islands. Similarly, closer home soothing eye wash was made from the flowers and used in the folk medicine of the Caribbean Islands. Traditional herbal eyewash made from the extract of the flowers was also applied to the affected eyes by the people of the Cuba, Puerto Rico, Jamaica and other islands.

It is a unique medicinal plant. Its leaves contain about 130 terpenoid indole alkaloids. There is good demand for these alkaloids in the world trade. United States of America imports about 1,000 tonnes of leaves while Germany, Italy and United Kingdom import about 1200 tonnes of roots to meet the requirement of alkaloids for drug manufacture (Halder and Rai, 2012). There is ample scope for this crop in India as the raw materials are used in the manufacture of drugs.

This herb contains at least two anti-cancerous alkaloids *viz.*, 'Vincristine' and 'Vinblastine' which are capable of inhibiting the growth of tumors in the human body. 'Vincristine' has proved more effective against childhood leukemia. The alkaloid 'Vinblastine', acts as cell cycle inhibitor, which has been found to be very effective in the treatment of testicular cancer. Alcoholic extract of periwinkle significantly lowers the blood sugar level in the body. It is also very effective in curing dementia caused due to insufficient blood supply to the brain. It helps in relieving muscular pain, depression of central nervous system and wasp's stings.

Other Uses

Periwinkle is gaining importance for insecticidal, nematicidal, anti-helminthetic antibacterial, antifungal and antiviral activities. All the parts of the plant have bioactive molecules like flavonoids besides alkaloids and as such they can be used as botanical pesticides. They are ecologically safe and biodegradable in nature.

Leaves of periwinkle extracted with various solvents have been found to be toxic to a wide range of phytophagous insects like gram pod borer, tobacco caterpillar, red cotton bug and termites. Its extract is also recorded to have strong inhibiting action against spore germination of *Fusarium moniliform*, *Alternaria alternate* and *Rhizoctonia solani*. Extract of periwinkle has also been found to be highly toxic to bottle guard mosaic virus and pepper vein mosaic banding virus in *Capsicum annum*.

Botanical Description

Periwinkle is an erect, branched herb which grows up to 0.5 m. The plants bear oblong-ovate, short stalked, opposite leaves, acute at the apex. Flowers are borne in axil in pairs with long cylindrical corolla tube. They are pink or white in colour. Fruits are follicle, containing up to 30 black seeds. The seeds are very small and light which lose their viability one year after their storage.

Cultural Practices

Climatic requirements: It can be grown in almost all kinds of climates ranging from arid, semi-arid and desert, coastal and even in hilly conditions.

Soil requirements: Although it grows on all types of soils but its best growth can be achieved in alluvial, well drained soils having sandy loam to loam texture and rich in organic matter. It can also be grown in salt affected soils, barren lands and sandy soils. However it grows best in partially shaded areas in acidic soils.

Method of planting and sowing time: Seedlings of periwinkle are planted in a bed having full or partial shade. It should not be planted during full sun since plants develop yellow leaves if grown under hot, bright conditions.

Prepare the bed in spring (March-April). Weed thoroughly and remove stones or other debris. Breakup the soil to 15 cm depth using rotary tiller. Work a 2.5 cm bed to improve the water holding capacity of the soil and nutrient content. Seedlings are then planted on a cloudy, cool day in big planting holes with same depth and diameter as the original nursery containers. Space the holes, 1.2 to 1.5 m apart for a traditional ground cover planting or 15-20 cm apart for faster coverage. Make sure the base of the plant is level with the soil.

Mulching: Spread 3.0 to 5.0 cm thick layer of mulch between the plants to deter weeds, help to keep the soil moist and cool. Leave about 3cm space between the mulch and the base of the plants to allow excess moisture in the soil to evaporate.

Irrigation: Water the plant to a depth of 5cm immediately after planting. Run the water into the soil at ground level rather than watering from overhead since moisture on the foliage create ideal conditions for a fungal disease.

Harvesting and yield: Crop is harvested after 12 months of sowing. Plants are cut at 7.5 cm above ground level and leaves, stems and seed are dried. Then roots are dug out, washed with water and dried in shade. On an average, the yield of leaves, stem and roots on dry weight basis is 4, 1.5 and 1.5 tonnes per hectare respectively under irrigated conditions. But the respective yield under rainfed situations is 2, 0.75 tonnes per hectare for leaves, stem and roots each.

References

Halder, J and Rai, A.B. 2012. *Vinca rosea* : An Herb for the future. *Agrobios Newsletter.* 11(4) : 61-63.

35

Safed Musli (*Chlorophylum borivilianum* Sant and Fernandez)

Common name: Safed Musli

Family: Liliaceae

Introduction

Safed musli is an important ayurvedic medicinal plant. The seedlings of musli are small and commonly found in forests during rainy season.The plant is observed in the valleys of Himalaya, Satpura, Vindhya, Aravali and in hilly areas of Bihar and Assam. It is also found in many parts of Gujarat, Rajasthan and in some parts of Maharashtra (Nibe *et al.*, 2008). Safed Musli, infact, is a major Indian medicinal plant used for the preparation of vital tonics. Dried powder from its fleshy tuber is medicinally important and saponins are the active ingredients.

The plant is a new addition to cultivation and has a vast potential in the international market because of its presumed aphrodisiac property (Jamwal, 2007). Apart from this, safed musli is highly effective for curing a number of human disorders and diseases. The same are described along with cultural practices for cultivation of safed musli.

Medicinal Values

The tubers of the safed musli are used as an ingredient in a large number of ayurvedic medicines. It is used as an energetic, a tonic, and as a appetizer. It is used as a medicine for diabetes and improves male potency. Tubers contain steroids sapoginine (1-2 per cent), proteins (10-20 per cent) and calcium to some extent with some water soluble minerals. As tubers are fat free so they are useful on Vata and Pitta. Safed musli is Vrishya, sheet, madhur, rasayana, slightly bitter, healthier and energetic in Ayurveda. It is also effective in fatigue and in blood purification. Safed musli has proved useful in diseases like renal calculus, dhupani, sangrahani, leucorrhoea. It is lactating, energetic to heart.

Daily use of about 3g of safed musli powder with milk is helpful in providing healthiness. It is highly effective for curing male impotency, natal and post-natal problems of women (Kour *et al.*, 2011).

Botanical Description

Safed Musli is known by many more names such as wonder crop, the crop of the 21st century, crop with golden roots, 'Kuber Ka Khazana' and 'Indian Viagra' (Jamwal, 2006). Dholi Moosli, White Musale, are the other names of safed musli.

Leaves of safed musli look like those of garlic but slightly yellowish and white flowers with six petals are arranged on the flowering stalks which emerge from the centre of the plant. About 20-25 flowers on the flowering stalk appear in July. The seed is very small, black and enclosed in the holes. There are about 10-12 seeds in a bole. The seed is very light in weight and near about 300 seeds weigh 1g. The number of tubers vary from plant to plant with average 5-30 per plant.

Cultural Practices

Climatic requirements: Sub-tropical climate is favourable for cultivation of musli. The plant can be grown up to an altitude of about 1500-1600m above mean sea level. Hot and humid climate is best suited for its growth and development.

Soil requirements: Safed musli is a root crop and can be grown in well drained, loamy sand and sandy loam soils. However, sandy soils are not found good for its cultivation because roots in such soils become long, thin and delicate. In short, it is concluded that light to medium textured, well drained fertile soil is the most suitable for cultivation of safed musli.

Land preparation: Land should be deep ploughed and harrowed 3-4 times to bring about proper aeration. After each harrowing (ploughing), planking is done. Broad beds of the size 90-120 cm and 15-30 cm height are prepared. Between two 90 cm wide path is left for intercultural operations and irrigation.

Propagation: It is usually propagated through tubers or roots but seed propagation is also used (Nibe *et al.*, 2008). The roots are used as planting material. Each root which has to be planted just have some portion of the crown attached to it, otherwise there will be less germination. To increase the germination, sprouted tubers are found economical. Roots each having 2- 5 g weight are considered to be the best for planting.

Planting: On the onset of monsoon, planting should be done preferably in June. However, if the monsoon rains are delayed, planting even can be done during the month of July or August. For planting one ha approximately 3.5q of tubers are required. At the time of planting soil must be moist and planting should be done at the spacing of 30 cm x 10 cm. Tubers should be placed deeply in the soil. If roots are seen naked in the fields, earthing should be done.

Weed control: The field should be kept weed free. Weeds are required to be removed by hand weeding. Care should be taken that no musli plant is removed with weeds. Two to three weedings are to be required in 30 -40 days.

Irrigation: To avoid the water stress during the growth period, irrigation should be given as and when required. During rainy season 2-3 protective irrigations are essential for the growth of tubers. However, water stagnation in the field be avoided.

Harvesting and yield: Generally leaves dry up after 100 days but the development of tubers goes on underground in the soil which get ready for harvesting after about 6-7 months (Jamwal, 2006 and Nibe *et al.*, 2008). So, digging of the roots should never be done when foliage of the crop has dried. Harvesting is resorted in the months of December-January when the matured leaves drop down with complete tuber formation. Before harvesting the field is properly irrigated for easy and safe dig out of the tubers. The tubers are dug out with the help of *kudal* in order to shun damage to the crop. Fresh tubers to the extent of 1.8 to 2.0 tonnes ha^{-1} or dried tubers 0.3 to 0.4 tonnes ha^{-1} are obtained.

Preparation of produce for marketing: Bunches of musli removed from the soil should be thoroughly washed with water. The washing should be done in bamboo baskets under water currents and then the tubers are peeled by pressing them and holding between 2 fingers. Such peeled tubers should be again washed with water and dried completely in the sun for 2 to 3 days. The hard dried tubers are ready for marketing.

References

Jamwal, J. 2006. Safed Musli (*Chlorophytum*) and Satawar. *The Kashmir Times*. 63 (261): 13 (Thrusday, September 21).

Jamwal, J. 2007. Medicinal/Aromatic plants and value addition. *The Kashmir Times.* 64 (338) : 13 (Thrusday, December 6).

Kaur, Navjyot, Kaur, Baljit and Singh, K. 2010. Medicinal plants of India. *Agrobios Newsletter*. 10(01) : 26-27.

Nibe, N. B., Dale, N. S., Umbarkar, R. B. and Vikhe, V.A. 2008. Safed Musali. An important medicinal plant. *Agrobios Newsletter*. 7 (2): 44.

36

Saffron (*Crocus sativus* L.)

Common name: Kesar

Family: Iridaceae

Introduction

Golden spice, saffron is known to the world from time immemorial, due to its fragrance flavour and colour. This magical spice is mentioned in many ancient literatures (Pallavi Nair, 2013). One such reference to the use of saffron in ancient Egypt, where it was used by Cleopatra as an aromatic and seductive essence

Saffron (*Crocus sativus*), called "Red Gold" of Jammu and Kashmir Himalayas is of much commercial and economic importance. Saffron, in fact, constitutes the second biggest cash crop after apple fruit in Kashmir (Anonymous, 2003). It is autumn flowering plant, rather a small herb with scented flowers, which has been cultivated in the vale of Kashmir since long. Saffron in Kashmir is legendary crop of the well-drained plateau of Pampore. According to a famous legend, a *naag* (snake) suffering from an eye problem went to a doctor in Padampur, now known's Pampore. He cured the *naag*; and to show its gratitude the *naag* gave the doctor a bulb of saffron which yielded the corms for cultivation. This is, how, the cultivation of saffron came into existence in this area. Although the exact date of saffron cultivation in Kashmir, is not precisely known yet the records dates back to about 550 AD near 400 centuries ago than its recorded cultivation in Spain by the Arabs. Indeed, it is believed that saffron originated in Asia Minor, Arabs who gave it the name Zaffron, and began cultivating it in some conquered parts of Spain. Thus, saffron a native of

Iran and Asia Minor, later on spread to several Asian and European countries. Pampore in Kashmir is the nerve centre of saffron production.

Currently, Spain produces around 70 per cent of the world's saffron, while India takes second place. In India, Jammu and Kashmir is the main contributor with respect to saffron cultivation. In Jammu region, saffron is grown in Kishtwar area traditionally. But in Kashmir it is now being grown in other places apart from Pampore. It is also grown in Kinnaur area of Himachal Pradesh in limited areas only.

Distribution

Apart from Pampore, cultivation of saffron has now been extended in the districts of Badgam and Anantnag particularly the Chrari-Sharief plateau and Mattan area, respectively. Thus, presently more than a lac people are engaged in saffron production,out of which about 50 per cent are women. In Kashmiri, the saffron is known as *Kong*, in Hindi as *Kesar* and in Urdu as *Zaffron*. In Kashmir, the saffron growers are generally referred to as *"Kong Zamindars."*

As far as total area under saffron cultivation is concerned, definite figures are not available. The land under saffron cultivation has shrunk from 10,000 ha to less than 4,000 ha. The area under saffron has further shrunk and now it is about 3,000 ha which is mainly attributed to rapid changes in the climatic conditions in the valley and organizational difficulties.

Medicinal Values

Saffron is a stimulant, warming in effect and dry in action, and as such assists in urinary, digestive and uterine troubles. The aromatic fragrance of saffron prompted the European herbalist, Gerard to mention during 16th century, "Moderate use of saffron is good for the head, and makes the senses more quick and lively, shake of heavy and drowsy sleep and make a man merry." Saffron is credited in pharmacy as colouring and flavouring agent, and also as an abortifacient.

In Asian countries, the pregnant women take saffron with milk for their good health and fairness of the new born. It was also used for body massage in winter because of its fragrance, stimulation and healing qualities. Besides, it inhibits skin tumors, improves arthritis, eye vision and menstrual discomfort as well as stomach upset and insomnia (Pallavi Nair, 2013). It is point to mention that Kashmiri saffron has 3 per cent more medicinal value than that of Iranian. Chemical constituents present in saffron are given in Table 36.1.

Saffron Formulations/Usage

The crop is valued for its dry stigmas obtained from flowers, which actually constitutes the saffron. The purple blossoms – arrogant flowers, some call them – are born by sun rise and die at sunset. It is lauded as a flavouring, digestive, sedative, dye, hangover cure, exhilarant, the present use of saffron is largely culinary. Madhur Jafery, a well-noted culinary expert in London once said, "While the taste is very subtle, saffron gives the dish an aroma at special occasions, I use it for its aroma which to me is headily sweet, and for its colour, which in India is thought very auspicious." Arabs, the greatest saffron's consumers use it to flavour lamb, chicken and rice dishes. In India, it is used for the colouring and flavouring of foods. The Kashmiri mostly employ saffron in various kinds of meat preparations, *kawah* tea and saffron *laddu* making (sweet dish) *vis-a-vis* sweet rice.

Table 36.1: Chemical Composition of the Saffron

Constituent	*Per cent*
Water	15.6
Starch and sugar	13.5
Essential oil	0.6
Oil content	5.6
Total N-free extract	43.6
Crude fibre	4.4
Ash	4.3
Colouring component	13.7

Source : Saraf, 2006.

Religious Importance

Held in reverence in the religious achelons by Jains, Hindus, Buddhists, saffron is also used in the form of perfume, strewn in Greek and Roman halls, theatres and baths. In Greek mythology, it was especially associated with the "hetaerae", a professional class of Greeks sprinkled with saffron when kings made their entry into the city. In ancient India a golden coloured water soluble fabric dye was distilled from saffron stigmas. Shortly after the demise of Mahatma Buddha, his priests made saffron the official colour for their robes.

Botanical Description

The saffron plant does not produce seeds and is highly sterile due to its triploid nature, it has three sets of chromosomes as compared to normal plants which have two sets of chromosomes. It is a bulbous perennial and grows from bulbs, each bulb giving rise to 2 to 3 flowers. Saffron is an autumn flowering plant, cultivated for a long time in the Kashmir valley and the Kishtwar region of Jammu. Owing to the sterility factor, the crop improvement

through the conventional breeding methods is not possible. The saffron plant, however, propagates through underground bulb-like structure called corms.

Cultural Practices

Climatic requirements: The cultivation of saffron calls for a specific type of conditions, a moderate *i.e.,* temperate to sub-temperate climate with well-drained sandy loam or loamy soil but it must be rich in available plant nutrients and adequate rainfall.

Soil requirements: In Kashmir, it is mostly grown on *"Karewas"*. *"Karewas"* are lacustrine deposits which are alluvial in nature and are free from stones. Soils of *"Karewas"* where saffron grows, are usually light yellowish (10 YR 6/4) to yellowish brown (10 YR 5/4) in colour with loam to silty loam and loam to sandy loam in texture (Gupta, 1992).

Generally, they are calcareous. However, had calcium carbonate leached from the surface and accumulated in lower horizons. The pH of the soils ranges between of 7.0 to 8.5 but never beyond 8.5. They are rich in available calcium, magnesium and potassium but poor in iron and zinc and have number of strains of *Azotobacter chroococcum* from these soils which fix atmospheric nitrogen to meet part of its requirement as most of the cultivators do not apply nitrogenous and other fertilizers.

Land preparation and planting: Land preparation for cultivation of saffron starts from April, by ploughing 4 to 5 times to a depth of 30-45 cm to bring the soil to fine tilth. This is followed by ploughing and leveling in the month of May. The healthy bulbs or corms free from fungal infection having a diameter of 3 cm or more are planted at a depth of 8 to 10 cm, usually in the month of August or September at a spacing of about 10 cm. Furrows are closed and the fields laid out into square beds. On an average 60 quintals of seed material (corms) is required to plant one hectare of land. The seed material is treated by dipping (after removing outermost fibrous covering of the corms) in 5 per cent copper sulphate solution before planting (Anonymous, 2006).

Flowering season: Usually, the flowering season of saffron is confined to about 3 weeks from mid-October to 1st week of November. The flowers are manually harvested daily during morning hours and stigmas with styles are trimmed immediately. Thus, during the month of October, the saffron fields are agog with flowers. The flowering season lasts for about a fortnight. Although the flowering season begins with the month of October yet it varies according to the location. It becomes earlier when the sun rays do not fall directly and the land is known as *"Khowet"* and gets late where the sun rays fall directly on the land throughout the day, which is known by the term *"Dachhun"*.

It is remarkable to note that the flowers of saffron bloom during night hours particularly between 1.00 a.m. to 5.00 a.m. The moonlit nights especially help the flowers to exhibit their brilliance and peculiar charm that the Creator has endowed with these flowers.

The flowers with their six lilac petals, being very sensitive to sunlight, have to be harvested prior to the sun rising. Sun-rays open the mouth of the flower and may make the entry of the bee into the interior portion of the flower, causing damage to the precious three red threads or stigmas nestling in each flower, known by the name *"Kong"*. So, the produce becomes virtually worthless unless picked up within an hour's time after their coming into full bloom. Flower picking soon after the dew disappears and before the sun gets hot becomes essential because flowers when exposed to sunlight too loose essential volatile oil necessary to impart fragrance.

Manures and fertilizers: Although most of the *"Kong Zamindars"* hitherto did not apply fertilizers yet these days it is recommended to add chemical fertilizers at the rate of 30 kg of N, 40 kg of P_2O_5 and 20 kg K_2O ha^{-1} so as to harvest higher yield of saffron. These fertilizers should be added in two splits, first *i.e.*, 15 kg N, 20 kg P_2O_5 and 20 kg K_2O at the time of second hoeing which falls during September. The second dose of fertilizers *i.e.*, 15 kg N and 20 kg P_2O_5 should be applied with third light hoeing. By doing this, the productivity at the Sher-e-Kashmir University of Agricultural Sciences and Technology, Kashmir's Research Station has been found to go up to 3.5 kg ha^{-1} as compared to the 2.0 to 2.5 kg ha^{-1} normally obtained in the saffron fields.

Pests and diseases: Saffron crop is not much affected by insects, except the damage which is caused by rats and moles. Birds, particularly sparrows, damage the flowers. The corms are, however, prone to disease like "Gladiolus Dry Rot" which shows various small black lesions on the corm, which coalesce into large black areas. The corms then rot and shrivel. "Gladiolus Scab" causes round sunken crates in the base of the corm.

Harvesting (picking of the flowers and drying): It is point to mention that it takes considerable skill to pick the flowers without doing any type of damage to the delicate and tender threads. The hand-picking of flowers is, infact, the most laborious and strenuous part of saffron cultivation. After picking previous days flowers, these are set apart carefully for drying. In Spain, the flowers are dried over a gas stove or on mesh screens over charcoal. In Kashmir, however, these are mildly sun-dried under the mellow autumn sun for 3 to 5 days. The direct sunlight is avoided to save them from rotting and losing their natural oil. Although it appears unbelievable yet generally about 1,80,000 – 2,00,000 flowers are to be handpicked to yield 1 kg of saffron. However, according to many *Kong Zamindars* about 2,00,000 – 2,40,000 flowers are needed to yield 1 kg of first grade saffron. Only an excellent harvest in a well-tended plot of about 1000 m^2 could yield about a kg. Flower picking is a job, involving whole of the families of the *Kong* growers. Kashmiri songs and *rauf* echoing in the air, men, women, children and hired labourers are seen strewn in the fields while collecting the delicate flowers in the wicker baskets. On the very day that flowers are picked, whole families sit up till late at night sifting the stigmas and separating them from these flowers. The stigma has an orange red tip which forms the true saffron or *"Shahi Zaffron"* called *Mongra* in the trade. The long white base of the stigma is of inferior quality and goes by the name of *Lacha*.

Mongra Zaffron, the premier quality from Kashmir, fetch over Rs. 20,000 for a kg. *Mancha Selecta*, the best quality saffron from Spain, sets the world standard (Singh, 2003). It sells as high as 1000 dollars a kg. The produce from Greece or Iran sells in the range of 300 dollars a kg.

Yield: The saffron yield estimates have not so far been recorded scientifically. However, 2 kg dried saffron ha^{-1} is the reported yield which can be harvested fifth year onwards of this crop.

Grading of Saffron: There are two grades of saffron viz; 'Mongra' and 'Laccha'. Mongra is the best quality saffron and consists of bright and red stigmas alone. The laccha grade belongs to the second quality which comprises both of stigmas and basal stylet with or without minor percentage of other floral wastes.

Grading standards of both of these grades are as follows:

Grade	*Colour*	*Floral Waste*	*Foreign Matter*
		(Maximum Content)	
Mongra	Deep Red	50 kg^{-1} or 5 per cent	50 kg^{-1} or 5 per cent
Laccha	Light Red	100 kg^{-1} or 10 per cent	100 kg^{-1} or 10 per cent

As per the Food Adulteration Act of India, the following parameters are required to be followed to fulfill the quality norms of the saffron (Kant and Misger, 2006).

Parameter	*Limit*
Total ash	Not more than 8 per cent by weight
Volatile matter	Not more than 14 per cent
Aqueous extract	Not less than 55 per cent
Foreign matter	Not more than15 per cent
Floral waste	Not more than 15 per cent on dry weight basis

Environmental Threats: No doubt, the Pampore saffron fields have acquired world fame, but these days, it is cause of concern that instead of giving saffron cultivation a boost, few decisions of the Government negate whole of the strides made since centuries. The Khonmoh industrial estate has taken up a lot of the saffron farms.

Establishment of Bharat Gas Petrol-cum-Pump, spreading over an area of 50 ha in saffron *Karewas* has resulted into the reduction of the saffron fields. In addition to this, the soap factories, the joinery mills and the Khrew and Wuyan cement factories have made the situation all the more abysmal. In this way, the land under saffron cultivation is fast shrinking.

The other reason destroying saffron fields in Pampore area these days, is the mushrooming growth of stone crushers, stone quarries being built-up in close vicinity of saffron growing fields. In Pampore alone there are about 30 to 40 stone crushers interspersed in the fields of saffron and on the edges of the roads. Dust created from the stone crushers and cement factories not only pollute the atmospheric air but also settles on the soils as well as leaves of the saffron plants. Accumulation of dust on the leaves of the saffron impairs the process of photosynthesis, thus, rendering the plant growth inactive and eventually lessening the efficiency of saffron production (Gupta *et. al* 1992). The flower quality also gets ruined.

The corms are prone to diseases like gladiolus dry rot which shows various small black lesions on the corm and later on coalesce into large black area. The corms then become rot and shrivel. "Gladiolus scab" causes round sunken crates in the base of the corm. The corm is also threatened by mice and birds particularly sparrows.

Adulteration: Sometimes the saffron is adulterated with other substances like turmeric, marigold or safflower petals, maize tassels and even bits of wax and molasses. Quality control on the part of farmers is undoubtedly; assured as they fear that even a minor step in the wrong direction on their part would rebound back. There is a general grouse among the *Kong Zamindars* that the middle men and wholesale dealers dye the floral parts of saffron to make them akin to saffron flowers for mixing them with real saffron.

Decline in saffron area and production: There has been a sharp decline in the area and production of saffron in Jammu and Kashmir during the last decade or so (Table 36.2). Productivity has also decreased. The reasons behind this declining trend include the conventional agro-techniques, poor nutrient input management.

Table 36.2: Area Production and Productivity of Saffron

Year	*Area (ha)*	*Production (kg)*
1994-95	4496	13425
1995-96	4496	14128
1996-97	5707	15952
1997-98	5361	17382
1998-99	4161	13021
1999-2000	3997	7554
2000-01	3827	4860
2001-02	3827	3950
2002-03	3830	5032
2003-04	2742	5154
2004-05	3063	4815
2005-06	2989	8852
2009-13*	3715	5611

*Area and production remained constant during these years

Source : Saraf, 2006, 2007, Malik,2013

The data given in Table 36.2 indicate that during the year 1994-95, the area under cultivation of saffron was 4496 ha which after a decade *i.e.*, during 2004-05 was reduced to 3063 ha and during 2005-06 further reduced to 2989 ha. However, with the efforts of scientists working in the "National Mission on Saffron", the area stands increased under saffron cultivation, which is now 3715 ha (Malik, 2013).

Setting up of National Mission on Saffron: The prestigious "National Mission on Saffron"' sanctioned by the central government during August, 2010 was set to achieve its target of doubling the saffron production from about 9 metric tonnes to 18 metric tonnes by introducing new technological interventions.

The implementation of the mission has started showing good results with the area brought under rejuvenation and replanting of good quality corms in 82 hectares in three public sector nurseries. The mission also envisaged efficient post harvest saffron recovery and quality saffron dried through the use of hot air driers.

The government *i.e.*, Department of Agriculture and Sher-e-Kashmir University of Agricultural Sciences and Technology, Kashmir have claimed that the productivity of saffron is expected to increase to 5 kg ha^{-1} from the present yield of about 3.0 Kg, after implementing of the National Mission on Saffron.

Improved production technology: The latest studies on production technology of saffron in Jammu and Kashmir revealed that maximum sunlight during March-April is essential for growth and development of corms and subsequent yield (Singh *et al.*, 2011). Further, it is added that adequate soil moisture from November to December is required for vegetative growth.

It has been estimated that generally about 1,50,000 flowers (Anonymous, 2006; Gupta, 2009) or 1,60,000 to 1,65,000 flowers are required to produce 1 kg dried saffron. About 40q healthy seed corm are required for planting of 1ha area. The appropriate time of corn planting is second fortnight of August instead of April-May. A suitable 5 years planting cycle has been recommended in lieu of 8 or 12 years.

Effect of climate change: As saffron crop is rain dependent, but untimely rains due to global warming, impact has been much pronounced since 1998-99 (Gupta, 2005) in Kashmir Valley.

Change of climate is, thus a major factor in reducing the saffron yield. It is because nowadays rains both of spring and winter season are erratic for the last fourteen to fifteen years. Hence, saffron growers are shifting to grow other crops like apple, walnut, pear, cherry etc. and Government of Jammu and Kashmir has now decided to provide irrigation facility to saffron crop (Anonymous, 2013).

Making of bore wells: Kashmir Valley's golden spice, saffron, witnessed a considerable decrease in production in 12 years (1997-2009). It was stated that 15.97 tonnes of saffron was produced in 1997 while only 5.61 tonnes was obtained in 2009 (Malik, 2013). This fall in production was attributed to erratic rainfall owing to change of climate (Gupta, 2009). In view of this, the Government of Jammu and Kashmir stated 70 bore wells of which 30 are made operational to irrigate saffron by the end of November 2013 (Anonymous, 2013).

References

Anonymous. 2003. "Zargar asks saffron growers to adopt modern technologies". *The Daily Excelsior*. 39 (305) : 5.

Anonymous. 2005. "Government to enact stringent law for quality control of saffron: Zargar". *The Daily Excelsior*. 41 (208) : 5 [Tuesday, July 26, 2005].

Anonymous. 2006. Saffron – The threads of gold. In : *Infotainment – The States Times*. 11 (62) : 5 [March 3, 2006].

Anonymous. 2013. 70 bore wells in offering for saffron fields. *The Tribune*. 133 (300): 4 Wednesday, October 30).

Gupta, R. D. 2005. Environmental degradation of J&K, Himalayas and their control Measures. Associated Publishing Company. Karol Bagh, New Delhi.

Gupta, R. D. 2009. Saffron-A wonderful medicinal and aromatic plant. *Indian Farmers Digest*. 42(11): 8-11.

Gupta, R. D. 2009. The fall of the golden spice. Down to Earth. June 15-30 issue, pp. 53.

Gupta, R.D., Kanwar, B.S. and Rana, K.S. 1992. Ecological considerations of the Sirmour Himalayas. In : Himalayan Environment – Man and the Economic Activities (Edited by JL Raina). Pointer Publishers, Jaipur (Rajasthan), India.

Gupta, R. D. 1992. Soil genesis in the Jammu and Kashmir Himalayas. In : Himalayan Environment – Man and the Economic Activities (Edited by JL Raina). Pointer Publishers, Jaipur (Rajasthan), India.

Gupta, R. D. 2003. Sufferings of saffron. *The Daily Excelsior*. 39 (87) : 6.

Kant, K. and Misger, F. A. 2006. Quality and grading of saffron. *Agrobios News Lettter*. 5 (7) : 36-37.

Malik, B. 2013. Area under saffron cultivation reduces by 70 per cent in Pampor. *The Tribune*. 133(298): 4 (Monday), October 28)

Pallavi Nair, K. 2013. Saffron : The Golden spice. *Agrobios News Letter*. 12(1): 38.

Saraf, S.A. 2006. Saffron – The pride of paradise on earth-I. *The Kashmir Times*, Jammu. 63 (358) : 13 [Thursday December 28, 2006].

Saraf, S.A. 2007. Saffron – The pride of paradise on earth-II. *The Kashmir Times*, Jammu. 64 (04) : 13 [Thursday January 4, 2007].

Singh, D.K, Misgar, F. A., Singh, A. P. and Singh, S. 2011. Production Technology of Saffron in Jammu and Kashmir. *Indian Farmers Digest*. 44(1): 43-44.

Singh, M. 2003. Saffron, the precious spice. *The Daily Excelsior*. 39 (81) : 4 [23 March, 2003].

37

Sesame (*Sesamum indicum* D.C.)

Common name: Til, tilee

Family: Pedaliaceae

Introduction

Among vegetable oil seed crops, sesame (*Sesamum indicum*) is considered one of the most preferred crop, especially under rainfed conditions even with low yield level. It is because of its higher price and very good quality of oil known as 'Tilee Da Tel' which becomes a good substitute for olive oil. Not only this, oil of sesame is effective in sore eyes and other human body's defects. Leaves and seeds of sesame also have number of medicinal values. Sesame is cultivated in many regions under rainfed conditions.

Medicinal Values

Seeds of sesame are laxative, demulcent, diuretic, nourishing, lactogogue and emmenegogue. They are useful especially in piles, dysentery and constipation taken in decoction or in powdered form. Decoction is also useful in cough and aphrodisiac. Ground sesame is given with butter in bleeding piles. If taken in large amount it brings about abortion., hence pregnant women should not take in large quantity (Singh, 2005).

Sesame oil is a good substitute for olive oil. Its drops are effective in soar eyes. It is given internally in gonorrhoea. Leaves are used in bowel infections, such as dysentery, cholera. Decoction made from leaves and roots make an excellent wash for hair. It serves as natural dye and promotes growth of hair. Seeds of sesame are highly rich in calcium. Oil is also used in cooking, salad preparation and sweets –'pugga' (Kharayal and Singh, 2015).

Botanical Description

Sesame is a small, annual plant indigenous to India and it extensively cultivated in many regions. It is a crop of rainfed areas. In Sanskrit it is called as 'Tila' or popularly as 'tilee'.

Plant is an erect, pubescent annual, growing up to 50 to 100 cm, branches from the base. The leaves are of 4 to 14 cm length, leaves opposite or upper ones alternate, the lower ones often 3-lobed. The leaf blade is linear. The inflorescence is an umbel. Flowers are tubular, 3 to 5 cm in size, pink to red or white. The fruit is a capsule, normally pubescent, rectangular in section and typically grooved with a short triangular beak (splits along two or more seams, when dry to release two or more seeds). Seeds are small, flat up to 5 mm length and of varying colours ranging from white, brown to black.

Cultural Practices

Climatic requirements: Being a rainfed crop, it requires warmer climate.

Soil requirements: Although it can be grown on a wide variety of soils yet this crop thrives best on sandy loam, loam and silty loam soils.

Sowing time and seed rate: Being a rainfed crop, sesame is sown with the onset of monsoon in the end of June or beginning of July. July sowing is always better than late sowing to get more yield (Kumar and Thakur, 2006).

Wherever possible, the crop should be sown in lines 30 cm apart with pora or drill. A plant distance of 15-20 cm is the best for branching varieties like 'Til No.1' and 'Brajeshwari'. Even when sown by broadcast, the crop should be thinned well to get good yields. A seed rate of 5 kg ha^{-1} is the recommended dose.

Manures and fertilizers: The crop is generally relegated to lands of low fertility where no manuring has been done. Contrary to this, the crop does not tolerate with excessive fertilization which causes luxuriant vegetative growth and renders it susceptible to diseases and lodging. However, about 30 tons of FYM should be applied before sowing preferably after last ploughing while preparing the field for sowing. To obtain good yield, fertilizing with 60 kg N, 40 kg P_2O_5 and 20 kg K_2O ha^{-1} is necessary.

Intercultural operations: When sown pure, one hoeing should be done to remove the weeds after three weeks of sowing. The crop should also be thinned at this stage.

Harvesting and yield: As the crop is susceptible to shattering, timely harvest is important to avoid losses, it should be harvested when the capsules turn pale yellow and then tied in small bundles. At least 2 shakings are necessary to get full recovery. Yield of the sesame varies from variety to variety. Generally, the yield of the sesame crop is found to range 5 to 8 q ha^{-1}. In case of Brajeshwari (LKT-4) average yield varies between 6-8 q ha^{-1} whereas, in Punjab Till No.-1 yields 5 q ha^{-1}.

References

Singh, S.S. 2005. Fate of the Medicinal Plant Resources of Jammu and Kashmir Directorate of Environment and Remote Sensing, Jammu.

Kharayal, B.S and Singh, Y. 2015. Diagnosis and integrated management of sesame diseases. *Indian Farmers Digest*. 47(8): 5-8.

Kumar, A. and Thakur, K.S. 2006. Production potential and economic feasibility of Sesame (*Sesamum indicum*) – based intercropping with legumes under rainfed conditions. *Indian Journal of Agricultural Sciences*. 76 (3): 183-189.

38

Soyabean
(*Glycine max* (L) Merr.)

Common name: Soybean

Family: Leguminosae

Introduction

Soybean, a leguminous plant product has been considered as one of the most wonder-nutritional gifts that nature has bestowed us, during the last half century (Wadhwa, 2009). It is native to East Asia and is now grown in many states of India but Madhya Pradesh ranks first in its production.

Soybean which is traditionally grown as a food crop in many states of the country, but now it is one of the most important oilseed crops. In the nineties, its consumption was strongly recommended to maintain good health and to assist in preventing and even treating certain diseases because of two main reasons. One, its being nutritionally very rich and the second its having certain non-nutritional components which have medicinal uses.

Medicinal Uses

Apart from having an appreciable amount of three macro-nutrients *viz.*, proteins, fats and carbohydrates, soybean contains vitamins and minerals that are required for good health and freedom from disease. Despite being a plant food, soybean is a rich source of proteins which have a biological value almost comparable to animal proteins and amino acid pattern virtually equivalent to proteins of milk and egg. Thus, it has an immense potential for narrowing down the protein gap in vegetarian diets of most people in the eastern countries (Wadhwa, 2009).

In addition to having good quality proteins, soybean has enough of fats which can well boast of possessing an excellent fatty acid profile. Oil from soybean contains only minimal saturated fats and almost zero content of cholesterol, whereas 85 per cent of it is made of polyunsaturated fatty acids. These fatty acids mostly belong to the linolenic acid group- the omega-3. Omega3's are anti-coagulant and as such prevent blood clot formation in the arteries that lead to brain strokes. Thus, oil of soybean is heart healthy and protects it from cardio-vascular diseases.

As most of the soybean carbohydrate is in the form of cellulose which is an edible fibre but does not get digested, so its calories are not too many and the glycemic index is also low. This keeps undercheck the glucose level of blood and reduces the insulin requirement of the body. Soybean is therefore, a valuable food to include in the diabetic diet.

Apart from being a rich source of nutrients soybean has some important phyto-chemicals which are non-nutritional and stems its disease fighting potential. The phyto-chemicals in soybean are isoflavones. They are usually referred to as phyto-estrogens because their chemical structure resembles with estrogen of the females. However, there is difference as they are much weaker than the natural hormones. Their low potency gives them a wonderful capability of extending estrogenic effects depending upon the requirement of the body. If the level of potent estrogens of the body is high, isoflavones of soybean block their action and prevent hormone driven diseases such as cancers. In the reverse situation, during menopause, these phyto-estrogens substitute the body's hormonal levels and ease symptoms like hot flushes and assist in the prevention of diseases like osteoporosis (Wadhwa, 2009).

Soybean has antioxidant properties which enhance its anti-carcinogenic effects. Regular consumption of soybean containing diet inhibits the development of breast cancer in females and prostate cancer in males. It plays significant role in preventing and treating chronic diseases such as heart ailments and menopausal syndrome (Sangeetha, 2011).

Harmful Effects

Some scientists believe that claim of soybean being a wonder food is based on scanty facts and half truths (Wadhwa, 2009). It contains some chemicals like phytic acid, trypsin inhibitors and nitrosamine which are highly toxic and can even be carcinogenic. Hence, it should not be used as staple food.

Other Uses

It is widely used in the industrial production of different antibiotics. Soybean flour is used to increase the shelf life of many products and improve the colour of pastry crusts. As its flour is free of gluten, it cannot replace all wheat or rye flour in bread making but can be used about 15 per cent to give dense bread with a nutty flavour and moist quality. Soybean is used for production of vegetable oil and margarine. In its pure form as a vegetable oil, it is often used in salad dressing. Soybean's lecithin acts as an emulsifier in some chocolate, breakfast cereals, ice cream, and sweets. Oil of soybean is also used in a wide variety of non-food products *e.g.* soap, biological detergents, plastics and CFC-free cooling agent, the derivative glycerin etc.

Botanical Description

Soybean varies in growth and habit. The height of the plant varies from 20cm to 2m. The pods, stems and leaves are covered with fine brown or grey heirs. The leaves are trifoliolate having 3-4 leaflets per leaf. The leaves fall before the seeds mature. The flowers are borne in the axil of the leaf and are white, pink or purple in colour. The fruit is a hairy pod that grows in clusters of 3-5, each pod is 38 mm long and usually contains 2-4 (rarely more) seeds, 5-11mm in diameter. Soybean occurs in various sizes and in many hull or seed coat colours, including black, brown, blue, yellow, green and mottled.

Cultural Practices

Climatic requirements: Soybean can be grown on a variety of climates varying from sub-tropical to sub-temperate. Due to its adaptability to varied agro-climatic conditions, it is gaining popularity.

Soil requirements: Although soybean can be grown on wide range of soils yet its best growth occurs on well drained sandy loam soils. The soil with pH 6.0 to 6.5 is preferred by the nodulating bacteria for efficient atmospheric nitrogen fixation.

Sowing time and seed rate: In India, soybean is a Kharif crop and can be sown with the first pre-monsoon showers (mid June to July). Late sowing results in poor plant stand. A seed rate of 75 kg ha^{-1} should be used to ensure good crop stand. Under rainfed conditions, the seed rate should be increased to 100 kg ha^{-1}. For late planting, 100-120 kg seed ha^{-1} is required (Singh and Bisht, 2014)

The crop may be sown in rows 45 cm apart. The plant should be spaced 10-15cm. Seedling is always preferred under optimum moisture conditions and seed should not be placed deeper than 3-4 cm in the soil.

Manures and fertilizers: Recommended doses of N, P_2O_5 and K_2O range from state to state *e.g.* under high rainfall conditions, the recommended doses for the soils of Himachal Pradesh are 20,60 and 40 kg ha^{-1}, whereas under low rainfall areas these are 20, 30 and 30 kg ha^{-1}.

Inter-culture and weed control: Two hoeings and weedings are required to be given in order to control weeds.

Harvesting and yield: The crop is harvested when the leaves turn yellow and fall off and pods change colour. Harvesting should be done in time to avoid losses through shattering

Yield varies between 2,500 to 3,400 kg ha^{-1} depending upon weather and soil conditions as well as tillage management (Ved Pradesh *et al.*, 2004).

References

Sangeetha C. 2011. Soybean : A miracle crop. *Agrobios Newsletter*. 10 (01): 13-14.

Singh, J.P. and Bisht, P.S. 2014. Soybean production technology for northern hill and plain zone. *Indian Farmers Digest*. 47(3): 29-30.

Ved Pradash, Bhattacharya, R. and Srivastva, A. K. 2004. Effect of tillage management on yield and soil properties under soybean (*Glycine max*) based cropping system in mid hills of north- western Himalayas. *Indian Journal of Agricultural Sciences*. 74(11): 573-77.

Wadhwa, K. 2009. Soybean: A wonder food. *Ayurveda*. 1 (12) : 38-39.

39

Spinach
(*Spinacia oleracea* L.)

Common name: Palak

Family: Amaranthaceae

Introduction

Like fenugreek and drum stick leaves, the leaves of spinach are also used all over the country as green leafy vegetable. Green leafy vegetables by virtue of having large amount of carotene (a precursor of vitamin A) not only protects our eyes but also promotes physical growth in children and helps in building up resistance to number of diseases.

Medicinal Uses

Succulent leaves of spinach and their tender stems, when boiled and seasoned, constitute an excellent cooling nutrition and demulcent dish. This nutritious dish has proved beneficial especially for those, suffering from anaemia due to the deficiency of iron.

Herbaceous parts are mildly laxative and used as emollient, poultice and decoction and infusion in fever, inflammation of lungs and bowls, difficult breathing biliary derangements and in urinary problems. Juice of leaves is a good gargle for sore throat. Raw juices are given in anaemia as spinach is very rich in iron. Poultice of leaves is effective in swelling scalds. It is best eaten raw. In cooked form, its oxalates, however, cause kidneys stones.

Recent studies indicate that spinach is a good source of chlorophyll and lipids, which are mostly present in its leaves. As such spinach possesses lot of antibacterial properties besides those of demulcent. The leaves of spinach are also used in poisonous condition, internally diuretic (Karunkar, 2013). Spinach being leafy is also an amazing food as it provides Vitamin C besides iron. Both these constituents are good for strong bones, teeth and hair.

Spinach is enriched with magnesium and vitamin E also, in addition to Vitamin A, thereby, making it very nutritious vegetable. It helps to prevent thickening and hardening of arteries, has anti aging and cancer preventing properties. It also prevents age related memory loss and cataract.

In the end, it is concluded that the health benefits of spinach are numerous and its ability to help prevent various diseases is well known. Spinach is at the top of many super food lists. Spinach does have an above average amount of iron in it, but has many other nutritional benefits as stated above. Apart from this, spinach also contains vitamin K, folate, manganese and potassium. It has antioxidants and vital nutrients making it a super food for the heart, brain, eyes and bones. It prevents colon cancer, osteoporosis and arthritis.

Spinach is cool, mild, relishing, causes the formation of urine. It activates the intestine. It is also beneficial in the diseases and eliminates the excessive bile in the body as well as destroys the excessive formation of bile in the body, purifies the blood and cures the stone, and strengthens the bones (Sharma, 2005). Spinach is a nutrition superstar, due to the occurrence of iron and other minerals like manganese, potassium. Moreover, it loaded with vitamins some of them are hardly found in other vegetables and foods (Mohindra, 2014).

Botanical Description

Spinach popularly known as palak, is a species of leafy vegetables, which is cultivated throughout India in all the agro-climatic zones. Its colour ranges from light to dark green and is available in two types; (i) the crinkle leaf and (ii) the smooth leaf variety. The edible portion of the plant is a compact rosette shape of fleshy leaves attached to a short stem. At flowering, it is 50-75 cm high with an erect stem and branching leaves. The fertile flowers are produced in groups at every joint very close to the stalk, while the barren flowers can be found in long terminal bunches. Spinach has two stages in its life cycle; the vegetative, rosette stage and the bolting, seed stalk stage.

Cultural Practices

Climatic requirements: Spinach is a cool season crop, which can tolerate frost better than other vegetable. High temperature, especially long day causes bolting and, reduces its market value. Hence, higher yields of spinach are obtained under short day and mild temperature.

Soil requirements: Spinach can be grown on wide variety of soils however; spinach thrives best on well drained soils having loamy texture. A pH range of 6.0 to 7.5 is most desirable.

Land preparation: Land should be ploughed twice or thrice and harrowed thoroughly to make it well pulverized. It should be then well levelled by planking.

Varieties: All green and Prickly seeded are the important varieties of spinach.

Sowing time: In the plains of Punjab, Himachal Pradesh and Haryana, it is sown twice a year *i.e.* from July to November and February to March. In sub-tropical areas of Jammu and Kashmir, it is sown from August to December, whereas, in intermediate or sub-temperate areas, the sowing time varies from August to September. Contrary to this, in temperate areas of Kashmir the suitable sowing times of Spinach varies from March to June and October to March.

Seed rate: Seed rate varies from 10-12 kg ha^{-1} in case of prickly seeded variety and in case of Shalimar green the seed rate is 5-6 kg ha^{-1}.

Manures and fertilizers: Farm Yard Manure @ 25 tonnes ha^{-1} must be added in the field before sowing the crop as a basal dose. Use of N, P and K @ 60-80 Kg, 25 kg and 25 kg, respectively, has been recommended. However, their application is discouraged in view of the "Organic production concept".

Irrigation: Irrigation at weekly interval is essential to have a good harvest.

Harvesting and yield: Harvesting should be done at frequent intervals when the foliage is tender and appropriate edible stage. Generally 4 to 5 cuttings may be taken. Yield of spinach varies from 100 – 200q ha^{-1}.

References

Karunakar, G.V. 2013.Vegetables are medicines too. DAV, *Ayurveda*. 1 (22): 18-19.

Mohindra, V. 2014. 7 vegetables you should eat to boost your disease fighting potential. *Ayurveda*. 1(26): 18.

Sharma, G.K. 2005. Let vegetables by your medicine. *Ayurveda for Total Health*, December 2005, p. 42.

40

Strawberry (*Fragaria virginiana* or *Fragaria chiloensis*) (L.) Mill.

Common name: Strawberry

Family: Rosaceae

Introduction

Strawberry, a member of rose family belongs to *Fragaria virginia or Fragaria chiloensis*. In the year 1950's, when wild strawberry was transplanted into home garden, it was given the genus name of *Fragaria vesca*. Initially this plant grew best in Northern countries where the cold winter kept the plants happier than those in the Southern regions. Because of their adaptability, strawberries are grown in all 50 states of United States and in all of Canada's provinces.

The fresh fruit is a rich source of vitamins and minerals as well as small amount of pectin which is needed for jam making. The fresh fruit contains 5 per cent sugar and 0.9 to 1.86 per cent acidity. The fruit is mainly consumed fresh but the best jam is made of strawberry. Other products that can be made from it are strawberry crush to serve as drink and preserve (Jamwal, 2012).

Nutritional Values

Strawberry shortcake is an all time favourable dessert and a frequent star on the sweet table, when strawberries are in season. A sumptuous desert, strawberry shortcake is composed of sponge cake, divided into two layers filled with sliced sweetened strawberries and whipped cream between the layers. As in present days, in India different varieties of strawberries are present in season at different times of the year so make this delectable dessert available the year around (Jamwal, 2008).

Medicinal Uses

The medicinal uses of strawberry have been known since long. For example during the 13th century, the French used it as a medicinal herb for numerous digestive discomforts. The roots, leaves and fruit of the Alpine strawberry were used as a digestive aid and skin tonic. The berry was prescribed for diarrhoea and digestive upset, while the leaves and the roots were supposed to relieve gout. The berry itself was rubbed on the skin to cease the pain of sunburn. The juice of the strawberry brightened discoloured teeth. The ancient Romans were staunch believers in curative powers of the strawberry. They believed that its use relieved melancholy and masked bad breath. Strawberry can cure inflammations, fevers, throat infection, kidney stones, gout, fainting spells and diseases of the blood, liver and spleen. John Gerard, a French herbalist touted the value of boiled strawberry leaves as a poultice. Of the fruit itself he says, "the ripe strawberries quench thrust, and take away if they be often used, the redness and heat of the face. Patients enjoyed the medicinal fruit treatment so much that they began eating them as food, accompanied by cream and wine". Strawberries have the highest total antioxidant power among the fruits and protect the body from cancer causing, blood vessel clogging free radicals (Suguna,2011).They contain high levels of Vitamin C, folate, fibre and potassium than most of the other fruits. Strawberries are filled with unusual phytonutrients that promote health (Gupta and Mahajan, 2012).

The fluoride content in strawberries whitens the teeth, prevents dental problems and gives strength to the gums, regulates electrolytes in the body. Presence of high fibre, folate, no fat and all the antioxidants like Vitamin C and flavonoids together form an ideal cardiac health pack as they effectively reduce the cholesterol. All other nutrients present in the berries make the heart muscles strong and help in proper functioning of the heart.

The copper content in the berries helps to reduce the bad cholesterol. It is helpful in production of red blood cells. These berries are virtually fat free. They suppress the appetite and stimulate metabolism. They also help in proper functioning of weight reducing hormones. Owing to presence of Vitamin C, strawberry protects from cough and cold, helps in lowering blood pressure, sharpens memory, cures arthritis and gout and protects eyes.

Botanical Description

Strawberry is unique in its ability to adapt well to a broader range of climate

than the most other fruits. Because of their adaptability, strawberries are grown in many countries of the world including India. Strawberries are unique in that their seeds do not serve to produce new plants. Strawberry plants multiply by sending out runners along the ground during the time that the fruit is developing. These runners develop roots and form the new plants. Because fruits of strawberry are delicate and highly perishable, they cannot be harvested with machine and are always handpicked.

Cultural Practices

Climatic requirements: As already stated that strawberries have good ability to adapt to a wider range of climate. They can be grown in tropical, sub-tropical and even sub-temperate and temperate climatic condition. However, strawberry plants cannot tolerate drought conditions.

Soil requirements: Well drained light to medium soil *i.e.,* loamy sand or sandy loam and sandy clay or silty clay having a pH range of 6.5 to 7.0 is most desirable. It can even be grown in soil pH range of 5.5 to 6.5 (Sharma, 2000).

Site selection and land preparation: As strawberry is highly perishable so the site selected for cultivation should either be near the town or city, and the marketing place. The field should be ploughed thoroughly twice or thrice and finally planked for making it well levelled. A dose of 50 tonnes of well rotten FYM or compost should be applied as a basal dose. Along with FYM, 100 kg each of superphosphate and potash, and 20 kg of ammonium sulphate ha^{-1} should also be applied as basal dose.

Planting: After land preparation, the raised beds of 90 cm width should be prepared. The distance from one bed to another should be such that a man can easily work and harvest the fruits. It should be 30 cm (Sharma, 2000).

The proper time for planting under North Indian conditions is the 3rd week of October. Before planting, runners should be treated with 2 per cent solution of Captan for 2 minutes.

Varieties: High yielding diseases resistant varieties suitable for the area should be planted. Varieties like 'Chandler', 'Pojaro', 'Selva', 'Fern' etc. introduced from California have been found very successful. About 55,000 runners are to be required ha^{-1}. The distance from plant to plant be kept at 30 cm.

Irrigation: In the beginning strawberry plants require irrigation till the plants are well established, weekly irrigation is sufficient. The drip or sprinkler irrigation is the best suited for the successful cultivation of strawberry, particularly where there is shortage of water for irrigation.

Manures and fertilizers : As strawberry responds well to manure and fertilizer application so the basal dose as stated above, 20 kg each of ammonium sulphate should be applied before flowering and at the time of fruit development. Fertilizers application should be immediately followed by irrigation.

Mulching: This is the most important cultural practice in case of strawberry. It helps to minimize water loss, creates congenial microclimates and reduces weeds growth and protects fruits from any kind of damage. Black alkathene of 200 gauge is good for mulching. The organic waste like paddy straw, sugarcane leaves or any other local material can also be used for mulching. Paddy straw has been considered best for mulching. Mulching should be done from December till the harvest of the fruit.

Protection from frost: Frost during December-January in North India may destroy the strawberry crop. Hence, protection from frost is necessary. It can also be done by covering the plants with thatch roofs which need to be removed during day time.

Plant protection measures: Although there is no major insect pest or disease that damages strawberry plants yet blotching, root decay and insects like spider and cut worm may cause considerable loss. Spraying of malathion, rogor and fungicides having sulphur compounds are found suitable. Neem based insecticides are now available which would be the safest sprays.

Harvesting and yield: The fruit starts ripening in about 100-110 days after planting. Harvesting should be done in the early hours of the day and only when colour of the fruit starts changing.

On an average each plant bears about 20 -25 fruits. The average yield of strawberry varies from 12.5 to 17.5 tonnes ha^{-1}.

References

Gupta, R. D. and Mahajan, N. 2012. 14 dietary tips to prevent cancer. *Ayurveda*.1 (21): 10-11.

Jamwal, J., 2006. More facts about strawberry. *The Kashmir Times*, Jammu.63(39): 13, Under Science and Environment column [February9, Thursday]

Jamwal, J., 2008. Strawberry devastated by rains. *The Kashmir Times*, Jammu.65(51): 15 [February 21, Thursday].

Jamwal, J. 2012.Strawberry cultivation is very remunerative. Personnel communication [189 Sector 1, Trikuta Nagar, Jammu, J&K State, India].

Sharma, D.D., 2000. Strawberry cultivation hope for unemployed rural youth. *Intensive Agriculture*.38(3-4): 3-4.

Suguna, 2011. The strawberry health. *The Daily Excelsior*. 47(134): 4 [May15, In Sunday Magazine].

41

Sugar Beet (*Beta vulgaris* L.)

Common name: Chukandar

Family: Chenopodiaceae

Introduction

Sugar beet, is cultivated in many parts of India, especially in gardens for its leaves and roots. In Punjab, this vegetable is grown for its enlarged roots which are blood red in colour and are eaten as salad. The roots before making into slices are boiled for about 30 minutes. These days, some tropical sugar beet hybrids are available for cultivation in Cauvery, Indus and Shubhra areas of the country. The duration of these tropical hybrids is 5-6 months depending upon hybrids and climatic conditions prevailing during crop growth period. Apart from using sugar beet as vegetable, it has also number of medicinal uses.

Medicinal Values

The medicinal properties of the sugar beet owe to an active principle betin in it. Juice of its white variety is good for liver. It cures eye inflammation also when applied to temples. When its juice is mixed with alum and oil, it is effective on burns. Beet root is a type of vegetable that helps in blood formation. It also helps to support proper spleen function besides that of liver (Pandey and Awasthi, 2012).

A decoction of roots at bedtime in the case of habitual constipation and in piles is good. Red beet is valuable in urine problems. Red beet decoction is applied externally to scalp for itching and dandruff in mixture with little vinegar. Raw or boiled, it makes a good salad and as a pot herb it is a good substitute for spinach.

Botanical Description

There are two main varieties of sugar beet viz; white and red. Sugar beet is a biennial sugar producing, root crop grown in sub-tropical, sub-temperate and temperate regions of number of countries (Bavalgave and Ingle, 2012). Now tropical sugar beet hybrids are gaining momentum.

It is a biennial, tall and succulent plant of 60 cm height and is usually grown for the fleshy root and young glossy leaves. The beet is a root vegetable with purple-green variegated leaves. The leaves (20–40 cm long) form a rosette from an underground stem. The inflorescence is tall (1.2-1.8 m), branched spike, with small, sessile flowers singly or in clusters. The fruit is a nut embedded in the base of the perianth of the flower. Each fruit contains a single seed, brown in colour, shape varying from round to kidney-shaped.

Cultural Practices

Climatic requirements: Sugar beet requires good sunshine during its growth period with a well distributed rainfall of 300-350 mm across the growing period. However, continuous heavy rains may affect the development of tuber and synthesis of sugar.

Soil requirements: Heavy soils retaining high soil moisture are found unsuitable for sugar beet production. Hence, well drained soils having loamy sand or sandy loam are suitable for the growth of sugar beet. It can also be grown in soils having sandy loam to clayey loam texture with porous soil structure (Bavalgave and Ingle, 2012). Optimum PH range is from 6.5 to 9.0 but it can also grow in saline and alkaline soil. The soils with good organic matter are more favourable for sugar beet.

Field preparation: Sugar beet being a root crop requires deep ploughing (45cm) followed by 2 or 3 ploughings to obtain a good soil tilth for favourable seed germination and proper root development. After proper leveling to ensure adequate drainage, ridges and furrows are formed at 50 cm apart.

Sowing time and seed rate: As sugar beet is a cold season crop so, it is sown during Rabi season - October to November. To maintain the required plant population of 1,00,000, use 5 pockets designer seeds. One packet contains 20,000 seeds weighing 600g (Bavalgave and Ingle, 2012). The recommended spacing is 50 x 20 cm. The designer seed is dibbled at 2 cm depth on the top of the ridges at 20 cm apart at one seed per hole.

Weeding and hoeing: The crop must be maintained weed free throughout the crop period. Although Pendimethalin @3 litre ha^{-1} is recommended to spray as weedicide after dissolving in 500 litres of water yet these days it is being discouraged owing to organic food production concept. Hence, hand weeding on 25, 50 and 75 days after sowing should be done. The earthing up operation coincides with the dressing of N-fertilizer.

Manures and fertilizers: Application of FYM @ 12.5 tonnes ha^{-1} with recommended doses of nitrogen phosphorous and potassium are 75:75:75 ha^{-1} in the form of nitrogenous, phosphatic and potassic fertilizers. Use of chemical fertilizers, however, should be limited due to their effect on the various food stuffs. Instead, use of bio-fertilizers and organic manures be encouraged.

Irrigation: Sugar beet is very sensitive to water stagnation at all stages of growth. First irrigation is crucial for the earlier establishment of the crop. It should be given after 5-7 days and 8-10 days after sowing in case of heavy and light textured soils respectively. The process should be continued, however, irrigation may be stopped at least 2 to 3 weeks before harvesting.

Harvesting and yield: Sugar beet matures in about 5-6 months. The yellowing of lower leaf whorls of mature plant and root brix reading of 15 to 108 percent indicate the maturity of root for harvest. The harvested root should be handled as gently as possible to remove soil and thrash to minimize the beet breakage and bruising to get quality sugar beet. The average yield of roots is 80-100 tonnes ha^{-1}.

References

Bavalgave, V. G and Ingle, A. B. 2012. Production technology of tropical sugar beet. *Agrobios Newsletter*. 10 (9): 28.

Pandey, H. and Awasthi, P. 2002. Winter vegetables: Treasures of nutrients and medicinal properties. *Indian Farmers Digest*. 45(2): 15-16.

Singh, S. S. 2005. Fate of the Medicinal Plant Resources of Jammu and Kashmir. Directorate of Environment and Remote Sensing, Jammu.

42

Summer Squash (*Cucurbita pepo* L.)

Common Name: Pumpkin, kashiphal, halwa kaddu

Family: Cucurbitaceae

Introduction

Pumpkin, also called 'Kashiphal' or *Halwa Kaddu*, is most common vegetable available in Punjab, Haryana, Himachal Pradesh and Jammu and Kashmir. Hitherto many of the people do not like to eat this vegetable except during fasts; either preparing 'Halwa' or 'Pakoras' using buck wheat flour or flour of sanghras. However, nowadays, its consumption has much popularized as a dish. It is attributed to its well known medicinal properties.

Medicinal Values

Pumpkins are considered the symbol of 'Thanks giving" in many countries. It is because they are rich in B-carotene, Vitamin A and anti-oxidant which prevent cancer diseases. One cup of cooked pumpkin provides 80 calories, fibre, iron, potassium. Potassium controls the blood pressure. When boiled it loses some of the nutrients and calories. Its seeds are good source of proteins. They are also high in unsaturated vegetable oil, Vitamin E and Vitamin B. Its cover is rich in fibres (Sawhney, 2005).

Recent study in China has given a new hope to diabetics by finding that compounds in pumpkin could potentially replace or at least drastically reduce the daily insulin injections (Anonymous, 2007). Rubbing of pumpkin fruit peduncle paste alleviates scorpion sting (Bala *et al.*, 2009).

The protective effect of pumpkin is thought to be due to antioxidants and D-chiro-inositol. It destroys the excessive humors of bile and phlegm in the body, very nourishing, semen builder and nourishing to the pregnant women and also clears away the constipation during that time.

Intake of juice of pumpkin kept overnight after extraction if taken in the morning kills all the worms. Juice mixed with honey has very cooling medicinal properties and is beneficial in cooling down the heat of the body, inflammation in the throat, impurities of the blood, boils, plethora, increase in temperature of the blood, blood coming out from the throat. Pumpkin helps to fight infections and improves vision, especially night vision and promotes healthy skin (Phartyal and Singh, 2014).

The crushed pulp of pumpkin wrapped in a piece of cloth if tied on the fore head, brings the temperature to normal in fever. The juice extracted from the leaves if applied on the affected part of the piles is beneficial. Eating *kheer* of pumpkin pulp develops the semen, it is also good for a pregnant women. It makes the heart strong (Sharma, 2005). Pumpkin like other vegetables (carrot, garlic, cauliflower, cabbage, beet root) has been found to check cancer and increase immunity in human body (Attri, 2013).

Botanical Description

Pumpkin is a vigorous plant, characterized by requiring much longer growing season, leaves large in size with lush green colour. Fruits are small to medium size depending up on the variety, round with flat ends or slightly pressed at poles, light green to light brown patches at maturity or deep orange yellow to light brown creamy patches at maturity.

Cultural Practices

Climatic requirements: Pumpkin requires somewhat warm climate, which must be free from the frost. However, unlike muskmelon or watermelon, it can tolerate cold climatic conditions to some extent.

Soil requirements: It can be grown on large variety of soils but soils having sandy loam and loam textures are considered to be the best for its growth. Soil reaction should be neutral (pH 6.5 to 7.5) or slightly acidic (pH 6.0 to 6.4), and well drained.

Sowing time and seed rate: Like other cucurbits, pumpkin is also sown twice in a year, during February to March and May to June. Seeds should preferably be sown in corner fields on raised beds 2 to 3 m wide with 0.5 m spacing between plants or even more up to 0.75 to 1.0 m. Seed rate varies from 5 to 6 kg ha^{-1}. Generally 2-3 seeds per hill are sown directly in the fields.

'Arka Surayamukhi' and 'Arkachandan' are the prominent varieties of pumpkin. In sub-tropical zone of Jammu region both these varieties are sown during June to July, whereas, time of sowing for sub- temperate/temperate zones are March and April- May, respectively. In Valley of Kashmir, the sowing time of 'Arkachandan' and 'Pumpkin Badami' varies from mid April to May.

Manures and fertilizers: Well decomposed manure or compost @ 25 tonnes should be added before sowing. It is wise to apply manure or compost @ 1-2 kg to each hill where the seeds are although to be dibbled at 2.5 cm depth. The application of organic manure is required to be supplemented with 100 kg N, 50 kg P_2O_5 and 50 kg K_2O ha^{-1}.

Irrigation and intercultural operations: Adequate irrigation should be given immediately after sowing and subsequently after every seven days. Fields are regularly hoed and weeded to make them free from all kinds of weeds.

Harvesting and yield: The fruits become ready for picking after about 120 to 150 days of sowing. Yields varies from 300- 340 q ha^{-1} in case of 'Arka Suryamukhi', whereas, the yield of 'Arka Chandan' varies between 300- 335 q ha^{-1}.

References

Anonymous, 2007. Pumpkins offer hope to diabetes. *The Kashmir Times.* 64 (192): 13.

Attri, B.L. 2013. Fruits and vegetables- Indispensable component of human diet. *Agrobios Newsletter.* 12 (4): 89-90.

Bala, S., Shukla, H.S. and Akram Mohd. 2009. Role of medicinal plants in human system. *Indian Farmer's Digest.* 42 (1): 28-30.

Phartyal, M. and Singh, P. 2014. Role of fruits and vegetables in diet. *Indian Farmers Digest.* 47 (5) 7-8.

Sawhney, D. 2005 Flavours of the month. *Ayurveda.* August Issue, pp.15-17.

Sharma, G.K. 2005. Let vegetables be your medicine. *Ayurveda.* December Issue, pp.38-42.

43

Sweet Flag
(*Acorus calamus* L.)

Common name: Bach, barian, safed bach

Family: Araceae

Introduction

A semi-aquatic perennial plant which is mostly found around the water bodies in the tropical and sub-tropical zones. It was most common in Manipur and Naga hills a few decades ago (Singh, 2005). However, these days it has become a rare species there. It is ascribed to its heavy exploitation and deforestation. Barian not only grows in tropical and sub-tropical areas but also grows in the temperate region of the country (Khanduri and Negi, 2010).

Being semi-aquatic or marshy herb in Jammu and Kashmir, it has a wide range of its distribution from border areas of Jammu plains to sub-temperate regions of Kud and Batote and is found along water courses and ditches (Singh, 2005).

In the Ayurvedic System of Medicine and in modern literature, it finds place along with other medicinal plants. It can, therefore, be cultivated on commercial scale. Tarai belt of Kathua and Jammu districts is eminently suited to barian as it already occurs there naturally under these situations.

The most important constituent of barian is a volatile essential oil of agreeable odour obtained from dried rhizomes yielding 1.5 to 3.5 percent oil. Aerial parts of plant yield a small amount of essential oil as roots, giving much higher percentage ranging from 2.5 to 3.5 per cent. The oil is used in Ayurvedic, Siddha and Unani System of Medicines. Apart from aerial parts, its roots are also used as medicines for preventing number of diseases. Briefly, the medicinal uses of this plant are as follows:

Medicinal Uses

Barian is an important medicinal plant with all parts being used. Digestive tonic prepared from its roots, is found effective in number of chest problems.

Effective in dyspepsia, flatulence and loss of appetite, it is provided to those who are suffering from fever. It is also found effective in hysteria and neuralgia. In combination with other drugs, it is given in colic and cough. Roots are simply chewed and it produces copious salivation and sensation of warmth in asthma. It is freely eaten in epidemics as it possesses properties to neutralize poisons.

Externally, it is used in chronic arthritis and rheumatism. For this purpose, roots are powdered and rubbed on the affected parts mixed with spirit. It is also diuretic and anthelmintic and used in many Ayurvedic formulations.

Rhizomes are found to be carminative in nature and can be used for treating nervous ailments. Also they are effective against stomach troubles, asthma, bronchitis, teething troubles in children. It possesses anti-carcinogenic property.

Botanical Description

Barian (bariyan) is a sweet scented marshy herb. In Sanskrit its name is *Vacha*, and in English sweet flag. The people of Jammu and Kashmir also call it by the name *Bareen* (Sharma, 2005). It is found both in temperate and sub-temperate zones.

Plant is a semi-aquatic perennial herb, in habit, with a long, indefinite, branched, cylindrical rhizome immersed in the mud and emitting numerous roots. The leaves are erect, yellowish-green, 2 to 3 feet in length, sheathing at their bases (which are pink), sword shaped, narrow and flat, tapering into a long, acute point. The flowering stem or inflorescence densely crowded with very small, greenish-yellow flowers arises from the axils of the outer leaves, but is longer and solid and triangular. The flowers are small and fragrant. The fruit is a berry, full of mucus, which falls when ripe into the water or to the ground, and thus dispersed, fruits are produced occasionally and propagated mainly by the rapid growth of its spreading rhizome.

Cultural Practices

Climatic requirements: Barian requires tropical and sub-tropical to temperate climate where temperature may vary from 10 to 38 degree Celsius and annual rainfall may range between 70 cm and 250 cm (Khanduri and Negi, 2010).

Soil requirements: Mixed red to red yellow sandy loam soils are considered best for the production with good yield. Red soils found along the slope of the hills and ridges are considered productive (Teli, 1995). Red yellow soils developed on moderate slopes are suitable to large number of crops including barian.

Field preparation: Field preparation for cultivation of barian is done by ploughing the soil twice followed by planking to bring fine tilth.

Manures and fertilizers: Well rotten organic manure either farm yard manure or compost @5 tonnes ha^{-1} is mixed in the soil after final ploughing.

Method and time of transplanting: The sprouted rhizomes of *Acorus calamus* are cut into small pieces and transplanted at a distance of 60 x 60 cm and 5 cm deep in a well prepared field. Transplanting is done during June - July and about 28,000 plants are needed ha^{-1} (Khanduri and Negi, 2010).

Irrigation: An important aspect of agro-technology is frequent and excess irrigation (2 days interval) except in rainy season.

Weeding and hoeing: Weeding and hoeing operations are resorted monthly to get better yield.

Harvesting and yield: Barian is harvested about 14-15 months after transplanting. As rhizome is commercial traded, digging of rhizome is generally done in the month of December-January. These are cut into pieces and dried in shade by spreading to connate volatile oil in the rhizome. Each Barian plant produces 1.5 kg green rhizomes with 140q green rhizome ha^{-1} or 35q dry rhizome ha^{-1}.

References

Khanduri, P and Negi, K.S. 2010. Scope in commercial propagation of *Acorus calamus, Allium sativum* L. and *Aloe barbadenses* Mill. to boost economy of the hill areas. *Agrobios News Letter* 9(5): 27-28.

Sharma, O.P. 2005. 151-Healing Plants of Jammu and Kashmir. Directorate of Social Forestry Project, Govt. of J&K.

Singh, S.S. 2005. Fate of the Medicinal Plant Resources of Jammu and Kashmir. Directorate of Environment and Remote Sensing, Jammu, Govt. of J&K.

Teli, B.L. 1995. Soil resources: Problems and planning in hills. In: Eco-degradation of Himalayas. Vinod Publishers and Distribution, Pacca Danga, Jammu, J&K, India.

44

Sweet Potato (*Ipomoea batatas* L. Lam.)

Common name: Shakargandhi

Family: Convolvulaceae

Introduction

Tubers form an important group of staple, serving as a secondary food for one fifth of the population, and third important group of food after cereals and pulses the world over. A number of tubers-tapioca or cassava, taro, yams and sweet potato are important food and raw material suppliers in the world. Possessing higher biological efficiency, tubers can substitute for cereals with high carbohydrates and calorific value. Besides, they are a potential source of alcohol, starch, glucose and have number of medicinal values. Sweet Potato is one of the tuber crops. Medicinal properties of sweet potato make it superior to all other tuber crops due to presence of antioxidants and anthocyanins.

Medicinal Values

"Talk about a nutritional power house! The more colour a veggie has the better it is for us, according to nutritionists- and sweet potato might be the brightest of all".

Potent oxidants found in sweet potato help to shield our hearts, sweet taste makes them delicious enough to eat for deserts.

Sweet potatoes are excellent source of vitamin A and vitamin C and a good source of vitamin B_6, fibre, copper and potassium. Sweet potato is the amazing food for anaemic patients. Always get firm, medium sized sweet potatoes with tapered

ends. Avoid ones with blemishes, sprouts or any sign of decay (Pandey and Awasthi, 2012). Some believe that they also lessen the effect of arthritis and asthma as they have anti-inflammatory properties. They are extremely rich in carotenes as well as Vitamin B_2 and Vitamin B_7 (biotin) besides Vitamin B_6 and Vitamin E. They also provide good amounts of manganese and iron and constitute a rich source of dietary fibre (Pandey and Awasthi, 2012). Sweet potatoes are absolutely loaded with nutrition and considered to be one of nature's most perfect vegetable (Mohindra, 2014). Packed with B-carotene and vitamin-A and one serving of sweet potato provides more beta carotene than 23 cups of broccoli.

Antioxidant properties of sweet potatoes can assist to get rid of free radicals linked with lot of the chronic illness like colon cancer and diabetes. Sweet potatoes act as an amazing food for anaemic patients.

Other Uses

The leaves of sweet potato are a rich source of nutraceuticals like beta carotene, anthocyanins etc. which can be used for making leaf *pakodas*. The soft and mealy textured varieties of sweet potato can be used for making 'gulab jamun' by mixing cooked paste with milk powder. Jam, pickles, and squash, other value added products are very relished by the people. It can be prepared by mixing mashed sweet potato paste with fruit pulp of banana and apple in the ratio of 70:30. The orange fleshed leaves are ideal for jam preparation.

Like potato chips, sweet potato, taro and yam tubers can be sliced, boiled, salted and sun dried and after deep frying they are made into edible chips. These fried chips are tasty and can be used as snacks. Such fried chips are available in markets of Orissa and West Bengal (Dutt, 2010). It has distinct sweet and savoury flavour which make such an excellent foil to many savoury dishes and they are fittingly paired with meat dishes they need a touch of sweetness (Anonymous, 2010).

Botanical Description

It is a warm season crop. Besides, growing of sweet potato in the fields on large scale, it is a very good vegetable for growing in kitchen garden.

Plant is a tuberous, perennial but usually grown as an annual. It has a thin stem, light green to purple in colour, with or without short soft hairs and which contain latex. The root system consists of fibrous roots and storage roots. As the plant matures, thick pencil roots that have lignification are produced. Other roots without lignifications are fleshy and thicken are called storage roots. Leaves are simple alternate, 5 to 15cm long, dense with bases. The leaf colour ranges from light green to deep purple, sometimes with a purple stain at the base. The shape of the leaves is variable; round, kidney shaped, heart-shaped, triangular, lobed and almost divided. The inflorescence is a cyme. Flowering behavoiur varies from those not flowering to with few flowers and with profuse flowering. Flowers funnel-shaped occur either singly or in clusters. The colour ranges from light pink to deep purple.

Cultural Practices

Climatic requirements: Sweet potato is warm season crop and is susceptible to frost.

Soil requirements: It can be grown on wide variety of soils. However, sandy loam soils are the best suited having pH range from 6.5 to 7.5.

Field preparation and sowing: The field should be ploughed twice or thrice to obtain fine tilth then it should be planked. Sowing is done during March to July. If cuttings of previous crop have to be done during March to April, and if the vines raised from the tubers, the crop is sown during May to July. Cuttings of 40cm length are planted on ridges about 45 cm apart. Middle portion of the cutting is buried about 10 cm deep leaving both the ends exposed. The planting can also be done on flat soil by keeping the same distance as in case of ridge sowing.

Irrigation: First irrigation is given immediately after sowing and subsequently after every 10 days interval.

Manures and fertilizers: Whichever quantity of organic manure is available must be added. Normally 5-6 tonnes of FYM or compost/vermicompost ha^{-1} is the recommended dose, which should be added and well mixed before sowing.

Intercultural operations: During crop growth at least 2 earthings are given to ensure proper root development.

Harvesting: Optimum time of harvest is indicated by the cessation of plant growth and quick drying of roots. For harvesting, vines are removed and the tubers are dug up open with the help of spade taking care that tubers are not damaged.

References

Anonymous. 2010. Once tasted never forgotten. *The State Times* 15(30): 6.

Dutt, S. 2010. Tubers for food security. The ICAR, New Delhi Research and Development Network is popular of tuber crops a lot. (Personnel Communication).

Mohindra, V. 2014. 7 Vegetables you should eat to boost your disease fighting potential. *Ayurveda*. 1(25): 18-19.

Pandey, H and Awasthi, P. 2012. Winter vegetables: Treasures of nutrients and medicinal properties. *Indian Farmers Digest*. 45(2): 15-16.

45

Tomato
(*Lycopersicum esculentum* Mill.)

Common Name: Tamatar

Family: Solanaceae

Introduction

Tomato, is said to be the native of South America. It was spread throughout the world following the Spanish colonization of the America during the 16th century and became popular in India also within the last seven decades (Katoch, 2012). Now a days the tomato is grown and eaten worldwide.

As a matter of fact, tomato is now a staple vegetable in every kitchen. These are consumed in different ways, including raw in salad, as an ingredient in many dishes and sauces, processed into ketchup or soup. Unripe green tomatoes can also breaded and fried, used to render salad or pickle. Tomato juice is sold as a drink, which can be used in cocktails such as the Bloody Marry.

Tomato being acidic in nature, enables it easy to be preserved in home canning whole or in pieces, as tomato sauce or paste. Tomato fruit is also preserved by drying, often in the sun and sold either in the bags or in jars with oil. It is worthwhile to mention that tomato is extensively used in Mediterranean cuisine especially Italian and Middle Eastern cuisines. Tomatoes are a key ingredient in pizza and pasta. Not only this, consumption of tomato is believed to benefit heart besides other body organs. It is attributed to lot of medicinal values of tomato, which are described in details as below:

Nutritional Composition

Tomato has been ranked first as a source of lycopene (71.6 per cent), second as a source of vitamin C (12.0 per cent), carotenoides (14.6 per cent), third as a source of vitamin E (6 per cent) and lipid soluble compounds (Pandey and Dhyani, 2015). The main antioxidant compounds in tomatoes are characterized by lipid soluble compound and constitute more than 80 per cent of total tomato carotenoid in fully red-ripe tomatoes. Other carotenoid in tomato constitute B-carotene and small amount of phytoene, phytofluence, L-carotene, γ-carotene and lutein. Tomatoes also contain moderate amount of water soluble phenolics and flavonoids. Tomatoes are also a good source of folate, potassium and vitamin E, soluble and insoluble dietry fibres (pectin, hemicelluloses and cellulose with a concentration ranging from 0.8 to 1.3 g/100g fresh weight).

Medicinal Values

Pulp and juice of tomato are digestive, a mild aperient. It promotes gastric secretion and is a blood purifier. It also serves as intestinal antiseptic, very useful in cancer of mouth and sore mouth. It is a rich source of Vitamin-C and a good substitute of orange. It stimulates liver and promotes flow of bile. But it is not recommended in gout, or in uric acid diseases. Peculiarity of this herb is that it is the richest source of vitamins and nutrients. They are the most extraordinary correctives for kidneys and natural stimulant and an effective agent for clearing poisons in the system.

Tomatoes contain lycopene, one of the most powerful natural antioxidant. In some studies, lycopene, especially in cooked tomato, has been found to help in preventing the ovarian cancer, in pre-menopausal women, digestive tract cancers (Mouth, throat, oesophagus, pancreas, colons and rectum), cardio-vascular disease, asthma, prostate cancer (Pandey and Dhyani, 2015; Mohindra, 2014).

Tomato also keeps the skin healthy because of lycopene which works as a sunscreen from within. It is recommended to have at least 16 mg of lycopene per day as it reduces the number of free radicals in the body and also helps to retain the moisture in the people who are used to sitting in air conditioned environments. One does not need to spend on expensive cosmetic treatments if one follows these simple steps to get healthier looking skin.

The acidity in tomatoes helps in reducing and cleaning up acne as they contain Vitamin A, Vitamin C and Vitamin B. If one has mild acne, cut a tomato in half and apply it on face. Recently it has been found that tomatoes are loaded with health benefits which include treating urinary tract infections, skin problems, diabetes and low hypertension and are also good for gut health and eye sight *vis-à-vis* anoxia (lack of appetite) cataract, night blindness, chronic dyspepsia (indigestion) etc.

Tomato contains numerous antioxidants, which combat several types of cancers *i.e.*, gall bladder, colon, rectum, ovarian, cervix, lungs, and breast and skin cancers. Since tomatoes are rich in potassium, clinical studies have shown that they have a positive effect on kidneys in many cases and for good renal functioning. Its regular use has proved to reduce the levels of LDL cholesterol and triglycerides in the blood which are the key culprits of cardiovascular diseases.

Botanical Description

Botanically tomato is a fruit, but it is considered as vegetable. It is better to say tomato a house vegetable grown all over India. Popularly known as Love Apple, is one of the most commonly known vegetables in the kitchen and homestead garden (Rana *et al.*, 2012). It is perennial with its natural habitat but annual in cold climates. The plant with short stems or vine-like with long, spreading stems grow up to 1–3 meters. The stems are covered with coarse hairs and the leaves are arranged spirally. The tomato plant produces bright yellow flowers, which can develop into a round fruit (berry) which is fleshy, smoothed skinned and can be red, pink, purple, brown, orange or yellow in color. Tomato flowers are pollinated by bees. On an average tomato weighs approximately 100 grams.

Cultural Practices

Climatic requirements: Tomato is a warm season crop. The plants cannot tolerate severe frost. Both high and low temperatures interfere with the setting of fruits. A temperature range of 20-27 degree Celsius is conducive for its successful growth. A warm sunny weather is found to be the most suitable for its proper ripening, colour development and higher yields. The tomato is one among the easiest grown vegetables in the kitchen garden provided abundant sunshine and adequate water facilities are there.

Soil requirements: Tomato can be grown on all kinds of cultivable soils but the soils having sandy loam texture are considered the best for its growth. The soils must be rich in soil fertility and organic matter content, and preferably well drained. Tomatoes grow best in soils having neutral soil reaction *i.e.*, in the pH range of 6.5 to 7.5.

Nursery and field preparation: Raised nursery beds are preferred to sow the seeds of tomato for raising plants in nursery. Land for transplanting seedlings is ploughed 3-4 times and planked to obtain fine tilth.

Sowing time and nursery raising: Seeds are sown in raised nursery beds about 5-10 cm apart and covered lightly with well decomposed compost.

a) Seed rate- 500-600g ha^{-1}

b) Sowing Time-varies from climatic zone to zone.

   i. Sub-tropical: Two crops of tomato can be raised in the sub-tropical climate *i.e.* during November/December and transplanted during February/March and in February and transplanted in March.

   ii. Intermediate/Sub-temperate: Sowing time varies from February-April anytime from March-June.

iii. Temperate: March-April and May-June are the times of sowing and transplanting, respectively.

Varieties: 'Pusa Ruby', 'Punjab Chhuara', 'Shalimar-1' are the suitable varieties which can be grown in the sub-tropical areas of Jammu. Contrary to this, 'Pusa Ruby', 'Shalimar-2', and 'Marglobe' are the varieties which have been recommended for the sub-temperate or inter-mediate areas. 'Shalimar-1', 'Shalimar-2', 'Roma' and 'Marglobe' varieties are recommended for growing in the temperate areas of Jammu and Kashmir.

Manures and fertilizers: Apply 25 tonnes of well rotten FYM and plough it into the soil. Apart from this, N, P and K@ 100-200, 15-50 and 60-100 kg ha^{-1} are recommended through chemical fertilizers. However, the application of chemical fertilizers should be discouraged owing to liking of organic food by people in the world being free from harmful chemicals/pesticides.

Hoeing and irrigation: Shallow hoeing and regular weedings are required to obtain better yield and quality fruits by ensuring clean cultivation. Plants are staked to shun fruit rotting. There is no doubt that tomato tolerates droughts to certain extent, however, irrigation at intervals of 7-10 days is desirable.

Harvesting and yield: Depending upon the mode of dispersal, tomato fruits may be harvested at various stages of maturity given below:

(a) **Green stage:** When the fruits are fully developed but are green and suitable for sending to a distant market.

(b) **Pink stage:** When some portion of the fruit is red pink and fruit is not fully ripe. It is most suited for local market

(c) **Ripe stage:** When the major portion of the fruit is red and softening begins. It may be picked up for table use

(d) **Full ripe stage:** It is suited for processing.

Yield of the crop varies from variety to variety. Generally it varies from 280-375 q ha^{-1}.

References

Anonymous.2013. Tomato for clear skin. *The State Times*. 18 (49): 5 (Tuesday, February, 19).

Katoch, O.R. 2012. Tomato produced in J&K. *The Daily Excelsior* (174): 6 (Tuesday, June, 26).

Mohindra, V. 2014. Vegetables-you should eat to boost your disease- fighting potential. *Ayurveda*. 1(26): 18-19.

Pandey, V. and Dhyani, S. 2015. Nutritional and medicinal properties of solanaceous vegetable crops. *Indian Farmers Digest*. 48(2): 24-28.

Rana, M.K., Brar, A and Adhikari, R. C. 2012. Tomato- A panacea for good health. *Agriculture Today*. 15(9): 44-46.

46

Turmeric (*Curcuma longa* L.)

Common name: Haldi

Family: Zingiberaceae

Introduction

Turmeric is one of the major spice crop grown in India since times immemorial. India produces nearly whole world's turmeric crop and consumes 90 per cent of it (Kailaimannan and Paramasivam, 2012). The crop is mostly grown in Tamil Nadu, Andhra Pradesh, Karnatka, Maharashtra and Odisha, which accounts for 95 per cent of the turmeric production in India. In the early years of twentieth century most of the allopathic medicines were based on herbs including turmeric. In fact, turmeric has been mentioned in early writings of the world in many languages (Bala *et al.*, 2009). The Unani practitioners have used it all over the world as well as other systems of mdicines. It was prescribed by many physicians owing to its medicinal values (Maurya and Yadav, 2014).

Medicinal Values

Turmeric is an important common flavouring spice of daily use. In India, it is used in the same way as saffron. Antiseptic properties of turmeric help in preserving the food. According to some physicians, the stringent and sour tasted turmeric is a time tested and a nourishing herb which not only gives natural gloss, royal glow and lustre but also imparts vigour and youthful vitality to whole of the body. Turmeric is, thus, a great tonic in general, aromatic, expectorant, blood purifier, skin tonic, carminative, pain reliever, germicidal, anti-flatulent, producer and enhancer of red

blood corpuscles, antibilious, protector of eyes, anti-inflammatory and imparts coolness to the system. Studies suggest that turmeric helps a lot in treating arthritis rheumatism (Anonymous, 2013). Taking of rhizome or its juice and dry powder, mixed in butter milk or plain water is found highly beneficial in intestinal problems. About 20 drops of the juice or raw turmeric mixed with pinch of salt, taken in empty stomach is an effective cure for expelling worms. A tea spoon of raw turmeric juice mixed with few drops of honey is taken daily for the treatment of anaemia. When dried turmeric powder is taken after mixing with a few drops of honey and juice of few bitter gourd leaves in morning for a few days, takes care of measles.

A teaspoon of turmeric powder with a glass of milk if taken twice or thrice daily in the morning prevents attack of asthma. Half a tea spoon of fresh turmeric powder, when taken after mixing in 30 ml of warm milk has been found effective in cough and cold. For treating sprains or swellings, turmeric paste mixed with lime and salt can be applied for good results. Application of turmeric powder to boils speeds up the healing process. Turmeric is very effective in treatment of skin disease like scabies. Applying paste of turmeric on the affected part of breasts eliminates pain. Rinsing mouth with turmeric water (5g of turmeric powder, 2 cloves and the dried leaves of guava in 200 ml water) gives immediate relief from pain. Applying the mixture of turmeric powder and lime at the place of insect bite nullifies the toxic effect. Though tests have been undertaken as a cure for alzheimer to cirrhosis of the liver, it is antiseptic for cuts and burns (Annonymous, 2009). Recent scientific studies confirm that turmeric can cure host of diseases, also it was found that turmeric restrain the growth of various types of cancer (Maurya and Yadav, 2014). Turmeric is used for the treatment of skin cancer or pre-cancerous skin condition. Turmeric is also found helpful in preventing the blockage of arteries which gradually lead to heart attack.

Botanical Description

Turmeric is a perennial, herbaceous herb reaching up to 1 m just like ginger with roots or tubers oblong-palmate. Its fragrance is sweet like the pollen of mango. The root has beautiful captivating golden complexion when taken out from the ground. The roots (rhizomes) are aromatic, with spicy fragrance, yellow to orange in colour, stimulant and a tonic.

Leaves are alternate, about 60 cm long, smooth, green, arranged in two rows, tapering at each end. Flowers are yellow white, in clusters of 3-5 in central spike, terminal, appear in July-August and are sterile.

Cultural Practices

Climatic requirements: The crop requires warm and humid climate.

Soil requirements: Loamy soils, rich in organic matter and available plant nutrients both major and micro, are the best for growing of turmeric.

Sowing time: March to April is the optimum time for its sowing.

Seed rate and sowing: The seed rate is 20-25 qha^{-1}. The seed rhizomes, each having one sound bud, are planted in shallow pits on the top of the ridges. Mother rhizomes are planted as such or split into two, whereas fingers are cut into pieces each 4.5 cm long. Row to row and plant to plant distance is maintained 60cm and 20cm, respectively.

Manure and fertilizers: Well decomposed FYM @ 10 tonnes ha^{-1} should be applied at the time of planting. Apply 60 kg N, 30 kg P_2O_5 and 60 kg K_2O ha^{-1}. Half of the nitrogen, full dose of P_2O_5 and K_2O should be applied at the time of planting. The rest of the N is applied as top dressing *i.e.* two months after planting.

Irrigation: The crop is watered immediately after sowing. As per need *i.e.*, depending upon type of soil, weather conditions like temperature, etc. watering is required.

Inter-culture operations: Weeding is done 3-4 times and earthing twice.

Harvesting and yield: The crop is ready for harvesting in about 8-10 months after sowing *i.e.* in December to March. The dry leaves are cut close to the ground and ploughed in between the rows or rhizomes are lifted with spade or digging fork.

The yield is about 300 q ha^{-1}. The cured produce is about 20-25 per cent of the fresh rhizomes by weight.

References

Anonymous. 2013. Arthritis: 5 foods that help reduce chronic pain. *The State Times*. 18 (251): (Monday, October 10).

Bala, S, Shukla, H.S and Akrum, M. 2009. Role of medicinal plants in the human system. *Indian Farmers Digest.* 42(1): 28-30.

Kailaimannan, R. and Paramasivam, M. 2012. Turmeric crop seenario in India. *Agrobios News Letter*. 10(9): 28-29.

Maurya, P. and Yadav, L. 2014. Medicinal uses of turmeric. *Agrobios News Letter*. 12 (10): 57-58.

47

Water Chestnut (*Ttrapa bispinosa* Roxb.)

Common name: Singhara

Family: Trapaceae

Introduction

Water chestnut is an annual aquatic herb. It is mostly grown in shallow water of perennial ponds tanks and lakes. It is also used as food for humans and animals in India, China, and South-east Asia. It is grown throughout Asia and tropical Africa in lakes and ponds and is often cultivated for its edible fruit. The medicinal values of the whole herb and fruit have long been recognized in folklore medicine as a cure for various diseases. Half mature kernel is eaten as such while mature kernel is used after boiling. Flour of water chestnut is used during fasts. In fact, Water Chestnut is an important food for many people. Processed dry nuts (kernels) are used not only during fasts but its flour is also used for making *abir* for 'Holi' festival and for preparing bread, *halwa* (pudding) and biscuits etc. (Maurya *et at.* 2004). In tribal areas, water chestnut is also used as a medicinal herb.

Medicinal Values

It is an important plant of Ayurvedic System of Medicine and is used in stomach related problems, genito-urinary system, liver, kidney, and spleen. It is bitter, astringent, stomachic, diuretic, febrifuge, and antiseptic. The whole plant is used in gonorrhea, menorrhagia, and other genital infections. These are used in the validated conditions in pitta, burning sensation, dyspepsia, haemorrhage, diarrhoea, dysentery, intermittent fever, leprosy, fatigue, inflammation, fractures, lumbago, pharyngitis, bronchitis and general debility. Edible portion (kernel) of water chestnut besides having carbohydrates (23.9 per cent), proteins (4.7 per cent) and fat (0.2 per cent) is also a rich source of mineral matter (1.1 per cent), Vitamin A and Vitamin C. In certain tribal villages, the flour of water chestnut is used for treatment of diarrhoea and abdominal pain.

Chemical analysis of the fruit and fresh nuts having considerable water content, citric acid and fresh fruit which substantiates its importance as dietary food due to low crude lipid, and mineral content as iron and manganese, potassium.

Other Uses

Flour called "Singhare ka Atta" is used in many religious rituals and consumed as a "phalahar" diet on the Hindu fasting days. Primary use of flour is as a thickening agent and to make batters for deep frying in Indian and Asian cooking. When used as a thickener, flour is usually stirred into water first before it is added to hot liquids and sauces (Amarya and Verma, 2013). This technique reduces the formation of lumps that may otherwise occur.

Water chestnut flour can be used to make pancakes, Indian fried bread-"poories" and even flat breads-"Chappaties". As a thickener and to butter fry chicken, cheese and vegetables flour is also used to make bread cakes and cookies. Fresh water chestnuts are utilized as appetizers and in variety of salads.

Botanical Description

Water Chestnut, commonly known as 'Singhara' or 'Paniphal', is grown throughout India. It grows abundantly in the lakes of Kashmir, India. It is an annual aquatic herb of great economic importance which is grown for its starch rich fruits. It has flexuose stem, ascending in the water; the submerged parts are furnished with numerous opposite pairs of green root-like spreading pectinate organs. Leaves are alternate, crowded on the upper part of the stem; 3.8–5cm long, rhomboid, somewhat truncate at the base, irregularly inciso-serrate, reddish-purple beneath; petiole dilated near the apex. Flowers are few, auxiliary, solitary, pure white. Fruits obovoid, angular, 2.2–5cm long, and broad, with a spreading flattened very sharp spinous horn at either side.

Cultural Practices

Climatic requirements: It prefers warm and humid climate. However, some varieties are also grown in the temperate zones of the world. The varieties grown in the temperate conditions cannot tolerate the conditions of tropical and sub-tropical zones, where it is cultivated on a commercial scale.

Soil requirements: It is widely grown in shallow water of perennial ponds and thrives best in ponds having heavy soil at the bottom. For successful cultivation of the crop, water depth in the pond should be in between 0.5 to 2.5 m during growth period.

Nursery raising: Nursery bed should be well puddled and seeds be sown in the ponds having at least 40 cm deep water. About 70-100 kg fully ripe nuts are sufficient to cultivate one hectare of pond. Seeds are broadcast during the month of December which takes about one month to germinate. Plants should be permitted to grow in the nursery till the onset of rainy season *i.e.* ending June to early July. Seedlings are transplanted in May/June in a perennial pond.

Transplanting: Crop is required to be transplanted in the month of July in ponds containing 120-150 cm water. Two or three plants should be tied up together and planted in the mud at the bottom of the pond at a spacing of 15-30 cm apart depending upon seedlings health and soil fertility. While transplanting, roots of the nursery seedling should be pressed in the mud by hands.

Maintenance of water level: Water depth in the pond should be between 0.5-2.5 m during different growth stages because excess water causes rooting in crop whereas, under shallow water conditions plants will not develop properly.

The optimum water level for pond should be 150-160 cm at branching stage, and 90-120 cm and 80-120 cm at flowering and maturity stages respectively.

Intercultural operations: Weeding of the pond during the month of July-August is essential to check the weeds.

Manures and fertilizers: Generally no manures and fertilizers are added, as the pond water itself is a good source of nutrients.

Harvesting and yield: Plant of Water Chestnut start flowering after one month of transplanting. The fruits are harvested from September to December. Generally two pickings are recommended. The first picking may be taken at 100 days after planting and other over one month thereafter. However, six to right pickings at 10-15 days interval from healthy plants can also be taken.

Normal crop yields about 250-300 q of fresh fruits ha^{-1} depending upon the variety, which after drying are reduced to about 80-100q. The dry weight of the fruits further reduces to 50-60q.

References

Amarya, S. and Verma, A. 2013. Not a nut- water chestnut. *Indian Farmers Digest.* 46(3) : 30.

Maurya, S.K., Singh, J.P and Awasthi, A. 2004. Water Chestnut cultivation. *Indian Farmers Digest.* 37(3): 29-30.

48

Watermelon (*Citrulus lanatus* Thunb.)

Common name: Tarbooz

Family: Cucurbitaceae

Introduction

Watermelons are native to Africa. The first recorded harvest of watermelon was nearly 5,000 years ago in Egypt, where watermelons were grown and regarded not only for their flavour but also for their beauty. It has been depicted in Egyptian wall paintings. Now over 1,200 varieties of watermelon are grown worldwide in 96 countries and in India there are about 400 varieties namely *Anarkali, Noorjehani, Sherbat-e-Anar* etc.

In India, the watermelon came to fore during the fourth century AD (Muthukrishnan, 2008). Susruta, the great Indian physician and surgeon, has mentioned in the Susruta Samhita that watermelons were cultivated along the banks of the Indus river and presently a number of varieties of watermelon are being grown along the banks of the river Ganga and the Yamuna. He called the fruit *Kalinda* or *Kalinga* (hence *Kalingad* in Marathi).

China ranks first in the world with respect to production with 40 m tonnes, followed by Turkey with 5 m tonnes, Iran with 3 m tonnes, United States coming fourth with 2 m tonnes. Agriculture Ministry, GoI has estimated the production to be at 1.5 m tonnes yet the actual production in the country is not known.

In addition to being a refreshing summer treat, watermelon is packed with lot of medicinal values and nutrition, which are given below:

Medicinal Values

It is a fruit which contains 92 per cent water rich in electrolytes (Na and K) which are being lost through sweat. Also, the fruit is very filling, has less calorie content and above all is very nourishing. It has no cholesterol and is an excellent source of vitamin A, vitamin B_6 and vitamin C and contains fibre.

A cup of watermelon cubes contains just 48 calorie, provides one-fourth of one's daily requirement of vitamin C. Its beta carotene content provides 11.1 per cent of daily needs of vitamin A. Lycopene content is well known for its antioxidant and anti-cancerous properties and is protective against prostrate, breast, lung and colon cancers as per the study conducted by All India Institute of Medical Sciences, New Delhi. Three or more servings of fruit daily may lower one's risk of age related macular degeneration, the primary cause of vision in adults.

Due to watermelon's moisturizing properties and contents of vitamin A, B and C *vis-à-vis* minerals like calcium, magnesium and potassium, it is very useful for removing blemishes and maintaining elasticity and texture of the skin (Rai, 2010).

Pulp is cooling and diuretic and fruit juice is used as antiseptic in typhoid fever. With cumin and sugar, juice is given in infection of urinary organs such as gonorrhea (Singh, 2005). Watermelon can be incorporated in daily diet in number of ways. Apart from having it on its own in chunks or in the form of juice, the fruit can be included in many recipes including slush, smoothies, *sherbats* and salad. If not relished as above, chop it up in a bowl and sprinkle sugar or drizzle honey on it. Watermelon juice with mint leaves is very refreshing on a hot day.

This versatile fruit is also well known beauty aid. It is widely used in beauty products like lipstick, lip gloss and shower gels. Watermelon essence is added to bath and body fragrance. Its juice protects the skin from the sun and clears the skin leaving it free from blemishes, fresh and lovely seeds are demulcent and diuretic.

Grate a watermelon, squeeze out the juice and apply it on face for 15 minutes. Wash with hot water and then splash cold water. Face will become fresh and clean. The fibre content in watermelon ensures toxins are scrubbed out leaving the skin healthy and glowing. Mixed with a pinch of gram flour, the pulp makes an excellent scrub.

Seeds are sold as "char magaz" after removing their testa and are as nutritive as almond. Fruit of the wild variety is bitter but superior medicinally. It is used as purgative.

In Indian history, the watermelon has very bitter overtones, as in 1739 after his sack of Delhi throne, the Persian Emperor gave a gift of camel load of watermelons to the Mughal Emperor Muhammad Shah as a return favour, for the greatest plunder in history. In many rural areas, watermelons are a part of the *Vastu Puja* done to remove ghosts before inhibiting a new house.

Cultural Practices

Climatic requirements: It requires relatively a longer growing period than muskmelon and takes about 85-110 days from seedling to maturity. It requires an optimum temperature of 21-26° Celsius with maximum of 35°Celcius and minimum 18° Celcius. It is highly susceptible to frost.

Soil requirements: It grows best in well drained sandy loam soils. Heavy soils are also satisfactory provided they are well drained and manured heavily. Unlike muskmelon, watermelon grows well even in moderately acid soils.

Sowing time: Under sub-tropical conditions *i.e.* North India, watermelon is sown during February-March. Thus, the season of watermelon availability extends from May to June upto July.

Seed rate and spacing: Four kg of neat and clean seed ha^{-1} is required. Row to row spacing is 2.0 to 2.5 m, while plant to plant it should be 0.6 to 0.75 m. Two to three seeds are sown per hill directly in the field after soaking in warm water for about 24 hours.

Manures and fertilizers: Apply 10-25 tonnes of well rotten farm yard manure (having brown or black colour, amorphous type and free from any kind of fungal mycelium) depending upon the availability. It should be broadcast in the fields about 10-15 days prior to sowing the seeds and thoroughly mixed with the soil. Apart from this, apply N, P and K ranging from 75 to 100 kg, 10 to 40 kg and 30 to 40 kg $ha^{-1,}$ respectively depending upon the status of the soils.

Apply whole of the phosphate, potash and one-third of nitrogen at sowing time and rest of the nitrogen as top dressing in two equal doses.

Intercultural operations: Shallow hoeing accompanied with weedings are required at regular intervals.

Irrigation: During summer, irrigate the crop every week. At the time of fruit maturity, irrigation should be given when it is absolutely essential, flooding should be avoided.

Hybrid cultivars: Due to extensive research at the Indian Institute of Horticulture Research and by Private Seed Companies, hybrid watermelons are taking over fields everywhere both in terms of high quality, high yield and disease resistance.

The colour of its skin varies from pale green to dark green with or without stripes. The colour of the pulp varies from light pink to even yellow.

The varieties with cream and cream yellow pulps have also been evolved in Japan. Like seedless oranges, Japanese scientists were the first to develop seedless varieties of watermelons. But the seedless varieties of watermelon (unlike those of grapes) have not reached a stage of commercial cultivation.

Harvesting and yield: When the melon is thumped with the finger, it gives a dead flat sound for ripe melon, whereas it gives a sharp ringing sound for immature melons. Yield of watermelon varies from 250-300 q ha^{-1}.

References

Muthukrishnan, R. 2008. The fruit for summer – watermelon. *The Kashmir Times.* 65 (144) : 4 [Sunday, May 25].

Rai, A. 2010. Watermelon-A boon for summer. *Ayurveda.* 1(17): 6-7.

Singh, S.S. (2005). Fate of the Medicinal Plant Resources of Jammu and Kashmir. Directorate of Environment and Remote Sensing, Jammu.

49

Wild Asparagus (*Asparagus racemosus* Willd.)

Common name: Satmooli, satavar, shatawar, shatavari

Family: Liliaceae

Introduction

In Sanskrit, it is called 'Shatavari', and in English and Hindi it is known by the word Satavar. There are three main species of Satawar viz; *Asparagus sparanjri, Asparagus offcinalis* and *Asparagus racemosus*. The first one is used as an ornamental plant; second as salad or vegetable and third is mostly used as a medicine. Ayurveda has called Satavar the "queen of herbs" and is the foremost herb recommended for female health (Gupta and Meena, 2012). Among the three ayureveda doshas; Vata,Pitta, Kapha, Satavar helps in balancing Pitta dosha.

Fresh roots and cladodes are the main parts of satavar, which are used in medicines. However, its roots are used as an ingredient in large number of Ayurvedic medicines. These are used as an energetic and a tonic (Jamwal, 2006) as well as in curing anaemia, leucoria and urinary infections. Their regular use also improves functional efficiency of all organs of body. Roots are used as aphrodisiac, diuretic, anti dysenteric (Joshi, 2013).

Chemical Composition

Fleshy roots contain sugar, mucilage and saponins. Fresh cladodes possess diosgenin, shatavari (four saponins), Shatavarin-IV is a glycoside of Sarsa, Sapogenin, Rhamnese and Glucose.

Medicinal Uses

People use *Asparagus racemosus* for upset stomach, constipation and stomach ulcers. It is also used for fluid retention, cancer, dementia, diabetes and general health tonic. Women use it for pre-menstrual syndrome, uterine bleeding and to increase breast milk production *i.e.* galactagaogue. Also used as an aphrodisiac. It is known to clean blood, nourish female reproductive organs, threatened miscarriage etc. It is also used in case of inflammation of sexual organs both in females and males. It helps in cases of nervousness, restless sleep and people with weak emotions. The fresh root juice when mixed with honey helps to control sensation pain in tumors. If taken with a cup of saffron milk, asparagus is good for post-menopausal women. It is beneficial in urinary stones and helps to increase urine output. Research has proved that it has antioxidant and anti-bacterial effects also.

Botanical Description

Satavar is a perennial, bushy, cosmopolitan, xerophytic plant in nature that can be grown as an ornamental plant. The plant is 1.0 to 2.0 m high, climber with thin branches, curved thorns and needle shaped leaves. The flowers are scented and fruits are pea-shaped, round/oval, which on ripening turn red in colour. The roots are tubular and occur in bunches.

Cultural Practices

Climatic requirements: The plant can be grown in sub-tropical and tropical climatic conditions with 50-100 cm annual rainfall. The temperature range where satavar can be grown varies from 10 to 40° Celsius. However, the plant cannot withstand frost.

Soil requirements: It can be grown on large variety of soils, but its best growth takes place in sandy loam or clay loam soils.

Time of sowing and seed rate: Nursery sowing is done during May-June and transplanting in August - September. 2.5-3.0 kg seed is recommended ha^{-1}. Hard seed needs 3-4 hours soaking in water before sowing. Two-three deep ploughings are required during land preparation.

Seeds are sown in polythene bags filled with sand/clay and FYM in 1:1 ratio. One to two seeds are sown per bag. Seeds germinate after 10-15 days after sowing. Thirty five to forty days old seedlings are then transplanted in the main fields at a spacing of 60 x 60 cm.

Manures and fertilizers: Compost or FYM@ 15-20 tonnes ha^{-1} should be added in the field at the time of their preparation. To obtain good yields of satavar, Urea, DAP (Diammonium Phosphate) and MOP (Muriate of potash) @ 80,100 and 225 kg ha^{-1} are applied respectively in splits.

Intercultural operations: Weeding/hoeing of the crop sown in June -July can be done during November - December.

Irrigation: In all 6-8 irrigations are required during the crop period. In winter irrigation at an interval of 1 month and in summer at 20-25 days intervals as per requirement is recommended.

Harvesting and yield: The digging of roots can be done during November-December (24 months after planting) when the crop is in dormancy. Generally, one and a half years old crop is the best for getting quality roots. Roots are thoroughly washed and are boiled for 45-60 minutes in boiling water and later on peeled and dried in sun for 3-4 days. Under normal conditions, an average yield of satavar is about 2-3 tonnes ha$^{-1.}$ On an average net returns to the extent of Rs. one lac per ha can be realized under ideal conditions.

References

Gupta, R and Meena, M.S.2012.Shatavari-A multitask herb. *Ayurveda*.1(21): 31.

Jamwal, J. 2006. Safed Moosli and Satavar. *The Kashmir Times*. 63(261): 13[September 21, Thursday].

Joshi, B. 2013. Satavari, an important medicinal plants for farmers. *Indian Farmers Digest*. 46(4): 32.

50

Wild Violet
(*Viola odorata* Linn.)

Local name: Banafasha

Family: Violaceae

Introduction

Being herbaceous, wild violet is found in temperate regions of northern hemisphere, divergent areas such as Hawai, Australasia and Andes as well as tropical/sub-tropical areas all over the world. In India, wild violet is mostly found growing in North-Western Himalayan states of Jammu and Kashmir, Himachal Pradesh and Uttrakhand. Besides these, it is also found occasionally in sub-mountainous areas of Punjab, Haryana, Himachal Pradesh, Uttrakhand and Jammu and Kashmir. In Jammu and Kashmir, the herb grows very well throughout the mountainous region of Jammu and abundantly in Kashmir valley ranging to an elevation of 4,200 m or so (Khajuria, 2005). Wild violet also called" Nuna Posh" which serve as a cooling agent, so it finds use in variety of herbal formulations.

Medicinal Values

The flowers dried and the leaves and whole plant as fresh is officially used. Flowers of wild violet are emollient, demulcent and are used in billousness and lung troubles. Also it is used as a colouring agent, perfumes and in making syrups. A glucoside, 'viola-quercitin', is found throughout the plant and especially in the rhizome. The rhizomes (the roots) are strongly emetic and purgative. The seeds are purgative and diuretic in action and are given in urinary complaints (Brahmvarchaswa, 2007). A modern homoeopathic medicinal tincture is made from

the whole fresh plant and is considered useful for spasmodic cough with hard breathing, and also for rheumatism of the wrists.

Botanical Description

Wild violet family comprises of 206 species widely distributed in temperate and tropical regions of the world. The sweet scented violet appears towards end of February and completes blooming by April. Leaves are heart shaped slightly downy especially beneath the stalks, rising alternately from creeping rhizome or underground stem. Blades of young leaves rolled up from either side into the middle or face of leaves. Flower stalk arise from leaf axils and bear single flower with a pair of scaly bracket, erect in position. The flowers are generally deep purple and full of nectar and attract honey bees. They bear 5 sepals and 5 unequal petals. Anthers are united into a tube.

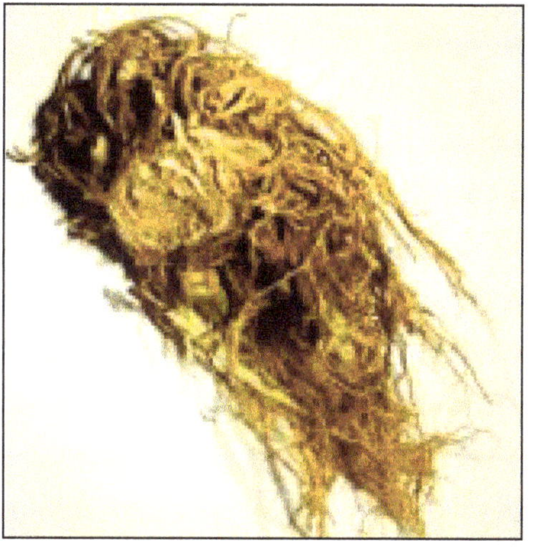

Cultural Practices

Climatic requirements: Wild violet can be grown in varying type of climates ranging from temperate, sub-temperate to sub-tropical type. Plants not only enjoy light shade but also thrive well in sunny locations.

Soil requirements: Wild violet grows on wide variety of soils. They require moist but well drained soils, rich in organic matter. An ordinary garden soil has been found suitable for successful cultivation of wild violet. It flourishes well in good medium soil (sandy loam, loam, loamy sand). Always avoid growing this herb in stiff clay soil. Deep digging is essential during autumn season.

Sowing/transplanting: Although its self seeding capability make it unnecessary, yet seeds after collecting can be sown during autumn season to spring season.

Choose the planting site with moist fertile soil and partial to full shade. It should be ploughed 2-3 times to make it well pulverized. Seeds are then sown @ 50 kg ha^{-1}. April in spring season or November in autumn season has been found to be the most suitable sowing time. It may take 4-5 years to see full bloom. The young plants are rooted runners each not less than 30 cm.

Transplant violet plants into field after the last frost date of the area. Space them 15 cm apart and they will tolerate little crowding. Water the plants but do not

soak planting site. Expect bloom during March-June. The plants may be healthy enough. To begin with, they are probably well and truly planted but after the first blooming they are allowed to spread.

Harvesting and yield: Spring is the best time for harvesting wild violet. On an average, the crop yields fresh herbage of 9.5 tonnes ha^{-1}.

References

Brahmvarchaswa. 2007. Botanical name, Hindi name and Family name of 304 medicinal plants. *Graham Vikas Jyoti*. July-September, pp. 48-53.

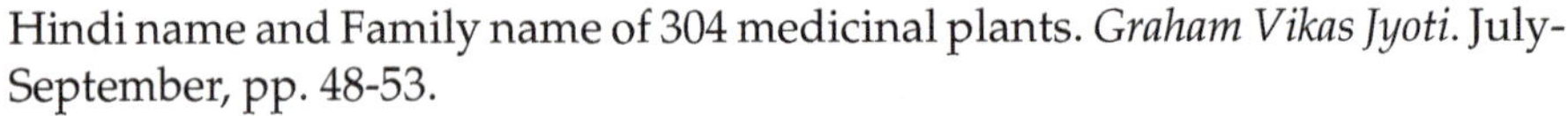

Khajuria, G.L. 2005. Protecting medicinal plants. *The Kashmir Times*. 62 (339): 13.

Glossary

Abortifacient: Anything (drug) used to cause abortion.

Abortion: Loss of reproductive structure (either flower or an ovule) during development.

Accessory: Fruit formed from the expanded dome-like receptacle of a single flower, covered with numerous as in case of strawberry.

Accessory species: A useful species of less value as compared to the principal species, which assist in the growth of the latter.

Achene: Dry, indehiscent fruit containing only one seed.

Acicular: Needle-shaped.

Active ingredient: That part of a herbicide or pesticide formulation which gives the substance its efficacy.

Acuminate: An acute apex or tapering to a point (referring to a plant or animal).

Acute: Apex formed by two straight margins, making an angle of less than 90 degrees.

Adjuvent: An inactive material which when added to a herbicide formulation enhances its efficacy.

Adventitious (bud or root): Refers to plant structures arising at unusual places on the plant usually on the stem or the roots.

Aerobic: Applied to the conditions in which organisms live only in the presence of oxygen and to conditions in which free oxygen is available.

Aggregate fruit: A category of fruit having many seed, developed from the fusion of several ovaries of one flower that are separated in a individual flower *e.g.* raspberry, strawberry.

Alkaloid: A group of basic organic nitrogenous compound of plant origin, bitter in taste and have pharmacological action on humans *e.g.* drugs (quinine, atropine, codeine) and poisons (atropine and strychnine).

Allergen: A substance capable of inducing an allergic response.

Alternate: Term used to leaves or buds of broad-leaved trees that arise first on one side of a twig or stem, then on the other.

Amenorrhea: An abnormal suppression or non-occurrence of menstruation.

Amoebicidal: A compound or chemical that kills amoebae.

Analgesic: A compound that relives from pain but maintains complete consciousness.

Angina: Heart disease indicated by acute constricting pains in the chest, spreading to shoulders, arms and neck.

Angiosperm: A plant that has true flowers and bears its seeds in fruits *e.g.* those in tropical zones, mostly are evergreen.

Annual plant: A plant that lives or grows for only one year or one growing season before dying, in contrast to perennial, which grows for more than one season.

Anthelmintic: Refers to a medicine that causes removal or destroys parasitic intenstinal worms.

Anthers: The pollen-bearing part of a stamen at the end of the stalk.

Anthrax: A bacterial disease caused by *Bacillus anthracis* in cattle and sheep and is often transmitted in humans. Skin and lungs are affected organs.

Anti-diabetic: A substance/medicine counteracting or controlling level of glucose in blood.

Antidote: Any remedy or substance administered to counteract or prevent the effect of a poison.

Anti-oxidant: Any substance that inhibits oxidation and thereby checks the deterioration of food, drugs and other materials.

Anti-febrile: A substance that subsides fever *i.e.* antipyretic.

Anti-inflammatory: A drug which reduces inflammation such as swelling, fever or pain.

Anti-leukemic: A drug that counters the disorder leukemia *i.e.* increase in leukocytes excessively coupled with low count of RBC and WBC's.

Anti-nephritic: Any substance having properties to counter or cure kidney diseases.

Anti-pyretic: Agent *i.e.* a drug that relieves or reduces fever.

Antiseptic: Any formulation or compound capable of preventing infection, sepsis or putrefaction of living tissues/skin *i.e.* a disinfecting agent.

Antispasmodic: A drug or formulation that relives, prevents or cures muscular spasms of stomach/intestine etc. as in epilepsy

Aphrodisiac: An agent, food, drink or compound that stimulates sexual desire.

Apiculate: A condition indicated by termination in a short, sharp flexible point in the leaf.

Arboreous: Refers to trees with well-developed trunk and crown.

Arid climate: Climate of a region that lacks sufficient moisture for crop production without irrigation. In cool region annual precipitation is usually less than 25 cm. It may be as high as 50 cm in tropical regions. Natural vegetation is xerophytic plants.

Aristate: An abrupt hard, bristle-like point or appendage *i.e.* in grasses on grains.

Astringent: A substance or component that checks the contraction of body tissues.

Auriculate: Having a pair of rounded lobes that resembles with human ear.

Axils: The angle formed by the junction of the upper side of a leaf stalk or branch with a stem or branch on which it grows.

Bactericide: Any compound or substance used to destroy bacteria irrespectively if pathogenic or not.

Bark: The exterior tissues covering a woody root or stem outside the cambium that protects the tree from injury caused by other plants, insects and other animals, disease and by fire.

Berry: Fleshy fruit developed from single pistil having no hard layers *i.e.* entire soft pericarp. The fruit is usually many seeded *e.g.* tomato or grape.

Biennial plant: A plant which completes its life cycle in two years. Such plants usually produce leaves and a well developed root system in the first year; stem, flowers in the second year and then die.

Biliousness: A conditions marked by liver dysfunction, giddiness, vomiting, headache, indigestion, excessive secretion of bile.

Biocide: Any agent that is able to kill living organisms.

Bio-fuel: Any fuel derived directly or indirectly from a renewable biological source. Bio-fuels can be solid, liquid or gaseous and include cellulosic ethanol, methanol or bio-diesel. Biomass or materials derived from biomass that can be used to generate energy.

Blade: Lamina, the flattened expanded portion of a leaf *e.g.* leaves of herbs or grasses.

Botanicals: Plant based materials used in herbal industry for making medicines or cosmetic formulations.

Bronchitis: An inflammation of the mucus membrane in tubes of lungs. It causes cough with thickened mucus.

Brown rot: Decay caused by a type of fungus that attacks cellulose rather than lignin, leaving a brownish residue.

Bud: Dormant and un-elongated stem composed of a short axis of meristem cells from which embryonic leaves, lateral buds, flower parts or all three arise.

Budding: A vegetative propagation technique used to produce fruit trees true to their parents having desired qualitative traits. The method consists of inserting a bud from a mother tree scion into a cut in the bark of a young seedling or clonally produced rootstock.

Bulb: A rounded, underground storage organ surrounded by fleshy scales which remain dormant during winters *e.g.* onion, garlic and other members of lily family.

Calculus: Concretions that occur in the form a stone in the animal body *e.g.* kidney stones.

Callus: Hard tissues which develop after a skin tissue is wounded and tends to cover the wound with un-differentiated mass of cells.

Calorie: The traditional unit of measurement for content of food energy and for daily energy requirement. The percent calorie content of protein, fat, cereal or sugar implies the proportion of the daily total calories coming from these dietary components.

Calorific value: It refers to the heat released from a substance in total combustion. Measured in kcal-1 or M cal kj-1 (1 calorie = 4.187 joules).

Calyx: The outermost whorl of a flower, consisting of sepals.

Capsule: Dry, usually many seeded dehiscent fruit composed of two or more fused carpels.

Carbohydrates: Organic compounds which are one of the primary sources of energy for cellular activities. Primarily made up of carbon, hydrogen and oxygen.

Carcinogen: A substance or agent having property to cause cancer.

Cardiac: Related to the heart *i.e.* cardiac disease.

Carminative: A substance that prevents formation of or promotes expulsion of gas from the alimentary tract, relieves flatulence.

Caudate: The tail-like appendage.

Caulescent: Having a well developed aerial stem, above ground.

Cellulose: The chief component of the cell wall of green plants. Cellulose is a carbohydrate *i.e.* polymeric compound composed of a long chain of glucose units and can only be digested by symbiotic bacteria growing in the guts of animals like ruminants and termites.

Chlorofluorocarbon (CFC): A family of compounds of chlorine, fluorine and carbon, entirely of industrial origin. CFC's include refrigerants, propellants for spray cans and for blowing plastic foam insulation, styrofoam packaging.

Chlorosis: Yellowing or blanching of foliage in plants caused by chlorophyll, or light deficiency or nutrient imbalance thereby resulting in decline of plant growth and vigour.

Cholera: A serious intestinal disease that can kill within few hours.

Chronic: Diseases or disorders that are of long duration or re-occur constantly either mild or acute.

Ciliate: A condition with fine hairs on the margin.

Cleft: Split or divided into two at the midrib or base of the leaf blade.

Climate: It refers to average weather conditions at a place or region of earth's surface over a long period of time.

Climber: A herbaceous or woody plant that climbs up trees or other support by twining round them or by holding on to them by tendrils, hooks, aerial roots or other attachments.

Colic: Severe pain resulting from abdominal contractions of muscles in the intestine walls, bile ducts, ureter, especially in babies.

Coma: A prolonged and abnormal deep sleep due to some injury or illness.

Conjunctivitis: Inflammation of the mucous membrane lining of the eyelids and interior part of the eyeball due to infection.

Constipation: A condition indicated by problem in bowels movement due to hardened faeces.

Contact dermatitis: Local allergic reaction provoked by exposure of skin to chemicals, allergens indicated by red itchy rashes.

Contraceptive: Any practice or device which prevents pregnancy by chemical or physical means.

Cordate: Refers to the leaves which are heart shaped with the petiole attached between the basal lobes.

Corm: A specialized part of a stem usually a short, enlarged base of a stem functioning as food storage. May be used as a propagule to propagate a plant vegetatively.

Crown: The uppermost (live) portion of a tree or the stand comprising of branches and foliage above the clear stem.

Cruciform: Refers to cross-shaped, *e.g.* sepals and petals of the members of mustard family.

Crude fibre: The components remaining after protein, soluble and easily decomposable carbohydrates and fats have been chemically removed from a plant sample by appropriate methods. Used in judging quality of animal feed.

Crude protein: The estimated protein content of plant material obtained by analyzing the nitrogen content of feed, assuming that all nitrogen is present as protein and that all proteins contain 16 per cent nitrogen. It is obtained by multiplying the percentage of the nitrogen content by 6.25.

Culm: The characteristic hollow and jointed stem of a bamboo or grass except at the swollen nodes.

Cuspidate: A sharp-pointed, abrupt tip on the leaves.

Cutting: A small detached portion of stem, branch or root placed in the soil or in other suitable medium in order that it may develop into a plant. Stem cuttings can be hardwood, semi-hardwood or softwood. They can be cut nodally or internodally.

Cyanogenic: Capable of producing hydro cyanic acid (HCN).

Cystitis: Inflammation of the urinary bladder caused by bacterial infection and indicated by painful urination.

Deciduous: A group of trees that shed or lose all of their leaves annually and are leafless for some time during the year.

Decoction: A concentrated liquid prepared by boiling medicinal plant parts (stem, leaves, roots, bark etc.) in water and extracting drug or active principle after straining the concentrate.

Decumbent: Stems lying along the ground with their ends upwards.

Dehiscent: A fruit or anther that splits naturally by sutures, pores, or caps etc.

Demulcent: A substance that sooths or relieves and protective in action; against irritation or inflammation of mucous mernbrane in the mouth.

Dengue: A viral disease in tropical climates indicated by fever and acute pains in joints, caused by mosquito bites.

Dentate: Having coarse, angular teeth directed outward or serrated margins.

Depilatory: A cream like formulation or product that causes removal of unwanted hairs.

Diabetes: A metabolic disorder affecting insulin production that causes excessive thirst and urination and increases level of blood sugar.

Diaphoretic: Drugs that promote perspiration as a result of stimulation of the sweat glands of the skin.

Diarrhoea: A condition of passing frequent and watery stool discharges.

Dicotyledon: An angiosperm having two cotyledons (seed leaves).

Dioecious: A flowering plant species or texon whose male and female flowers (or strobilli) are on different trees.

Disinfectant: A physical or chemical agent that frees a plant, organ or tissue from infection *e.g.* bacterial contamination.

Diuretic: Agent that increases urine flow by acting on the kidneys.

Drought: A prolonged period marked with absence of precipitation long enough to make the ground and even water courses dry and severe drought in the summer results in forest fires.

Drupe: Fleshy indehiscent (not regularly opening) fruit with the seed contained in a woody stone.

Dysentery: A disorder cuased by intestinal infection which causes evacuation of mucous and blood in the stool.

Dysmenorrhea: Abnormal pains during the menstruation period. The pain may be either spasmodic or continuous.

Dyspepsia: Refers to painful indigestion indicated by nausea or vomiting.

Ecobolic: A drug that causes contraction of the uterus leading to abortion or promoting parturition.

Eczema: A non-contagious itching; inflammatory skin eruption characterized by rough skin, blisters and postules which often itches or bleeds.

Elephantiasis: A chronic enlargement of the cutaneous and sub-cutaneous tissues, common in the tropics and caused by nematode parasites.

Emetic: A drug or an agent having the power to empty the stomach by vomiting.

Emmenagogue: Refers to a drug which have the power of stimulating the menstrual flow.

Emollient: A substance applied externally to soften or soothen the skin.

Enteritis: Acute or chronic intestinal (small) inflammation.

Environment: The sum of all external biotic and abiotic conditions and influences that affect the development and survival of an organism in a given area.

Enzyme: An organic (protein) catalyst, produced by a living organism which speeds up a specific biochemical reaction. Enzymes are necessary to make almost all processes within the cells fast enough to sustain life.

Epidemic: Relating to a condition which is marked by the rapid development of a population (plant, animal and virus) to an extremely high level which is harmful for the maintenance of the ecosystem.

Erysipelas: An acute skin inflammation caused by *Streptococcus,* results into raised red patches on skin accompanied by fever.

Ethnobotany: A branch of botany that seek an understanding of how and in what way people use plants and nature in their local environment. Ethnobotany is an interdisciplinary approach including botany, anthropology, history and chemistry.

Evergreen: Belonging to a group of trees which do not shed all of their leaves every year but go through a gradual replacement of the oldest leaves each year. Instead of being bare in winter, they have leaves throughout the greater part of the year.

Exanthema: A disease resulting in eruptions of skin, such as measles or scarlet fever.

Exotic: A plant species, animals, fungi or micro-organisms which have been introduced outside their natural habitat as a result of human activity. Exotic species are also known as alien species, foreign species, introduced species, non-indigenous species, and non-native species.

Expectorant: A medicine promoting secretion of bronchial mucus and curing cough.

Extract: A pharmaceutical preparation containing active constituents of a drug in a solvent and often in concentrated form with a suitable solvent, evaporating the solvent, and adjusting to prescribed standards.

***Ex -situ* conservation**: The preservation of a plant or plant parts outside its natural or original habitat *e.g.* a gene bank.

Farmyard manure (FYM): The well decomposed excreta of domestic animals, mixed with straw or other litter. It is used as organic manure to soil to improve soil physico-chemical properties.

Febrifuge: A drug used to reduce fever.

Fertilizer: A product (organic or inorganic) that is used to enrich the soil fertility and to increase or improve crop yield or plant growth by supplying essential plant nutrients.

Fibrositis: Inflammation of fibrous tissue as a result of injury.

Filariasis: A disease caused by parasitic worms especially in tropical climates spread by flies or mosquitoes.

Flavonoid: A group of organic compounds which imparts colours in fruits and flowers. Also often used along with mordant for the dyeing of fabrics.

Flora: The plant life (including trees and shrubs) growing in a specific area or a region.

Flower: Strictly, a reproductive structure bearing pistils or stamens or both and usually sepals and petals in angiosperms only. The so-called flower of conifers is the male or female strobilus.

Fodder: Plants or plant (leaves or sometimes flowers and pods) parts eaten by browsing or grazing animals *e.g.* fodder trees including species of *Acacia, Leucaena, Prosopis* and many others. Often fodder is collected and stored for future consumption.

Fruit: A mature, developed or ripening ovary is called fruit.

Galactogogue/Lactagogue: An agent that induces or increase the secretion of milk.

Gastritis: Inflammation or irritation of the stomach living.

Gastrointestinal: Pertaining to the stomach and intestines.

Genito-urinary: Relating to the genital and the urinary organs or functions.

Germicidal: A substance that kills germs and pathogenic micro-organisms in general.

Germination: The emergence of a normal seedling from a seed or resumption of growth of a seed or spore. Usually recognized by rupture of the seed coat or spore wall and appearance of the radical, hypocotyl or thallus.

Gingivitis: Inflammation of the gums.

Glaucoma: A disease characterized by increasing fluid, intraocular pressure causing gradual loss of sight.

Glossitis: Inflammation of the tongue.

Gonorrhea: A venereal disease that causes inflammation of the mucous membranes of the urethra or vagina.

Gout: A condition of defective uric acid metabolism. It is characterized by pain in joints and increased uric acid in the blood.

Grafting: A method of vegetative propagation by placing a small shoot from one (scion) in close contact with another (root stock) so that the cambium layers from both come in contact and the shoot grows as part of the main plant. This is normally used to obtain high quality fruit from hardy, well established plants (rootstock).

Gregarious flowering: The general flowering of species which do not flower annually, over considerable areas when all or most of the individuals of certain species do not flower annually; in some cases followed by the death of the plant *e.g.* bamboos and strobilanthes. Opposite to sporadic flowering.

Gum: Complex polysaccharide formed by cells in reaction to wounding or infection.

Hemolytic: An agent capable of destroying red blood cells.

Haemorrhoids: An enlarged and swollen blood vessels or vein in the anal region or alimentary tract.

Haemostatic: An agent that stops flow of blood or prevents bleeding.

Hectare: A unit of area measurement equal to 10, 000 square meters.

Hepatitis: Inflammation of the liver which can lead to liver cancer or failure caused by a virus.

Herb: A plant whose stem is always green and tender and height is usually not more than one meter. It can be annual, biannual and perennial depending on its life span.

Hernia: Protrusion of an organ through its cavity wall. It may occur in any part of the body, but is especially associated with the abdominal cavity.

Herpes: A skin disease caused by a virus with patches of distinct blisters and high fever.

Home garden: A land-use practiced in and around individual's house with a definite fence, in which several tree species are cultivated together with annual and perennial crops (stratified canopy), often with the small livestock. There are different forms of such gardens *e.g.* village forest gardens, compound gardens, kitchen gardens.

Indigenous: Native to a specified area, not introduced. An indigenous tree is one that grows naturally within a specific environment or within certain pre-determined boundaries. *See also exotic.*

Insecticide: Any chemical or biological preparation used to kill or disrupt the development of insects.

Internode: A region of the stem between two successive nodes.

***In-situ* conservation**: To conserve the species or population inside their natural habitat or host plant.

Irrigation: The controlled application of water to arable lands to meet water requirements not satisfied by rainfall.

Jaundice: A condition with yellowing of the skin, mucus membrances and secretions, as a result of excess bile pigments (bilirubin) in the blood.

Lactation: The formation and secretion of milk by mammary glands.

Laryngitis: Inflammation of the larynx resulting in horse voice or painful cough.

Laxative: A drug which tends to stimulate or facilitate bowel movement.

Layering: The production of adventitious roots from a prostrate main stem or a branch in contact with the ground, to produce a separate plant after separation from the mother plant.

Leprosy: A chronic disease of the skin and nerves characterized by loss of pigment and lumps on skin.

Leucopenia: An abnormally low number of leukocytes (white blood cells) in the blood.

Leucorrhea: A white or yellowish mucopurulent vaginal discharge.

Loam: A soil composed of relatively equal proportions of sand and silt and lesser clay so that the properties of the soil are not dominated by any one of them.

Mange: A contagious skin disease in mammals by itching and hair loss caused by parasitic mites, communicable to humans.

Manure: Animal excreta with or without a mixture of bedding or litter.

Meningitis: A disease caused by bacterial or viral infection that causes inflammation of the membranes enveloping the brain and spinal cord.

Menorrhagia: Abnormal excessive menstrual flow.

Mesocarp: The middle layer of the fruit wall, often the fleshy edible portion.

Migraine: A recurring and painful headache, often accompanied by vomiting, giddiness, and disturbed vision, contractions of voluntary muscles especially of the jaw.

Monocotyledon: Angiosperms having one cotyledon (seed leaf).

Monoecious: Of flowering plant population or species, having functional male or female flowers borne separately on the same plant *i.e.* plants bearing unisexual flowers, both staminate and pistillate on the same individual.

Multiple: A category of fruit derived from the fusion of an entire inflorescence, as in the pineapple.

Malaria: An acute, usually chronic, disease caused by protozoa belonging to the genus *Plasmodium* and transmitted by *Anopheles* mosquito. It is characterized by intermittent fever, anemia, and debility, and in its acute form, by chills, high fever, and profuse sweating at regular intervals.

Neuralgia: An acute pain along the course of a nerve either on head or face.

Nursery: An area set aside for raising young plants for eventual planting out and has ordinarily both seedling and transplant beds. Nurseries are either permanent or temporary.

Nursery bed: An elongated strip of area in a nursery where seed is sown or into which transplants or cuttings are raised.

Nutraceutical: A product isolated or purified from foods (including from specific forest-based foods) that is demonstrated to have a physiological benefit or provide protection against chronic disease. Nutra-ceuticals are usually sold in medicinal forms, not as foods and are generally considered part of the vitamin and pharmaceutical market.

Oblique: Slanting, neither parallel nor at right angle.

Oblong: About two or three times longer than broad; rectangular with rounded corners.

Obovate: Same as ovate, except the petiole attached at the narrow end at the base.

Obtuse: Apex formed by two lines that meet at more than a right angle.

Ophthalmia: Inflammation of the eyeball or the conjunctivitis.

Opposite: Two leaves at a node.

Organic fertilizers: A compound containing nutrient elements derived from organic sources without any synthetic additions *e.g.* FYM.

Out planting: Transferring of nursery stock to the planting site as compared with transplanting in the nursery.

Oval: Broadly elliptic, having rounded shape *e.g.* egg.

Ovate: The shape of the longitudinal section through an egg, broad leaf rounded at the base and tapering towards the end.

Palmate: A leaf having five or more lobes with midribs arising from one point *e.g.* maple leaf.

Palpitation: A rapid, strong or irregular heartbeat or throbbing of the heart due to agitation or exertion or illness.

Panicle: A loose flower cluster *e.g.* in cereal crops.

Papillate: Pimple-like hairs or pimple-like protuberance.

Parallel: Several to many veins of same size running side by side with same distance between them *e.g.* in grasses.

Parasiticide: Any agent or formulation that destroys parasites.

Parturition: The act of giving birth to a child.

Perennial: A plant that grows or lives for more than one year in contrast to an annual, which grows only for one year. Some resprout from a root system or re-seed themselves every year.

Peristalsis: Wave like muscular contractions of the muscles of intestine.

Pest: Any organism (insect, fungi or virus, weed or avian or mammalian *i.e.* bird, rodent) capable of causing injury or damage to the plants or other life forms.

Piles: A synonym for hemorrhoids.

Pityriasis: A skin disease in which the epidermis sheds thin scales as dandruff.

Pneumonia: Refers inflammation or congestion of the lungs, caused by bacterial or viral infection, air sacs filled with pus and become solid.

Pod: A type of dehiscent dry fruit with one to many seeds, formed from a single carpel *e.g.* in members of leguminosae family.

Postpartum: Following childbirth.

Poultice: A soft, moist mass of material (bran, flour, herbs etc.) applied to sore or inflamed part for supplying heat and moisture or acting as a local stimulant.

Pre-treatment: Physical or chemical treatments applied to seed in difficult to germinate species, prior to sowing in order to increase the rapidity or completeness of germination *e.g.* cold water soaking, hot water or acid treatments etc.

Propagation: Regeneration of new plants by means other than seed *e.g.* by cuttings.

Prophylactic: A drug or measure to prevent a disease.

Protein: A functional organic macro-molecule assembled from amino acids linked with peptide bonds; a product of gene expression. Provide a defense mechanism in plants.

Pruritus: Localized or generalized itching of skin due to certain body disorders.

Psoriasis: A non-contagious skin disease characterized by red, itchy patches and white scales.

Purgative: Drugs which are strongly laxative in action.

Pyorrhoea: Purulent inflammation of gums and tooth sockets associated with discharge containing pus.

Quinones: Organic compounds based on aromatic groups of benzene.

Restorative: A drug having the ability to restore heath, strength and weakness.

Rheumatism: A disease characterized by painful inflammation of muscles, joints, *e.g.* rheumatoid arthritis.

Rhizome: A horizontal stem or root like appearance, lying on or under the soil, which produce adventitious roots and sends up shoots.

Ringworm: A contagious fungal disease affects the skin, hair, or nails of humans and domestic animals.

Rosette: Cluster of leaves at or near the base of the plant.

Rubefacient: A substance applied to counter irritants to the skin, producing blisters and redness of skins.

Runner: A slender horizontal stem with elongated internodes and produce roots at nodes that touch the ground.

Sandy soil: A soil which is composed of very fine in-organic particles, such as sand grains.

Saponin: A toxic compound characterized by its ability to form emulsions and soapy lathers found in many plants.

Scabies: A contagious skin disease caused by a mite that burrows in the horny layer of the skin and causes itching and small red spots.

Scurvy: A nutritional disorder caused by deficiency of vitamin C; characterized by extreme weakness, spongy and bleeding gums and bumps under the skin.

Sedative: Any drug or an agent that induces sleep or promote calmness.

Seed: Small, usually rounded body, sometimes with a wing to aid wind dispersal, that develops from the fertilization of the ovule and from which a new plant develops after germination. In simple words seed is a ripened ovule.

Seed dormancy: A condition of mature viable seed in which germination is considerably delayed even though external conditions favour germination.

Seed germination: The first stage of growth of seed, when the protective seed coat breaks down and water enters the seed. This happens only under the favourable environmental conditions.

Seedling: Young plant usually with height of one meter either grown from seed in a nursery, prior to transplanting or out planting.

Septum: An internal partition within a fruit.

Shrub: A plant having persistent woody stems but lacks main stem and a relatively low growth habit and generally produces branches from the base instead of a single bole. It differs from a tree by its low stature and non-arborescent form. Usually less than 16-20 ft. at maturity.

Soil: The uppermost weathered layer of the earth's crust formed by the action of soil forming factors (climate, relief, parent material, time and living organisms). Soil comprises of weathered mineral particles, dead and living matter, air space and soil solution. The following sub-division is recognized:- (i) Surface soil- the more or less completely weathered surface layer, rich in soluble material and containing a relatively higher proportion of organic matter and fine earth; also called top-soil. The zone of aeration and intense root and microbiological activity. (ii) Subsoil - the layer immediately below the surface soil which is incompletely weathered and contains much less of soluble nutrients, organic matter and fine earth.

Spasmodic: Periodic sharp attacks marked by spasms.

Sporadic flowering: Flowering in a small proportion of a population of a species with pronounced periodicity *e.g.* in bamboo.

Stimulant: Substance that promotes the level of physiological or nervous activity.

Stomachic: Refers to drugs for increasing appetite or facilitating digestion.

Succulent: A plant in which the tissues have an unusually high vacuole to cytoplasm ratio, thus very large cells.

Syphilis: A chronic bacterial disease, spread by sexual contact and caused by *Treponema pallidum*.

Tannin: An astringent substance (usually polyphenol), available naturally in the bark (and/or wood) of different species of trees (*e.g.* most of acacia and mangrove species), used in converting hide to leather. They bind protein and provide a defense mechanism in plants.

Temperate climate: Climate which is warm in summer and has a definite winter season. Determined in India more by altitude than latitude.

Tetanus: An acute infectious disease caused by bacteria and characterized by rigid spasms of voluntary tissue.

Tonic: A drug or an agent taken to improve vigour or general well being of an individual.

Transplant: A seedling grown from seed, after it has been moved one or more times in a nursery *i.e.* transplanted in the nursery, in contrast to a seedling planted out direct from the seedbed.

Transplant nursery: A nursery into which seedlings are transplanted before eventual field planting. In India it is called pricking out.

Tree: A large woody perennial plant having a single well defined stem (bole or trunk) and a more or less definite crown and root system, usually attaining a height of at least 2½ metres.

Tropical climate: A climate which is not very hot, without a definite winter season.

Tuber: Refers to an underground shoot or stem, generally irregular in shape, which acts as a store of nutrients *e.g.* in potato.

Tuberclosis: A highly contagious disease caused by bacterium *Mycobacterium tuberculosis*. It is characterized by the presence of nodules in the tissues especially lungs.

Tumor: Abnormal swelling of the body parts either solid or fluid filled caused by abnormal growth of tissues but without inflammation.

Typhus: An infectious disease caused by the *Rickettsia* characterized by purple rashes, headache, fever or delirium and mostly transmitted by lice, mites or rat fleas.

Ulcer: Any open sore other than a wound on external or internal surface of body characterized by tissue disintegration or pus formation.

Varicose: Swollen twisted or lengthened veins especially.

Variety: Any clone or product of breeding given a separate name having enough desirable characteristics to be commonly cultivated. Often an exclusive geographic range or an assemblage of cultivated individuals with useful and re-produceable characters.

Vegetative propagation: A form of asexual reproduction (natural or artificial) in which plant parts produce new plants, which are genetically identical to the parent *e.g.* rooting of cuttings, layering, grafting, budding and micropropagation and fertilization is not involved. Reproduction of trees by coppice or root-suckers is vegetative reproduction.

Vermicide: A substance which is used to kill worms.

Vermifuge: Substance that kills or expels intestinal worms.

Vesicant: A blistering agent, or drug that produces blisters on the skin.

Vesicatory: A chemical or an agent capable of causing blisters.

Villous: Shaggy or covered with slender, soft, villi or hairs.

Wart: Refers to a skin tumor caused by virus infection i.e. a small rounded growth on skin of an animal or on the plant surface.

Waterlogged: A condition of land where the groundwater stands at a level which is detrimental to plant growth. It may be due to over irrigation, seepage or poor drainage.

Weeding: Removal or suppression of un-wanted or competing vegetation including herbaceous growth by cultural, physical or chemical means.

Woody perennial: A plant with lignified stems that continues to grow from year to year.

Yellow fever: A tropical epidemic disease caused by mosquito-borne viral infection which affects liver and kidneys, causing fever and jaundice and often causes deaths.

Yield: Plant parts harvested and the quantity of it.

Zonation: An arrangement of different plant communities in more or less definite zone, following one another in a constant order due to a continuous change in the habitat factors.

Index

B

T

www.ingramcontent.com/pod-product-compliance
Ingram Content Group UK Ltd.
Pitfield, Milton Keynes, MK11 3LW, UK
UKHW021322190726
13845UKWH00003B/26

9 789351 309949